uantum Many-Body Physics in a Nutshell

Quantum Many-Body Physics in a Nutshell

Edward Shuryak

PRINCETON UNIVERSITY PRESS · PRINCETON AND OXFORD

Published by Princeton University Press, 41 William Street, Princeton,
New Jersey 08540
In the United Kingdom: Princeton University Press, 6 Oxford Street,
Woodstock, Oxfordshire OX20 1TR

press.princeton.edu

Library of Congress Control Number 2018950316

ISBN 978-0-691-17560-7

British Library Cataloging-in-Publication Data is available

Editorial: Eric Henney and Arthur Werneck
Production Editorial: Ali Parrington
Text and Jacket Design: Leslie Flis
Jacket image: Courtesy of Brookhaven National Laboratory
Production: Jacqueline Poirier
Copyeditor: Cyd Westmoreland

This book has been composed in Scala

Printed on acid-free paper. ∞

Typeset by Nova Techset Pvt Ltd, Bangalore, India
Printed in the United States of America

10 9 8 7 6 5 4 3 2 1

Contents

Preface

Quantum Many-Body Physics in a Nutshell contains a carefully selected introduction to the theory of multiple physical systems connected by certain common phenomena. Its goal is to present the maximal amount of physics—both phenomena and theoretical tools—in the simplest possible way. According to the quotation above, even this optimal path will require a certain level of effort from the reader. I hope the level of presentation is accessible to the average graduate student, or even a curious undergraduate, with the appropriate background.

While I have taught many physics courses, over nearly 50 years, most of the material for this book was accumulated over the past decade at Stony Brook from lecture notes on graduate quantum and statistical mechanics, many-body theory, nuclear physics, and several courses on special topics. This book is basically written for a serious reading with self-education in mind, but it can also be used for teaching, forming, say, a one-semester introductory course.

One aim of the book is to introduce a wide a range of phenomena, from trapped gases at nano-Kelvin temperatures to quark-gluon plasmas at temperatures reaching giga-electron-volts. Yet the scale itself is irrelevant; what is emphasized instead is *the unity of physics*: the many common features, tools, and ideas used for all of these phenomena.

Another aim is to show the reader how the main technical tools work. The central tool is the *Feynman path integral form of the density matrix*, continued into imaginary (Euclidean) time defined on a Matsubara circle, which provides a natural finite-temperature extension of quantum mechanics and field theories (QFTs). We start in the simplest setting possible, in the $(1 + 0)$ dimension, namely, the quantum-mechanical motion of a particle at finite temperature. In this simplest setting, we introduce the Feynman diagrams, one by one. A pedagogically important fact is that all of them are finite and simple to calculate.[1] There are renormalizations but no divergences.

After this pedagogical introduction, we turn to the simplest QFTs, the scalar field theories, first with real and then complex fields. We do so both in relativistic and nonrelativistic versions. Eventually the gauge theories, QED and QCD, appear as natural extensions.

Most applications—the electron gas, nuclear matter, liquid ^{3}He, and quark-gluon plasma—all require generalization to Fermi fields and path integrals. For all of them we go through similar learning paths: start with weak coupling, evaluate key lowest order perturbative diagrams, and then perform their resummation or sometimes use the renormalization group. Then last comes a discussion of various pairing phenomena near the Fermi surface.

All physics topics considered here are discussed in much more detail in other, specialized books. Specifically, books on many-body theory include the electron gas, superfluidity, and superconductivity; books on nuclear physics address nuclear matter; books on finite-temperature field theory discuss the quark-gluon plasma (and some include lattice gauge theories). Most of these books are not really written as textbooks, providing instead full expositions of the material, intended for experts in the field. In this book we include all of these topics, keeping the level of presentation in each chapter more or less at the same level. Hopefully, each chapter can be explained during about two lectures, and the book can be read in independent study, with a reasonable effort, providing an introductory exposition of all these phenomena and theoretical tools.

The term "quantum many-body systems" in the title implies that we will not discuss any finite systems, such as atoms or finite nuclei. We will only discuss the macroscopic limit (e.g., an electron gas or nuclear matter). Our smallest systems—trapped quantum gases or drops of quark-gluon plasmas—still include no fewer than thousands of particles. Indeed, only macroscopic systems can display true phase transitions, which will be discussed. Of particular importance for us will be the phases of matter possessing certain nonzero quantum *condensates* of various boson fields (or fermion pairs). These phenomena, known collectively as spontaneous symmetry breaking, reappear in multiple settings of modern physics.

This book has no problem sets, as most textbook have. Instead I suggest a exercises imbedded in the text where they are needed. Being educated myself via the Landau-Lifshitz *Course of Theoretical Physics*, I recall rather frustrating experiences, when not just the problems but also their solutions, given at the end of each section, looked intimidating. (Only later did I learn that in these books, the "problems" usually contained the essence of some famous paper or somebody's dissertation. They were not really meant to be solved by the reader, but admired.) My exercises are different: Their purpose is to illustrate the ideas or tools under consideration by an application. I tried to make them as

---

[1] Feynman himself, who invented his famous diagrams for QED, for reasons unknown (to me) never used them—not in his famous book on path integrals in quantum mechanics nor in his other book on statistical mechanics.

simple as possible. All are straightforward, and the methods discussed should lead to a solution, occasionally using tools like Maple or Mathematica for plotting, doing algebra, or evaluating integrals. No tricks need to be invented. These exercises are intended to build self-confidence. My idea of a good exercise is one, that a reader can say after completing it: "Wow, I did not expect that this famous subject is actually that simple!"

uantum Many-Body Physics in a Nutshell

1 | Introduction

1.1 Prerequisites and Textbooks

The content of this book is sufficiently general and traditional, as to merit a place in the standard curriculum of physics graduate programs. Yet it is hardly to be found in most universities. In fact, it does not even have a single widely accepted name, being called a "many-body theory," "statistical field theory," and the like.

Bits and pieces of this topic can be found in lectures on atomic, condensed matter, nuclear, and high-energy physics, which all may discuss very similar and generic phenomena. A famous Einstein quotation notes that the number of good ideas is so small that the same phenomena pop up repeatedly in many different settings. And indeed, as we will see, one can collect all these pieces together, revealing the logical and historical paths that are common to these settings. Using the examples at hand, we will be demonstrating the surprising unity of physics in its many fields.

Quantum macroscopic phenomena famously include *superfluidity* and *superconductivity*. Their discoveries go back to the low-temperature frontier that was first explored a century ago. Their studies in the 1940s–1970s have resulted in amazing experiments and applications. Since 1990 we have witnessed the discovery of Bose-Einstein condensation in trapped atomic gases and other amazing applications, such as strongly coupled fermionic systems, perhaps with quantum computers to come in the near future.

Past decades have also witnessed remarkable progress in understanding strongly coupled quantum field theories (QFTs). The vacuum we live in—the quantum chromodynamic (QCD) vacuum – is a kind of superconductor, with a gap induced by quark-antiquark pairing. It is also a *dual superconductor* with Bose-condensed color-magnetic monopoles. Experiments on heavy ion collisions reveal the corresponding phase transition to the normal phase, known as the quark-gluon plasma, possessing quite unusual properties of a dual plasma, with a complicated interaction between electrically and magnetically charged quasiparticles.

Yet the major theoretical tools we will discuss are not so new. Feynman's diagrams, invented for quantum electrodynamics (QED) in the 1950s, became the main tool not only of QFTs but also of statistical mechanics (many-body theory). Weakly coupled Bose and Fermi systems were understood first, followed by realistic electron gases and nuclear matter. The Bardeen-Cooper-Schrieffer (BCS) theory explains not only the superconductivity of metals but also a large number of other phenomena as well. Along the way, it has been entirely reformulated, and we will explore it within the framework of renormalization group flow, another unifier of a huge number of phenomena.

Naturally, in the past decades many more discoveries have been made, and new directions have now taken central stage. A big event in the 1990s was cooling of the trapped quantum gases (both Bose and Fermi varieties) to temperatures so low that quantum condensates spontaneously appear. The so-called Feshbach resonance had allowed experimentalists to tune the main interaction parameter—the scattering length a—to any desired value, making it an ideal laboratory for many-body physics.

A much larger experimental effort in the nuclear physics community is the quest to produce and study quark-gluon plasma via high-energy heavy ion collisions. Performed mostly using the two largest devices of modern physics—the Relativistic Heavy Ion Collider (RHIC) at Brookhaven National Laboratory and the Large Hadron Collider (LHC) at CERN—they reveal rather unusual properties of such matter. Another objective is to understand the phase transitions and structure of the QCD vacuum state we live in. At the end of the book we will discuss some open issues and current ideas on them.

And last but not least, the exponential growth in computer power made numerical evaluation of the Euclidean path integrals practical in late 1970s. First-principle studies of quantum many-body systems (e.g., liquid helium-II) became possible. This trend swept through QFTs, in particular to lattice gauge theories, and these studies now occupy many of the most powerful supercomputers in the world. These simulations are reproducing the hadronic spectrum and quark-gluon plasma equation of state. Due to these massive computational efforts, these results now complement the experimental effort and even allow researchers to study situations that would be impossible for real experiments.

A word about prerequisites and the use of other sources. It is assumed that the reader is familiar with standard quantum mechanics and statistical mechanics. However, it is *not* assumed that the reader has taken even introductory QFT courses: That is why the early sections contain an "alternative introduction" to the Feynman diagrams. Yet extra knowledge is always helpful: As a good self-education QFT book from the *Nutshell* series, I recommend that by A. Zee [1]. I will mention textbooks on particular subjects, as we proceed.

The classic books by Feynman on path integrals [2] and statistical mechanics [3] are the foundation of the the early chapters in my book. So the reader is advised to go through the first chapters in the Feynman books. Strangely enough, Feynman's books do not contain Feynman diagrams: Filling this gap will be one of our first tasks. Diagrammatic applications in statistical mechanics were covered in the classic pioneering book by Abrikosov, Gorkov, and Dzyaloshinski [61]. Other recommended books are Fetter and Walecka [5] and Negele and Orland [6]. These are all excellent books. However, they were

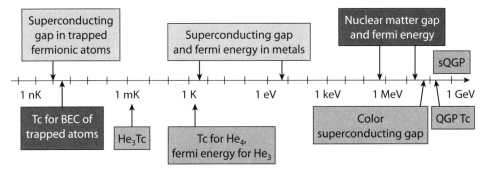

Figure 1.1. The logarithmic temperature scale and certain phenomena to be considered. Each vertical tick is an order of magnitude, and in total the temperature changes from nano-Kelvins to Giga-eV, by about $9 + 4 + 9 = 22$ orders of magnitude. Note that the frontiers of high-energy physics (on the right) and that of low temperature physics (on the left) are at roughly equal distances from room temperature (300 K) in the middle of the plot.

all written for theorists and focus mostly on technical tools rather than on phenomena, so they are not suitable for a first reading. They are also dense, to the extent of being difficult to read, and somewhat dated (e.g., they do not use path integrals, now standard). For a reader interested in a particular application, it would be more useful to look at these books after going through this one.

1.2 Physical Phenomena and Theoretical Tools

Let me start with a map in Figure 1.1, which indicates the location of some of the systems to be discussed in this course on a temperature scale.[1]

A proximity in scale by no means implies similarity of physics. In fact the absolute scale does not matter at all. The first thing physicists do when tackling a problem is introduce the appropriate units to make the problem scaleless. For example, two most extreme cases, the *trapped Fermi atomic gases in the strong coupling regime* at temperatures of the order of nano-Kelvins $T \sim nK$ at one end, and the quark-gluon plasma at the other end, are in fact rather similar. They both are very strongly coupled yet conformal (scale-independent) types of matter, displaying similar kinetic properties, the understanding of which remained challenging until recently.

For example, Fermi systems we will discuss include (in the order to be addressed) the electron gas, liquid ^{3}He, nuclear matter, trapped atomic gases, and quark-gluon plasmas. For all these cases we will see several recurring motives:

1. We use the same set of theoretical tools in all cases, namely, Euclidean path integrals, defined on the Matsubara circle, with the circumference $\beta = \hbar / T$ and with zero temperature corresponding to $\beta \to \infty$.

---

[1] As in most modern texts, by T I mean the energy, setting the Boltzmann constant $k_B = 1$ and freely combining the Kelvin K and electron-volt eV scales. The standard suffixes are n = nano = 10^{-9}, m = milli = 10^{-3}, k = kilo = 10^3, M = mega = 10^6, G = giga = 10^9.

2. We start by defining the action (Lagrangian). It always consists of a quadratic part (harmonic oscillator $V \sim x^2$ in quantum mechanical examples and the kinetic energy for QFTs) and the more complicated interaction part.

3. We will work out the lowest-order diagrams, with one or two loops, for the total (free) energy.

4. Then we will study the "mass operator" Σ and/or the "polarization operator" Π, correcting the zeroth-order propagators of the theory.

5. We then resum some series of the diagrams, using either geometric series or more sophisticated tools like the renormalization group equation.

6. Because the resummed ("dressed") Green functions possess either shifted or even entirely new poles, we will find that matter supports different propagation modes. Some modify the original particles, making them *quasiparticles*, with the same quantum numbers but new properties. Some are completely new (e.g., *plasmons* in electron gas or *zero sound* in liquid ^{3}He).

7. Furthermore, Bose-Einstein condensation of bosons (or Cooper pairs) of various types effectively destroys particle number conservation. This leads to the so-called *anomalous Green functions* describing the appearance or disappearance of particles. Using such diagrams, known as Gorkov formalism, we will describe multiple superconducting phases and the superfluidity of liquid ^{3}He, with their rich phenomenology.

We will show that superfluid and superconducting phases possess certain topological excitations, the *quantized vortices*, giving rise to complicated phase structures, the *topological matter*, of magnetized (or rotating, or both) matter. Furthermore, the vortices are only one member of the topological soliton family. We will briefly discuss the *skyrmions*, and (more important for this book) *Dirac monopoles* in three dimensions. Topological solitons in four dimensions are called *instantons*;[2] we will show how they lead to fascinating effects in quark-gluon plasmas and the QCD vacuum.

The most complex matter we will discuss near the end of the book is the vacuum of QCD, which has multiple condensates. One is due to quark-antiquark pairing, producing the so-called quark condensate $\langle \bar{q} q \rangle \neq 0$, at the surface of the Dirac sea. It is responsible for a gap in the quark spectrum—an effective quark mass—which is a significant (if not dominant) part of the mass of nucleons (and thus of our own mass). Another condensate is even more unusual: it is a dual superconductor made of bosonic objects with a magnetic charge, the color-magnetic monopoles. It leads to the dual Meissner effect, expelling the electric field into flux tubes and causing "quark confinement." One may ask whether the discussion of these fascinating physical phenomena in this introductory volume is premature. My answer is that, after one goes thorough all the needed theoretical tools and sees how they describe several traditional condensed matter applications, it should not be difficult to understand them.

---

[2] The solution was found by Polyakov and colleagues in Belavin et al. [179] and called "pseudoparticle," but the name "instanton"—suggested by 't Hooft and meaning "existing for an instant"—became the standard term.

To complete this general introduction, let me praise our tool of choice, on which the whole formalism will be based. It is the *the Feynman path integral* discussed from the beginning in the so-called *Euclidean time formulation*. As we will soon see, (1) it is the simplest and the most natural way to introduce and evaluate the Feynman diagrams; (2) it provides the most natural basis for the semiclassical approximations; and—last but not least—(3) it opens the door to direct first-principle numerical simulations of the path ensembles. We will discuss some computer-based approaches, which range from studies of few-body quantum-mechanical problems and the properties of liquid ^{4}He to numerical simulations of lattice gauge theories, which keep busy a fair fraction of the most powerful supercomputers available.

1.3 Feynman's Path Integrals

The concept is due to Feynman and originated from his studies of QED. The meaning of these path integrals in a quantum mechanical setting is described in his book [2], and in statistical mechanics in [3]. While we will be exploring the main formulas in this book, the reader is encouraged to read a few chapters of the Feynman volumes, where the concept is explained in much more detail.

Let me start by noting that all standard texts on quantum mechanics pay some tribute the so-called *matrix formulation* due to Heisenberg, which operates on abstract states in Hilbert spaces, usually a representation of some closed operator algebra. Then the standard presentations move to the Schrödinger formulation, based on the notion of *wavefunctions*. In stationary (time-independent) problems, the wave functions are the eigenstates of the Hamiltonian

$$\hat{H}|n\rangle = E_n|n\rangle. \tag{1.1}$$

Hence their dependence on time is simple, via[3] the exponential factors e^{-iE_nt}.

The coordinate operator defines another set of states $|x\rangle$. It is rather singular, assuming that a particle is at the point x and nowhere else, so the operator should be treated with care. The stationary wave functions—the main element of the standard textbook Schrödinger formulation of quantum mechanics—are the projection coefficients of one set to another, namely,

$$\psi_n(x) \equiv \langle x|n\rangle. \tag{1.2}$$

Another concept, central for both quantum mechanics and statistical mechanics, is the *density matrix*. Let us discuss it in the simplest setting of single-particle motion in a time-independent one-dimensional potential $V(x)$. First define *the amplitude of the particle's propagation from the point x to the point y, during certain time t:*

$$G(x, y, t) \equiv \langle y|\exp(-i\hat{H}t)|x\rangle = \Sigma_n \psi_n(x)\psi_n^*(y)\exp(-iE_nt). \tag{1.3}$$

---

[3] Here and in most places below we use units in which $\hbar = 1$.

The first equality in this equation is the definition of the amplitude as a matrix element of the exponent of the Hamiltonian. The second equality is obtained by inserting a complete sum of the projectors over all stationary states

$$\hat{1} = \sum_n |n\rangle\langle n|$$

into the definition. We start with two sets of stationary states $|n\rangle, |n'\rangle$ and use the definition of the wavefunction (1.2). Because in its eigenbasis the Hamiltonian operator has nonzero elements only on the diagonal, the last expression has only a single—not a double—sum over states $n = n'$.

Exercise *First we calculate the amplitude for the free motion of a particle, using the usual plane waves $\psi_p(x) \sim \exp(ipx - iE_pt)$, $E_p = p^2/2m$. The sum over n is thus the integral over momentum p, and the answer is*

$$G(x, y, t) = \sqrt{\frac{m}{2\pi it}} \exp\left[\frac{im(x-y)^2}{2t}\right]. \tag{1.4}$$

This expression tells us that a free quantum particle spreads diffusionally, with the distance traveled $|x - y| \sim \sqrt{t}$. (A rather confusing imaginary i in the exponent will soon be gone; see below.) We will later use this expression for normalization of the motion in nonzero potentials. Show that the expression in the exponent of equation (1.4) is none other than the classical action for a straight path going from x to y and taking time t.

The Feynman formulation of quantum mechanics is based on the celebrated path integral representation for this amplitude, namely,

$$G(x, y, t) = \int_{x(0)=x}^{x(t)=y} Dx(t)\exp(-iS[x(t)]). \tag{1.5}$$

From classical mechanics we are familiar with the classical paths that satisfy the Newtonian equations of motion. According to Feynman, however, any quantum system is described by an ensemble of all possible paths, which are not required to satisfy any equations, only the boundary conditions specified as the limits on the integral. By this definition, the path integral should be taken over all paths $x(t)$ with the ends fixed: they all start at point x at time $t = 0$ and end at point y at time t.

Note that the weighted phase factor for a path includes a phase related to its action $S[x(t)]$, not a Hamiltonian. For a particle of mass m moving in a static potential $V(x)$ this factor is, as defined in classical mechanics courses,

$$S = \int dt \left[\frac{m}{2}\left(\frac{dx}{dt}\right)^2 - V(x)\right], \tag{1.6}$$

with a minus sign for the potential.

The definition of the path integral itself is as follows. Let us split the time interval t into a large number N of small steps, each of a small duration $a = t/N$. The "discretized path" can be defined as the set of all intermediate positions of the particle, $\{x_i\}, i = 1, \ldots, N$.

One can evaluate the N-dimensional integrals using x_i as variables. The continuous limit of the path integral is naturally defined as the limit of the discretized path integral under the following condition

$$N \to \infty, \quad a \to 0, \quad t = aN = \text{fixed}.$$

Exercise *Derive expression (1.4) for a free particle by evaluating the path integral as N integrals over intermediate positions X_i. In case of difficulty, consult [2], where the evaluation is done in detail.*

In both of his books Feynman shows in detail how one can do path integrals for potentials V containing only linear and quadratic powers of the coordinate. In such cases all integrals over the intermediate coordinates x_i are Gaussian and can be done explicitly. Among such cases are the practically important harmonic oscillator and motion in a magnetic field.

Unfortunately, Gaussian path integrals are the only ones that allow analytic evaluation. Any non-Gaussian path integrals should in general only be calculated either perturbatively (as expansions near Gaussian ones) or numerically. We will discuss actual numerical evaluation of path integrals in the many-body setting in section 12.3.

1.4 Quantum Statistical Mechanics of a Particle

Applications of path integrals in their original formulation is difficult because of the interferences of oscillating weight factors. To get rid of such oscillations, we can analytically continue $G(x, y, t)$ into imaginary (also called Euclidean) time:

$$\tau = it. \tag{1.7}$$

In this case, the paths enter the sum with the new (and nonoscillating) weight $\exp(-S_E[x(\tau)])$, where the *Euclidean action* is defined as

$$S_E = \int dt \left[\frac{m}{2} \left(\frac{dx}{d\tau} \right)^2 + V(x) \right]. \tag{1.8}$$

Note that the relative sign between the kinetic and potential energies has now changed, and the expression looks like a Hamiltonian. This is equivalent to a motion in a flipped $V \to -V$ potential.

The highest weight in the resulting path integral has the "lazy path" $x(\tau) = x_{\text{min}}$, corresponding to the particle sitting at the minimum of $V(x)$ at all times. Because the potential can always be shifted, let us assume that $V(x_{\text{min}}) = 0$ and is positive ($V(x) > 0$) elsewhere. If so, our simplest (lazy) path obviously has zero action.

As emphasized by Feynman, the quantum particle is not just sitting at one location, the bottom of the potential, as a classical particle would do if put there. But the lazy path is just a single path, and the quantum particle is tempted to try many other paths!

The behavior of a quantum particle is a compromise between the *action* (which tries to enforce classical behavior, minimizing the action) and the number of possible paths, or their *entropy* (which counts these possible paths).

Let us now recall the definition of the amplitude, written as a sum over states, and rewrite it in Euclidean time:

$$S(x, y, \tau) = \sum_n \psi_n^*(y)\psi_n(x)e^{-E_n\tau}. \tag{1.9}$$

The oscillations are now changed by decreasing exponentials, so that at large values of τ, the integral is dominated by the lowest (ground) state.

Note that this amplitude satisfies the (analytically continued to τ) Schrödinger equation

$$-\frac{\partial S(x, y, \tau)}{\partial \tau} = \hat{H}_x S(x, y, \tau), \tag{1.10}$$

because both the l.h.s. and the r.h.s. simply add a factor E_n to the sum. Note also that as $\tau \to 0$ in (1.9), the exponent drops out, and the remaining sum becomes (by completeness of the stationary wave function set) just $\delta(x - y)$, which can be used as the initial condition for the equation.

Next we define the set of all periodic paths by putting $x = y$. For these paths the Euclidean time interval can be closed into a circle. Let us call its circumference β, $\tau \in [0, \beta]$ related to the temperature via[4]

$$\beta = \frac{\hbar}{T} \tag{1.11}$$

The path integral over the periodic paths thus defined is the quantity

$$\rho(x, T) = \sum |\psi_n(x)|^2 exp(-E_n/T), \tag{1.12}$$

which is nothing other than the probability of finding a quantum particle at point x in a heat bath with temperature T.

Furthermore, let us integrate over the initial (=final) value x, using the normalization condition of the states $\int dx|\psi_n(x)|^2 = 1$. What is obtained is in fact the basic quantity of statistical mechanics, the *statistical sum*

$$Z = \sum_n e^{-E_n/T} = \int Dx(\tau)e^{-S_E}, \tag{1.13}$$

for which we have now found a path integral definition.

Indeed, the r.h.s. is the path integral over all periodic paths, with the duration parameter $\tau \in [0, \beta]$ defined on the so-called Matsubara circle with the circumference β as defined in (1.11). This expression for Z in terms of the path integral opens many new application in quantum statistical mechanics. This fact has been realized in several key papers from the 1950s, such as [7–9].

---

[4] Unlike in the previous expressions, where $\hbar = 1$, here I remind the reader about the presence ot this term. In the classical limit of $\hbar \to 0$, the circumference $\beta \to 0$.

Example: (from [3]): The transition amplitude for a quantum particle in a harmonic oscillator potential

$$V = \frac{m^2 \Omega^2}{2} x^2 \tag{1.14}$$

has the following form:

$$G_{\text{osc}}(x, y, \tau) = \left(\frac{m\Omega}{2\pi\hbar\sinh(\Omega\tau)} \right)^{1/2} \exp\left[-\left(\frac{m\Omega}{2\hbar\sinh(\Omega\tau)} \right)((x^2 + y^2)\cosh(\Omega\tau) - 2xy) \right]. \tag{1.15}$$

Exercise *While it can be derived from multiple Gaussian integrals, show instead that equation (1.15) satisfies both equation (1.10) and the natural initial condition at $\tau \to 0$.*

Exercise *For $x = y$ the density matrix is the probability of finding the particle at x at temperature T. Show that the expression in the exponent in (1.15) for the harmonic oscillator is the action of a classical path for inverted potential $-V$, rolling toward the maximum $x = 0$, starting and ending at x and taking time $\beta = \hbar/T$. Such paths, called "fluctons," can be defined for other potentials and even QFTs. They are a basis for multidimensional semiclassical approximations. The reader can find more on this topic in [10].*

The diagonal elements of the amplitude S are the probability of finding a particle at point x: so setting $y = x$ and $\tau = \beta$, we find that the harmonic oscillator at *any* temperature has a Gaussian distribution of particles

$$P(x) = \sqrt{\frac{m\Omega}{2\pi\hbar\sinh(\hbar\Omega\beta)}} \exp\left(-\frac{x^2}{2\langle x^2 \rangle} \right), \tag{1.16}$$

where the (temperature-dependent) width is given by

$$\langle x^2 \rangle = \frac{1}{2m\Omega} \coth\left(\frac{\Omega}{2T} \right). \tag{1.17}$$

This expression, which we will meet again later in the book, has two important limits. At small $T \to 0$, the width corresponds to the quantum mechanical ground-state wavefunction $\psi_0(x)$ of the oscillator. In the opposite limit of high $T \to \infty$, it corresponds to the classical thermal result $\langle x^2 \rangle = \frac{T}{m\Omega^2}$. Let us rewrite this expression once again to elucidate its physical nature. Since for harmonic oscillators the total energy is just twice the potential energy, which is related to mean $\langle x^2 \rangle$, we also have an expression for the mean energy of the oscillator at temperature T. Check that it can be put into the following "physical" form:

$$\langle E \rangle = \Omega \left(\frac{1}{2} + \frac{1}{e^{\Omega/T} - 1} \right). \tag{1.18}$$

Now we can see the meaning of the two terms in parentheses: they are the energies corresponding to (T-independent) zero-point quantum oscillations (familiar from quantum mechanics courses) plus the energy of the thermal excitation (familiar from statistical mechanics courses). Note that we automatically get the correct Planck (or Bose) distribution from the transition amplitude in Euclidean time: we will see similar "miracles" many more times below.

Let me now outline the rather straightforward transition, from quantum mechanics to the QFT context, which we will need later. The path integral over $(1+0)$-dimensional time-space, over all particle paths $x(\tau)$, is directly generalized to the path integral over the *field histories* in $(1+d)$ time-space dimensions. So the functional integrals change their integration variable

$$x(\tau) \to \phi(\tau, \vec{x}), \tag{1.19}$$

preserving its physical meaning. The weight is still given by the Euclidean action of the corresponding QFT. Most importantly, the same Euclidean time setting on a thermal or Matsubara circle can still be used for the partition function.

Let us jump a bit ahead and outline three major applications of the Euclidean path integrals. They can be evaluated by

1. perturbative series,
2. semiclassical approximation, and
3. direct numerical methods.

In this book we will discuss (1) in significant detail, presentation of (2) will be mostly missed (see [10] if interested), and (3) will be covered for a few basic applications.

1.5 A Few Gaussian Integrals

Two basic one-dimensional Gaussian integrals (defined for $A > 0$) are:

$$I_1 = \int_{-\infty}^{\infty} dx\, e^{-Ax^2} = \sqrt{\frac{\pi}{A}} \tag{1.20}$$

$$I_2 = \int_{-\infty}^{\infty} dx\, e^{-Ax^2} x^2 = \frac{\sqrt{\pi}}{2} A^{-3/2}. \tag{1.21}$$

Note that the second can be obtained from the first by a derivative over A, and their ratio gives the r.m.s. average $\langle x^2 \rangle = 1/(2A)$.

Let us generalize the integral to several dimensions, with coordinates x_i, $i = 1, \ldots, N$. A generic integral has some quadratic form A_{ij}

$$I_3 = \int_{-\infty}^{\infty} \Pi_{n=1}^{n=N} dx^n e^{-\sum_{i,j} A_{ij} x^i x^j}. \tag{1.22}$$

Without loss of generality, the form can be taken to be symmetric and can be diagonalized. If all eigenvalues of A are positive, $\lambda_n > 0$, the integral is well defined. Since it simply factorizes, the result is the following product of I_1 integrals, so

$$I_3 = \frac{\pi^{N/2}}{\Pi_{n=1}^{n=N} \sqrt{\lambda_n}} = \frac{\pi^{N/2}}{\sqrt{\det A}}. \tag{1.23}$$

Using this result, we can easily get an expression to be used below for a *correlator* of two coordinates:

$$\langle x_i x_j \rangle = \frac{\int_{-\infty}^{\infty} \left(\Pi_{n=1}^{n=N} dx^n \right) x_i x_j e^{-\sum_{i,j} A_{ij} x^i x^j}}{\int_{-\infty}^{\infty} \left(\Pi_{n=1}^{n=N} dx^n \right) e^{-\sum_{i,j} A_{ij} x^i x^j}}. \tag{1.24}$$

In the diagonal basis of the form, $A_{nj} y_j = \lambda_n y_n$, this integral factorizes and is simply given by the ratio of two basic integrals:

$$\langle y_i y_j \rangle = \delta_{ij} \frac{1}{2\lambda_i} \tag{1.25}$$

In general, coordinates x_i can be defined as a superposition of the diagonal ones $x_i = \sum_j C_{ij} y_j$ with some coefficients C_{ij}. Therefore, a correlator of two general coordinates can be written as

$$\langle x_i x_{i'} \rangle = \frac{1}{2} \sum_{ii'j} \frac{C_{ij} C_{i'j}}{\lambda_j} = \frac{1}{2} \left(A^{-1} \right)_{ii'} \tag{1.26}$$

the matrix element of the matrix A^{-1} *inverse* to the original matrix.

This is the essence of the perturbative construction: The propagators representing correlations of two quantum paths/fields are nothing other than the inverse of the relevant quadratic form in the action. The Euclidean time helps to avoid unwanted zeros in this inversion.

1.6 Another Take on the Path Integral for the Harmonic Oscillator

As explained above, the path integral for the harmonic oscillator is defined using the (Euclidean) action

$$S = \int d\tau \left(\frac{m\dot{x}^2}{2} + \frac{m\Omega^2 x^2}{2} \right), \tag{1.27}$$

where $\dot{x} = dx/d\tau$. We can recognize a certain Gaussian form, since $S \sim x^2$. Feynman originally used the discretization $x(\tau) \to x(\tau_n = n\beta/N)$ and approximated the derivative by a difference

$$\dot{x}(\tau) \to \frac{x(\tau_{n+1}) - x(\tau_n)}{a},$$

in which case we have a nondiagonal matrix.

Let us now try another approach. The differential operator of kinetic energy is simple enough to be diagonalized. It acts as written on both paths, but using periodicity in time, we can integrate the kinetic term in parts, rewriting the relevant differential operator as the one acting on one of the paths $x(\tau)$:

$$\hat{A}_{\text{osc}} = -m\frac{d^2}{d\tau^2} + m\Omega^2, \tag{1.28}$$

so that now the exponent contains $\int d\tau x(\tau) \hat{A}_{\text{osc}} x(\tau)$. Note the minus sign of the second derivative, which comes from integration by parts.

This operator plays the role of the quadratic form in section 1.5. But unlike the matrix form there, we now have a differential operator. It acts in the Hilbert space of functions, which has infinitely many dimensions rather than a finite number of them. Yet the same approach can in fact be used: the matrix or the differential operator may and should be diagonalized!

The eigenfunctions and eigenvalues of this particular operator are in fact easy to find: they are Fourier harmonics

$$y_n = \exp(i\omega_n \tau), \quad \lambda_n = \frac{m}{2}(\omega_n^2 + \Omega^2), \tag{1.29}$$

which are discrete because of the periodicity in τ:

$$\omega_n = \frac{2\pi n}{\beta} = 2\pi T n. \tag{1.30}$$

Note that the integer n runs over all integers, from $-\infty$ to ∞, including zero. The decomposition of the paths, periodic in τ with the period $\beta = \hbar/T$, is thus just what is called the Fourier decomposition.

Following the same logic as in section 1.5 on Gaussian integrals, we arrive at the following expression for the correlator of coordinates at different times:

$$\langle x(\tau)x(\tau')\rangle = \frac{T}{m}\sum_n \frac{\exp[i\omega_n(\tau - \tau')]}{\omega_n^2 + \Omega^2}. \tag{1.31}$$

We can recognize the coefficients C_{ij} from equation (1.26) in the numerator, related the original and the diagonal coordinates, as the Fourier harmonics, with the inverse eigenvalues in the denominator. The only differences from the formula we had before is that the sum over n is now infinite and the harmonics are complex, so one of them appears complex conjugated.[5] However, these differences do not matter: what is important is that by acting with the operator on this expression, we get the eigenvalue in the numerator, canceling the denominator. The resulting sum is

$$\sum_n \exp[i\omega_n(\tau - \tau')] \sim \delta(\tau - \tau'), \tag{1.32}$$

because Fourier harmonics form a complete set on periodic functions. So the expression above *is* indeed the inverse of the differential operator.

This correlator plays a very important role in perturbation theory and has a special name: the *propagator*. Note that it is a function of the time difference, because the action has no explicit time dependence anywhere, so all time moments are equal. Thus we can set $\tau' = 0$ and call the propagator a function of one variable, $G(\tau)$. The calculation we just made,

$$m\left(-\frac{d^2}{d\tau^2} + \Omega^2\right)G(\tau) = T\sum_n \frac{-(i\omega_n)^2 + \Omega^2}{\omega_n^2 + \Omega^2}e^{i\omega_n\tau} = \delta(\tau), \tag{1.33}$$

shows that $G(\tau)$ is nothing other than the *Green function* of classical equations of motion. Indeed, the operator acting on one of the arguments of the Green function returns the delta function, or unit operator. The mathematical meaning of the Green function is just

---

[5] It is the same for the quadratic form if we had not assumed there that the C_{ij} are real.

the *inversion* of the corresponding differential operator, a direct generalization of the inverse matrix in section 1.5.

For clarity of the presentation, let me add that in this case the Fourier transform corresponds to the multidimensional Gaussian integral as follows. The functions we integrate over, the analog of the original coordinates x_n, are the set of all periodic functions $f(\tau)$. These can be decomposed into the other set, the diagonal coordinates y_n, with the coefficients being the Fourier coefficients $\tilde{f}_n$. The relation between the two sets is given by

$$\tilde{f}_n = \int d\tau f(\tau) e^{i \frac{2\pi}{\beta} n \tau} \tag{1.34}$$

with the inverse relation being

$$f(\tau) = \frac{1}{\beta} \sum_n \tilde{f}_n e^{-i \frac{2\pi}{\beta} n \tau}. \tag{1.35}$$

You should not be intimidated by the fact that one of the formulas has an integral and the other a sum (although an infinite one): both representations—in time or in frequencies—provide mathematically identical descriptions of all paths we need to sum over in the path integral.

At this point it is also useful to discuss the limit of large $\beta \to \infty$ or vanishingly small temperature $T \to 0$. Physically, this is the case when one wants to go from statistical mechanics back to quantum mechanics or QFTs at zero temperature. In this case the periodicity condition is no longer restrictive, and we can use Fourier transforms with *continuous* frequencies. The formulas for the direct and the inverse transforms in this case are

$$\tilde{f}(\omega) = \int d\tau f(\tau) e^{i\omega\tau}, \quad f(\tau) = \int \frac{d\omega}{2\pi} \tilde{f}(\omega) e^{-i\omega t}, \tag{1.36}$$

and the expression for the propagator would have an integral over continuous ω instead of the discrete sum over n.

Completing this somewhat lengthy discussion of the propagator $G(\tau)$ for a harmonic oscillator, let us find its explicit form. It can be done either by performing the summation in (1.31), which I leave as an exercise, or by solving the equation for the Green function.

Let me do it by using some educated guesses. The first is to realize that at the point $\tau = 0$ (same as $\tau = \beta$ on the circle), the function $G(\tau)$ must have a jump in the first derivative: The second derivative will then produce $\delta(\tau)$, as required by the equation. The second consideration is that, starting from the opposite point on the circle, $\tau_{\text{opposite}} = \beta/2$, the function should be smooth and symmetric in both directions. The third is that the solutions of the equations of motion without the r.h.s. are just exponents of the type $\exp(\pm\Omega\tau)$. Therefore the needed symmetric function must be $\cosh(\Omega(\tau - \beta/2))$ times some constant C. The constant can be found from the derivative jump required or from the fact that at $\tau = 0$ the function is just the average of the coordinate fluctuation squared $\langle x(0)^2 \rangle$ (see l.h.s of (1.31)), which we already know from (1.17). Thus the explicit form we are looking for is

$$G(\tau) = \frac{\cosh(\Omega(\tau - \beta/2))}{2m\Omega\sinh(\Omega\beta/2)}. \tag{1.37}$$

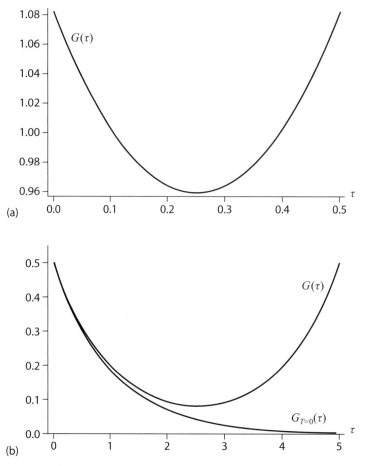

Figure 1.2. The correlator of coordinates $G(\tau)$ for harmonic oscillator, for $m = 1$ and two values of $\beta = \hbar/T = $ (a) $1/\Omega$, (b) $5/\Omega$. Panel (b) also shows the limit $T = 0$ (1.38).

For the small (zero) temperature T limit, the circumference of the circle goes to zero, and point $\tau = 0$ can be considered as the middle of the $[-\infty, \infty]$ line. In this case the propagator takes a very simple form:

$$G(T \to 0, \tau) = \frac{1}{2m\Omega} e^{-\Omega|\tau|}. \tag{1.38}$$

Note the modulus of the time in the exponent: this function is indeed singular at the origin and decreases in both directions, $\tau \to \pm\infty$.

It will be important later to get some feeling for which T values are "low" or "high": thus I have plotted the propagator $G(\tau)$ in Figure 1.2. Panel a for the case $T/\hbar\Omega = 1$ can in fact be considered a high T case, as $G(\tau)$ depends on τ weakly, deviating from 1 by only several percentage points. Panel b corresponds to the case $T/\hbar\Omega = 0.2$: it can be considered a low T case, as $G(\tau)$ follows the $T = 0$ line, falling significantly—by an order of magnitude—until the midpoint.

2 | Feynman Diagrams in Quantum/Statistical Mechanics

Feynman diagrams are taught as the main topic of any standard QFT course. However, a standard approach suffers from at least two problems. One is that this is done directly in Minkowski spacetime. As a result, to derive Feynman diagrams one needs to motivate and introduce the celebrated Feynman's $i\epsilon$ prescription—leading to a particular combination of advanced and retarded Green functions, known as the *Feynman propagator*. This step was necessary but historically rather painful for Feynman himself, and it is still hard to explain to newcomers trying to understand QFT. Students still suspect something is wrong with that approach and keep asking what would happen if one used instead the (classically motivated) advanced Green functions.

We will avoid this problem by defining the vacuum Feynman rules as the low-temperature ($T \to 0$) limit of the thermal theory. The thermal theory is naturally defined in Euclidean time, and so there are no zeros of the operators, and thus no special prescriptions (or related logical jumps) will be needed. In fact this approach is also historically accurate, following how Feynman originally got his idea to use his celebrated propagator in the first place. Only this Green function has good analytic continuation, from the Euclidean to the Minkowskian theory.

The second famous set of problematic QFT issues is the divergences of some diagrams and related *regularization* procedures, which are often confused with the *renormalization* of certain physical quantities, in higher order in perturbation theory. Our introduction of the Feynman diagrams, in the quantum/statistical mechanics setting and (1+0) dimensional setting, will clearly show that those issues are in fact unrelated. We will suffer *no* divergences and thus have no need for any regularization, and yet we will have the notion of mass and charge renormalization, order by order, in its full glory.

We deal first with the quantum mechanical problem of a weakly anharmonic oscillator, the simplest problem (the "hydrogen atom" of this book) in which the Feynman method can be fully developed.

In section 2.1 I start by reminding the reader how the problem is treated in standard quantum mechanical textbooks. Then, in section 2.2, Feynman diagrams for the same problem are introduced, and the two perturbative expansions are compared. Section 2.3 proceeds to show that there is no need to calculate the disconnected diagrams. Section 2.4 introduces the propagators in frequency representation and provides the first examples in which the summation of certain diagram sequences can be made.

2.1 Anharmonic Oscillators à la Quantum Mechanics Textbooks

Before introducing the Feynman diagrams, let us solve the same problem in the traditional way, following the methods from quantum mechanical textbooks. (Otherwise, the reader would not be able to appreciate dramatic improvements and simplifications that Feynman diagrams bring to our theoretical toolbox.)

The first thing physicists usually do is to select the appropriate units, in which the problem at hand looks as simple as possible. Three mechanical units—for time, length, and mass—can be selected at will. We define our units for the harmonic oscillator by setting three main parameters to be equal to 1:

$$m = 1, \quad \Omega = 1, \quad \hbar = \frac{h}{2\pi} = 1. \tag{2.1}$$

In such units, the oscillator Hamiltonian (as usual, the hats indicate the operators) and its spectrum are

$$\hat{H} = \frac{\hat{p}^2 + \hat{x}^2}{2}, \quad E_n = \frac{1}{2} + n, \quad n = 0, 1, ... \tag{2.2}$$

The anharmonic oscillator has some perturbations, usually written as powers of $\hat{x}$ with some coefficients. The textbook approach to anharmonic oscillators relies on the observation that, by introducing excitation and annihilation operators $\hat{a}^+, \hat{a}$ à la Heisenberg, one can calculate matrix elements of any power of $\hat{x}$ by directly multiplying the matrices. So, in the stationary diagonal basis $|n\rangle$, the anharmonic Hamiltonian can be viewed as an infinite matrix with simple nonzero matrix elements. Taking a large part of this matrix, one can diagonalize it numerically or analytically.

The coordinate operator in the harmonic basis has simple nonzero matrix elements next to the diagonal

$$\langle n|\hat{x}|n-1\rangle = \langle n-1|\hat{x}|n\rangle = \sqrt{\frac{n}{2}}. \tag{2.3}$$

Let the **anharmonic oscillator** we consider have the following two additional terms in the potential:

$$\Delta V(x) = g x^3 + \lambda x^4. \tag{2.4}$$

If the coefficients g, λ are considered to be small, we can calculate corrections via some version of the perturbation theory. We will start with the so-called old-fashioned perturbation theory and evaluate corrections to the ground state energy E_0 of the order

$O(\lambda)$ and $O(g^2)$. Any quantum mechanics textbook contains the appropriate expression for this:

$$\Delta E_0 = \langle 0|\Delta V|0 \rangle - \sum_{n>0} \frac{|\langle 0|\Delta V|n \rangle|^2}{E_n - E_0} + O(\Delta V^3). \tag{2.5}$$

Let us start with the cubic term. It is obvious from parity that $\langle 0|x^3|0 \rangle = 0$, and thus only the second term contributes. This by itself fixes the sign of the shift: it must be negative, as the numerator is a square and the denominator is always positive. In the sum over states excited by the operator $\hat{x}^3$ the highest state is obviously $|3 \rangle$, because the nondiagonal jumps appear up to three times. Another intermediate state contributing is $|1 \rangle$, since two jumps can be upward and one jump can be downward. Their respective contributions are

$$\Delta E_0 = -\frac{9}{8}g^2 - \frac{1}{4}g^2 = -\frac{11}{8}g^2. \tag{2.6}$$

Now we can similarly calculate the effects of the quartic perturbation. The first term is nonzero and positive, as $\langle 0|x^4|0 \rangle \neq 0$. The second term has contributions from two states, $n = 2, 4$. The result is[1]

$$\Delta E_0 = \frac{3}{4}\lambda - \left(\frac{9}{4} + \frac{3}{8}\right)\lambda^2 = \frac{3}{4}\lambda - \frac{21}{8}\lambda^2. \tag{2.7}$$

It all looks nice and elementary. Yet suppose we want to continue the calculation to higher orders and generate many terms of the perturbation series. In this case we find that the methods proposed in standard quantum mechanics texts are no longer of any use: none of them includes expressions for any higher orders! In fact, there is no closed form for the nth order corrections anywhere!

(An aside for the theoretically inclined reader: determining the properties of perturbative series for an anharmonic oscillator is a famous problem in mathematics. The series themselves have been generated for hundreds or more orders. Attempts to sum them up did not work, because they are divergent or asymptotic series with factorially growing coefficients. Furthermore, they lead to unphysical imaginary parts, which can only be canceled in the so-called trans-series. These include not only the powers of the couplings but also powers of two more functions, $\exp(\text{const}/g^2)$ and $\log(1/g)$, known collectively as the "nonperturbative terms." Unique formulation of the trans-series for path integrals remains an unsolved problem. Relations between perturbative and nonperturbative parts of the series is the subject of resurgence theory, which is also a current hot research topic in physics and mathematics.)

---

[1] The simplest way to get matrix elements of powers of $\hat{x}$ is to write a piece of it, on paper or in Mathematica, as a finite matrix and multiply it by itself the needed number of times.

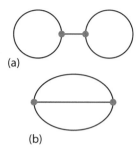

(a)

(b)

Figure 2.1. The diagrams of the order g^2: dots are triple vertices, connecting three propagators. Their popular names are, "the dumbbell" (a) and "the sunset" (b).

2.2 The First Feynman Diagrams

The Euclidean action for the anharmonic oscillator is defined by

$$S = S_{\mathrm{osc}} + \Delta S = \int_0^\beta d\tau \left[\frac{(\dot{x})^2 + x^2}{2} + g x^3 + \lambda x^4 \right], \tag{2.8}$$

where S_{osc}, as before, is the quadratic part.

The strategy is to expand the exponent term by term, in powers of ΔS:

$$Z = \int Dx e^{-S} = \int Dx e^{-S_{\mathrm{osc}}} \left[1 + \left(-\int d\tau \Delta S \right) + \frac{1}{2!} \left(-\int d\tau \Delta S \right)^2 + \cdots \right]. \tag{2.9}$$

Since the coefficient of the nth term comes from the expansion of an exponent, it simply is $1/n!$.

Then we evaluate the path integrals for each subsequent term. All are Gaussian integrals, with certain extra powers of coordinates in the numerator. So the main idea of the Feynman method is that there are some simple and general rules for such integrals. And indeed, just as for the multidimensional Gaussian integrals we discussed before, each of them can be expressed in terms of a certain number of the correlators of two coordinates, $\langle x_i x_j \rangle$. These correlators are given by the propagator, the inverse of the original quadratic form of the oscillator, and the Green function for the differential operator, which we already know. The rules to calculate them are the famous Feynman rules: one should draw all possible diagrams in which all vertices (stemming from the nonlinear terms) are connected by the propagators.

For the cubic perturbation, the vertex (the coupling g) is coupled to three outgoing lines, shown by the dots in Figure 2.1. The lines are the propagators (correlators or Green functions, which are just different names for the same entity). Denoting the times of the dots as τ_1, and τ_2, we write the following expressions for the first and the second diagrams

$$\text{"dumbbell"} = C_a g^2 \int d\tau_1 d\tau_2 G(\tau_1, \tau_1) G(\tau_1, \tau_2) G(\tau_2, \tau_2) \tag{2.10}$$

$$\text{"sunset"} = C_b g^2 \int d\tau_1 d\tau_2 \left(G(\tau_1, \tau_2) \right)^3, \tag{2.11}$$

where we have included the combinatorial coefficients C_i in front. These coefficients are obtained from the multiplicity of pairings of the coordinates that lead to this particular diagram.

For the dumbbell diagram (Fig. 2.1a), we can select one out of three x values at each vertex, which go into a line connecting the two vertices in 3×3 ways, times $1/2$ from the expansion of the exponent to second order, which gives $C_a = 9/2$. In the sunset diagram (Fig. 2.1b), the first $x(\tau_1)$ can be paired with $x(\tau_2)$ in three ways, the second in two, and the third has to be content with the only partner left. The same $1/2$ from the second-order expansion is there as well, so $C_b = 3!/2 = 3$.

Let us start with the zero temperature limit first, just for simplicity. In this case Euclidean time is defined from $-\infty$ to ∞, and the Green function is simply $G_0 = \exp(-|\tau_1 - \tau_2|)/2$, as derived before. Therefore a "bubble" of the dumbbell diagram, where both ends have the same times, is just a number $G(\tau, \tau) = 1/2$.

The double time integrals can be rewritten as the integral over the time difference $\tau_1 - \tau_2$ and the absolute (mean) time $(\tau_1 + \tau_2)/2$: the former is convergent, while the second is not, since there is no dependence on the absolute time anywhere. Remember that T is small but not really zero, so we recall from (1.11) that the circumference of our time circle

$$\int_{S^1} d\left(\frac{\tau_1 + \tau_2}{2}\right) = \beta. \tag{2.12}$$

In relative time we can put one of the times to zero, say, $\tau_1 = 0$, and calculate those elementary integrals (not forgetting that τ_2 can be both negative and positive). The results, including combinatorial factors, for the two diagrams separately and in total are

$$\delta Z / Z_0 = 1 + \left(\frac{9}{8} + \frac{1}{4}\right) g^2 \beta = 1 + \frac{11}{8} g^2 \beta. \tag{2.13}$$

Both contributions are from the second-order expansion in perturbation ΔS, and thus both are obviously positive.

In contrast, for low $T \to 0$ the excitations must be negligible, and the partition function must be dominated by the ground-state energy, $Z \sim \exp(-\beta E_0)$. Putting corrections to Z in the exponent, we can interpret both corrections found for the partition function as a shift of the ground-state energy by

$$\Delta E_0 = -\frac{11}{8} g^2. \tag{2.14}$$

Note now that not only is the total answer to second order the same as in the old-fashioned quantum mechanics calculation we made above, but there is also a term-by-term correspondence: the dumbbell diagram with one intermediate line corresponds to the contribution of the $n = 1$ intermediate state, and that from the sunset, with three lines, to the contribution of the $n = 3$ state.

The same logic applies to the quartic $O(\lambda)$ perturbation terms, with diagrams shown in Figure 2.2. The first double-bubble diagram gives the $O(\lambda)$ contribution, and two others, with two vertices shown by dots, $O(\lambda^2)$. The combinatorial factor in Figure 2.3a is $C_a = 3$, because the first x can be paired to any of the three remaining ones. $C_b = 6 \times 6 \times 2 \times (1/2)$

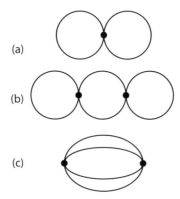

Figure 2.2. The diagrams of the order (a) λ and (b, c) λ^2.

and $C_c = 4!(1/2)$, where the factors of $1/2$ are from exponent expansion, as before. Calculating the relevant integrals for Figure 2.3b,c with $\int d\tau G^2(\tau)$ and $\int d\tau G^4(\tau)$, we find

$$\frac{Z}{Z_0} = 1 - \frac{3}{4}\lambda\beta + \left(\frac{9}{4} + \frac{3}{8}\right)\lambda^2\beta. \tag{2.15}$$

Note now that the exponentiated result again agrees with the ground state shifts found in the old-fashioned way. Furthermore, the b and c diagrams correspond to the contributions of the $n = 2, 4$ excited states, respectively.

Let us summarize what we have accomplished so far. We calculated the lowest-order corrections to the ground-state energy of the anharmonic oscillator in two ways, the old-fashioned one, and using Feynman diagrams. The results agree, even term by term. Yet I am going to claim that Feynman perturbation theory is better. Why?

The **first answer** comes from the fact that we can use the same expressions in statistical mechanics at finite T. All we need to do is to substitute the small-T Green functions we used above by their finite-T version (1.37). The standard textbook formulas for the partition function would require evaluating corrections to all levels separately, which is much more complicated.

Exercise *Calculate the lower-order corrections to the free energy of an anharmonic oscillator at finite temperature from the same diagrams (Figures 2.2 and 2.3).*

Recall that the bubble, $G(\tau, \tau) = \cosh(\beta/2)/2\sinh(\beta/2) = 1/2 + n_T$, is now the thermal average of the number of quanta from statistical mechanics plus $1/2$ from the ground state in quantum mechanics. Work out both the low- and high-T limits of the result. Check that the former agrees with the calculation we just did. Interpret each term in the high-T limit.

The **second and the main answer** is that it is completely straightforward to continue calculation of corrections to higher orders. Let me put it in the form of the following exercise.

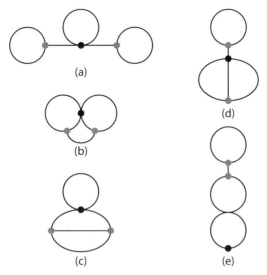

Figure 2.3. The diagrams of the order λg^2.

Exercise *Calculate corrections to the ground-state energy of the order $\sim g^2\lambda$ given by the five diagrams shown in Figure 2.3. The first one (panel a) corresponds to the following analytic expression:*

$$g^2\lambda C_a [G(0,0)]^3 \left[\int d\tau G(\tau,0) \right]^2 = g^2\lambda \times \left(3 \times 3 \times 4 \times 3 \times \frac{1}{2} \right) \times \frac{1}{8}. \tag{2.16}$$

Similar expressions correspond to the other diagrams. The contributions (diagrams a–e) are

$$g^2\lambda \left(\frac{27}{4} + \frac{5}{24} + \frac{3}{2} + \frac{3}{2} + \frac{27}{16} \right) = g^2\lambda \frac{559}{48}. \tag{2.17}$$

2.3 Disconnected Diagrams

Disconnected diagrams are those which can be split into two or more independent parts. For example, the first-order double-bubble diagram $O(\lambda)$ in Figure 2.2a appears twice in the next order $O(\lambda^2)$.

We have not included it in the calculation of the ground-energy shift: this omission was not a mistake but the consequence of a generic feature we will now explore. Since there are no relations between the parts of the disconnected diagrams, their contributions include multiple repetitions of the same diagrams. These, in general, can be summed up. In fact the result is direct exponentiation of the correction, since

$$1 + d + \frac{d^2}{2!} + \cdots = e^d. \tag{2.18}$$

Note that although we recalculated above a correction to Z, what we really want is a correction to E_0/T, (or at finite T, to $\log(Z)$ or to the free energy $-F(T)/T$). So all terms

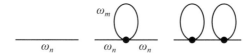

Figure 2.4. The sequence of diagrams producing corrections to the propagator.

that just exponentiate in Z appear linearly in the ground-state energy (or F). A practical conclusion from this consideration is very simple: we should ignore all disconnected diagrams altogether.

2.4 Diagrams in ω Representation and Their Resummation

In section 1.5, when we derived the Green function, we obtained its decomposition into eigenmodes of the quadratic part of the action. At finite T the expression is a sum over the Fourier harmonics, or the so-called Matsubara frequencies:

$$G(\tau_1, \tau_2) = \frac{T}{m} \sum_n \frac{e^{i\omega_n(\tau_1 - \tau_2)}}{\Omega^2 + \omega_n^2}. \tag{2.19}$$

In the zero-T limit the sum becomes an integral over the continuous ω variable, as we have already discussed.

Suppose we use this expression for the propagator and put it into the diagrams. It is easy to see what then happens with all the time integrals. Each time corresponds to a specific vertex, which has, say, k lines. The time dependence is only in the exponents of the propagators, and the integral over the exponent is just the (discrete in this case) delta function:

$$\int d\tau e^{i\tau \sum_{i=1}^{i=k} \omega_i} = \beta \delta_{0,\Sigma}, \quad \Sigma = \sum_{i=1}^{i=k} \omega_i. \tag{2.20}$$

The meaning of this result is that an *energy conservation* condition holds: at each vertex, the frequency sum must vanish ($\Sigma = 0$).

As an example of how this representation helps, consider the sequence of diagrams shown in Figure 2.4, giving corrections to the propagator $\langle x(\tau_1)x(\tau_2)\rangle$. Suppose we single out a particular frequency ω_n at one end. It is clear that each bubble has its own frequency ω_m, which comes and goes, but the energy conservation forces all intermediate propagators to have the same frequency ω_n as the original one. Therefore, the only sums to be done reside in the bubbles. The corresponding value for the bubble we already know well from (1.17), because it is just the mean square quantum/thermal fluctuation of the coordinate in the harmonic oscillator:

$$bubble = \lambda T \sum_m \frac{1}{\Omega^2 + (2\pi Tm)^2} = G(\tau, \tau) = \langle x^2 \rangle. \tag{2.21}$$

The Matsubara sum simply provides another way to calculate it.

Some further thinking about combinatorial weights will convince you that the sum of the terms for the sequence of diagrams looks like

$$\frac{1}{\Omega^2 + (2\pi Tn)^2} + \left(\frac{1}{\Omega^2 + (2\pi Tn)^2}\right)(bubble)\left(\frac{1}{\Omega^2 + (2\pi Tn)^2}\right)$$

$$+ \left(\frac{1}{\Omega^2 + (2\pi Tn)^2}\right)(bubble)\left(\frac{1}{\Omega^2 + (2\pi Tn)^2}\right)(bubble)\left(\frac{1}{\Omega^2 + (2\pi Tn)^2}\right) + \cdots.$$

Since this sum is just a geometric series, it can be directly summed up:

$$\frac{1}{\Omega^2 + (2\pi Tn)^2} \frac{1}{1 - (bubble)\left(\dfrac{1}{\Omega^2 + (2\pi Tn)^2}\right)} = \frac{1}{\Omega^2 + (2\pi Tn)^2 - (bubble)} \qquad (2.22)$$

The physical interpretation of this sum, in the spirit of quantum field theories, is that the bubble represents the lowest-order *renormalization* of the classical oscillator frequency Ω by quantum/thermal oscillations. It is our first example of the "mass operator."

Let me emphasize the lesson here: renormalization of some physical constants, order by order, has nothing to do with the divergences of QFT diagrams. In fact, none of the diagrams in quantum and statistical mechanics applications will ever be divergent, and yet their resummation is interpreted as renormalization.

Let us slightly generalize this result to a sequence of diagrams, in which one line connects an iteration of an arbitrarily complicated "mass diagram" M, instead of a simple bubble one. The initial line we will call the bare propagator G_0, and the sum is the "dressed" propagator $\mathbf{G}$. The resummation statement is

$$G = G_0 + G_0 M G_0 + G_0 M G_0 M G_0 + \cdots = \frac{1}{G_0^{-1} - M}, \qquad (2.23)$$

based solely on the geometric series.

3 | Real Scalar Fields and the Renormalization Group

In this chapter we generalize what we learned in the $(1+0)$-dimensional case to fields defined in $D = (1 + d)$ spacetime dimensions. As we will see in Section 3.1, this generalization is natural and very straightforward. Furthermore, note that in the special case of zero temperature, $T \to 0$, $\beta \to \infty$, the Euclidean formulation treats time and space coordinates in an identical way.

However, for a nonzero temperature there is always a difference, because, unlike space, the Euclidean time is defined on a circle of finite circumference β. Thus the "paths" (or rather, field histories) are always periodic with period β, and their Fourier transforms have discrete Matsubara frequencies $2\pi n/\beta$ with integer n. Basically, the only modification of the QFT Feynman rules for nonzero temperature would be the use of Matsubara sums instead of the frequency integrals.

We discuss how to calculate them in Section 3.2. The procedure will be based on the Sommerfeld-Watson trick for the sums, rewriting them as certain integrals over carefully defined lines in the complex plane. The trick naturally leads to the expression containing three terms, interpreted as the contributions of the *positive energy* on-shell particles, *negative energy* on-shell particles (or *holes*), and the *virtual (off-shell) quanta*. The first two contributions come from the "heat bath". They depend on T, and in the zero temperature limit $T \to 0$ disappear, leaving only the last one. After a contour rotation this last contribution coinsides with what the original Feynman rules from QFT courses prescribe.

We calculate directly some of the simplest diagrams for weakly anharmonic scalar field, and then, starting in Section 3.4, we discuss application of the renormalization group method for evaluation of the critical indices for $d = 3$ dimensional systems with second-order phase transitions. I introduce the crucial notion of the *order parameter* and the Landau theory of symmetry breaking in Section 3.4. This mean-field theory will be further supplemented in Section 3.5 by the notions of *critical indices*, lattice discretization of the theory, and *Kadanoff scale doubling*. The limit of the doubling will result in a nontrivial limit known as the *fixed point* of the renormalization group. From it we can derive the values of the critical indices using *Wilson's epsilon expansion* (Section 3.6).

3.1 Path Integrals for a Scalar Field

We are now ready to make our first generalization, moving from the playground we used in Chapter 2 (quantum-mechanical examples or $(1+0)$-dimensional paths) to the first QFT example in $(1+3)$ dimensions. Let us start with the simplest case: a free (noninteracting) real scalar field.

Its real time (Minkowskian) action has the form

$$S_M = \int d^4x \left[\frac{1}{2} \left(\frac{\partial \phi}{\partial t} \right)^2 - \frac{1}{2} \left(\frac{\partial \phi}{\partial \vec{x}} \right)^2 - \frac{m^2}{2} (\phi)^2 \right]. \tag{3.1}$$

With some abuse of the notation, the first two terms with field derivatives are called the kinetic term. Note that this combination of derivatives, corresponding to $p^\mu p^\nu g_{\mu\nu}^{\text{Minkowski}}$ with the relativistic Minkowski metric, is Lorentz invariant. Since d^4x is Lorentz invariant as well, the whole action—as it should for relativistically invariant theory—is a Lorentz scalar, the same in all frames. The last term is called the mass term: below we use the units in which $m = 1$.

Shifting to the Euclidean time $it \to \tau$, we rewrite the action as

$$S_E = \int_0^\beta d\tau \int d^3x \left[\frac{1}{2} \left(\frac{\partial \phi}{\partial \tau} \right)^2 + \frac{1}{2} \left(\frac{\partial \phi}{\partial \vec{x}} \right)^2 + \frac{1}{2} (\phi)^2 \right]. \tag{3.2}$$

Note that instead of flipping the sign of the time derivative term, I did so for all other terms instead. The Lorentz invariance now becomes the rotational invariance in D dimensions, which is, however, broken at finite T, because the time τ, as before, is defined on a circle of circumference $\tau \in [0, \beta = \hbar/T]$.

Standard QFT courses explain that a noninteracting field can be easily quantized. Indeed, using the spatial Fourier transform $\phi(\tau, \vec{x}) \to \tilde{\phi}(\tau, \vec{k})$, we can rewrite this action as a sum of independent harmonic oscillators:

$$S_E = \int_0^\beta d\tau \frac{d^3k \, V}{(2\pi)^3} \left[\frac{1}{2} \left(\frac{\partial \tilde{\phi}_k}{\partial \tau} \right)^2 + \frac{1}{2} (k^2 + m^2) \tilde{\phi}_{\vec{k}} \tilde{\phi}_{-\vec{k}} \right] \tag{3.3}$$

with frequencies

$$\omega_k = \sqrt{m^2 + k^2} \tag{3.4}$$

corresponding to the relativistic energies of the field quanta with momenum $\vec{k}$.

Our extensive study of the Gaussian integrals and harmonic oscillators in Chapter 2 led us to write down the propagator (Green function) for this theory. It is again a standard sum over all eigenmodes of the corresponding differential operator, namely, plane waves.

The only new element here is the sum over momenta:

$$\langle \phi(\tau, \vec{x}) \phi(\tau', \vec{x}') \rangle = G(\tau - \tau', \vec{x} - \vec{x}') = T \sum_n \int \frac{d^3 k \, V}{(2\pi)^3} e^{i\omega_n(\tau-\tau') + i\vec{k}(\vec{x}-\vec{x}')} \frac{1}{\omega_k^2 + \omega_n^2}. \tag{3.5}$$

Indeed, the exponents are eigenfunction of the operator in question, the denominator is the corresponding eigenvalue, and the sum over n includes all Matsubara frequencies $\omega_n = 2\pi T n$, as for a single oscillator. Note that the normalization is that of the sum over momenta, to be done in a standard normalization box with the 3-volume V, corresponding to a sum over all standing wave states in that box.

If $T \to 0$, the Matsubara sum can be approximated by the integral over continuous ω: the result in this case is the so-called massive Euclidean propagator for a particle of mass m. By taking the integral, we find its explicit form

$$G(\tau, \vec{x}) = \frac{m}{4\pi^2 \sqrt{x^2 + \tau^2}} K_1(m\sqrt{x^2 + \tau^2}) \tag{3.6}$$

in terms of the Bessel function K_1. As expected, the answer depends only on the four-dimensionl distance.

After the explicit function form for the propagator is obtained, we may try to analytically continue it into the real-time Minkowski world by simply flipping the sign of $\tau^2 \to -t^2$. If the distance is spacelike, $x^2 - t^2 > 0$, nothing really changes. But if it is timelike, $x^2 - t^2 < 0$, the square root gives an imaginary unit i. The Bessel functions of imaginary arguments can be rewritten as oscillating Bessel functions.

A historically important puzzle appears at this point. What we calculated is the correlator of the fields at two spacetime locations. The result we found—known as the Feynman propagator—turns out to be nonzero even *outside the light cone*. However, classical fields, e.g. in electrodynamics uses different Green functions, which follow the causality argument literally and vanish outside the future light cone. This mismatch provoked a lot of thinking on Feynman's part and remains a notorious stumbling block in teaching QFT. The Feynman propagator in QFT is not the usual retarded Green function of the classical equations of motion; this fact stems from the basic idea that quantum particles and fields take all paths, not only the casual ones inside the future light cone. This conclusion looks strange, but it is true, and we have to live with it!

More specifically, in the real-time formulation, the denominator of the Green function (3.5) is $m^2 + k^2 - \omega^2$. Since it contains a singularity on the integration contour for energy ω, its proper definition requires a rule defining the integral uniquely. Historically, Feynman came up with a certain $i\epsilon$ prescription, which is nothing but the analytic continuation from the Euclidean formulation we gave above. Since we have used the Euclidean formulation from the beginning, Feynman's prescription appears naturally. So in this book, this formulation is unproblematic. Finally, some practical advice: calculate as much as possible in the Euclidean formulation and carry out analytic continuations only when needed and only at the end of the calculation.

Exercise *Derive the zero-mass limit of the propagator from the Matsubara momentum sum, and compare it to the limit of the result given above.*

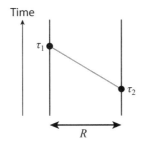

Figure 3.1. The setting for the calculation of the interaction potential induced by a scalar exchange. The propagator is shown by the intermediate line.

To gain some confidence that we are on the right track, let us consider the following example problem. Consider two point sources coupled to the massless ϕ field and placed a distance R from each other (see Figure 3.1). A diagram connecting two points on their worldlines at times τ_1 and τ_2 produces the following expression:

$$\int \frac{d\tau_1 d\tau_2}{R^2 + (\tau_1 - \tau_2)^2} = \frac{\pi t_{\text{total}}}{R}. \tag{3.7}$$

The denominator of the l.h.s. integral is the massless version of the propagator, which the reader should have obtained from the previous exercise. The total time t_{total} appears here, since the propagator depends only on the relative time. At small but nonzero temperature, $t_{\text{total}} = \beta = 1/T$.

The exponentiated set of such diagrams (recall the discussion of disconnected diagrams in Section 2.3) gives $Z \sim \exp(-V(R)/T)$, and from the exponent we can read off the effective potential induced by the virtual field exchange:

$$V \sim -1/R.$$

What we have just derived is the statement that an exchange of the massless scalar between two static sources leads to a Coulomb-like attractive potential.

Exercise *Using the nonzero-mass propagator at zero temperature, calculate the form of the potential induced by the exchange of a massive scalar quantum.*[1]

These calculations of the propagator, and the forces it induces, can be extended to $T \neq 0$. Let us do the Matsubara sum first. In particular, consider the bubble diagram—a loop in Green function, which is supposed to give us a correlation of two fields at *the same spacetime point* $\tau = 0$, $\vec{x} = 0$, or $\langle \phi^2 \rangle$. We find the already familiar expression (1.18)

$$G(0,0) = \int \frac{d^3 k\, V}{(2\pi)^3} \frac{1}{\omega_k} \left(\frac{1}{2} + \frac{1}{\exp(\omega_k/T) - 1} \right), \tag{3.8}$$

now supplemented by a new element, the momentum integral. The first (T-independent) term inside the parentheses produces an obviously divergent integral at large momenta

---

[1] This exercise is important: an exchange by a massive scalar called the σ-meson between two nucleons is the attractive part of the nuclear force, responsible for the existence of nuclei, thus atoms, and thus ourselves.

$k \to \infty$. The second (T-dependent) term is convergent because of the thermal cutoff at large k.

This observation indicates a general trend: the UV divergences that we will later find will always be T independent. It is my opinion that their removal—the regularization of the theory—belongs to basic QFT courses and is not considered in this book, since the medium is not involved.

So, we will only pay attention to the T-dependent heat bath or "matter" part of the results, rather than the vacuum problems. In the particular case of the bubble, for the massless field it is just the obvious power of T times a constant, which is readily found to be

$$G(0, 0) = \frac{T^2 V}{12}. \tag{3.9}$$

This is our first result for a relativistic finite-T field theory.

3.2 Matsubara Sums Lead to the Particle/Hole Interpretation of the Diagrams

Before we proceed to interacting fields and actual diagrams, let us work out some general method for evaluation of the Matsubara sums. The idea is based on the so-called Sommerfeld-Watson method in mathematics. Its first step is to express the sum in question as a contour integral

$$S = \sum_n F(n) = -i \int_C dz F(-iz) \frac{1}{e^{2\pi z} - 1}, \tag{3.10}$$

where the contour C surrounds the imaginary axis of the complex z variable. Note that the second factor introduced by hand is a certain function that has poles at all integer points on the imaginary axis

$$z_n = i \times n,$$

and that all of them have the unit residuals. The equality of the sum and the integral is proven by using Cauchy's theorem and including the contributions from those poles, one by one. For example, for $n = 0$ the last factor is $\approx 1/(2\pi z)$ at small z and the integral a over a small circle around zero gives F(0), etc.

Step two of the method includes motion of the right-hand-side and left-hand-side parts of the C contour away from each other, both following the direction along the imaginary axis (see Figure 3.2). While moving them to the right and to the left, respectively, to $z \to \pm\infty$, we can again use Cauchy's theorem and pick up singularities, now of the function $F(-iz)$ itself.

The *right-hand* side of C is simpler to deal with, because the function in the second parentheses decays exponentially to the right $\sim \exp(-2\pi \text{Re}(z))$: thus the distant $z \to \infty$ part of large contour contributes nothing. The only contributions come from singularities of the function F itself, located in the right halfplane. In particular, poles of F will pick

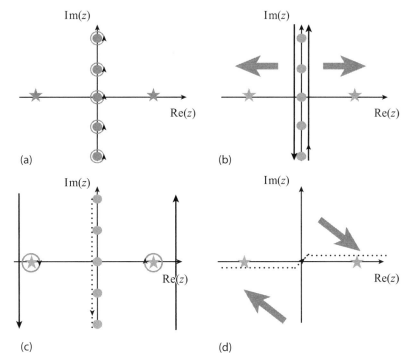

Figure 3.2. Moving the Sommerfeld-Watson contours in the complex z plane. (a) The series of circles around poles at $z = in$ with all integer $n = -\infty, \infty$. (b) These circles are replaced by two lines enclosing the imaginary axis, which are then moved left and right, as indicated by large arrows. (c) Moving these contours leaves small circles around the singularities of the function F, shown by two stars. The dotted line indicates the remaining integral of $F(z)$ without the Sommerfeld-Watson function. (d) This line can then be rotated to the real axis as shown, explaining the Feynman $i\varepsilon$ prescription for going around the singularities on the real axis.

up the following contributions:

$$S_{\text{right}} = \sum_{\text{Re}(z_i) > 0} \frac{2\pi \text{Res}(F(z_i))}{e^{2\pi z_i} - 1}. \tag{3.11}$$

Here Res stands for the residuals of the poles of the function F.

The left-hand side of C cannot be treated in the same fashion, because for $z \to -\infty$ the exponent decreases to the left and does not provide a cutoff. Fortunately, we can get around this difficulty by using the following identity:

$$\frac{1}{e^{2\pi z} - 1} = -1 - \frac{1}{e^{-2\pi z} - 1}.$$

So the left-hand contribution can be split into (1) just an integral over the imaginary axis, with a flipped direction and without any Sommerfeld-Watson function, and (2) one with a modified exponent that is now increasing to $z \to -\infty$ and thus can be dropped near $-\infty$.

Thus the left-hand integral produces a contribution

$$S_{\text{left}} = \sum_{\text{Re}(z_i)<0} \frac{2\pi \text{Res}(F(z_i))}{e^{-2\pi z_i} - 1} + \int_{-i\infty}^{i\infty} dz F(-iz), \tag{3.12}$$

which differs from equation (3.11) in the sign of the exponent, plus a plain integral over the imaginary axis stemming from the -1 term in the identity above. Note that the left side of C was going downward along the imaginary axis: by flipping its direction, we now have an integral from $-i\infty$ to $i\infty$.

These three terms each have a specific physical meaning: the term S_{right} corresponds to contributions of all the positive-energy particles belonging to the heat bath: thus it has a thermal Bose distribution. The term S_{left} similarly accounts for negative energy states or antiparticles. The remaining integral in (3.12), without a distribution function, accounts for *virtual particles*, the only ones appearing in the vacuum of a QFT.

Note that three such terms appear for each summation over the Matsubara frequencies. The number of summations is, as is easy to see, the number of loops in the diagram, N_{loops}. So this is the number of thermodynamical weights—Fermi or Bose— that appear in the diagrams.

Note further that usually there are more particle lines in the diagram than the number of loops, $N_{\text{lines}} > N_{\text{loops}}$. Therefore, many lines appear without thermal weights. This seems puzzling. Yet it is actually true— the physical meaning of it will become clear a bit later, as we proceed.

Example: Consider the function corresponding to the Green function (propagator), namely,

$$F(n) = \frac{1}{(2\pi Tn)^2 + k^2 + m^2}. \tag{3.13}$$

Changing $n \to -iz$ changes sign of the first term in the denominator and determines two poles on the real axis at the following locations:

$$z_{\pm} = \pm \frac{\sqrt{k^2 + m^2}}{2\pi T}. \tag{3.14}$$

Plugging these poles into Sommerfeld-Watson functions, at the right and the left, we find the same result, namely, a Bose statistics occupancy factor, for a particle and antiparticle, with energies given by $\omega_k = \sqrt{k^2 + m^2}$. The residual of the pole is $1/(2\omega_k)$, which is to be added into the momentum integral:

$$\int \frac{d^3k}{(2\pi)^3} \frac{1}{\omega_k} \frac{1}{\exp(\omega_k/T) - 1}.$$

The remaining integral over the imaginary axes $\int dz F(-iz)$ can be rotated clockwise to the real axis, provided no singularities are crossed. If these (as in this example) are on the real axis, then they should be passed from above if they are in the right semi-plane $\text{Re}(z_i) > 0$ and *from below* if they are in the left semi-plane (see Figure 3.2).

Returning to the historical remark made after (3.6), I reiterate: this clockwise rotation is the origin of the famous Feynman $i\epsilon$ prescription for propagators, which creates difficulties for any QFT student trying to understand them. The explanation of why the contours should be placed differently in the right and left halfplanes has just been explained by the rotation of the countour from the imaginary to the real axis.

Exercise *Check that for the propagator, the direct summation over Matsubara n and the Sommerfeld-Watson method lead to the same answer for a finite-T propagator.*

3.3 Weakly Interacting Scalar Field

Let us now introduce nonlinear terms into the action, for example, a quartic one,

$$\lambda \phi^4,$$

as we did for the anharmonic oscillator. All the diagrams we did for anharmonic oscillator in Chapter 2 are still the same. The only new element is that the Matsubara sums are for fields complemented by the integrals over momenta.

Example: The double-bubble lowest-order diagram (see Figure 2.2a) provides the lowest-order correction to the free energy. The evaluation of it is very simple, since the bubble has already been calculated. The resulting correction to the free energy $F = -T\log(Z)$ is

$$\Delta F = 3V\lambda \left(G(0,0)\right)^2 = \lambda \frac{3}{12^2} VT^4 \tag{3.15}$$

(where the factor 3 is the combinatorial factor we discussed before in Section 2.2).

Example: A bubble sitting on a line, or the mass operator of the lowest order. Note that it has a combinatorial coefficient of 12, which conveniently cancels the numeric factor in the bubble. Thus, resumming the geometric series of such diagrams, one gets a propagator dressed by bubbles, which includes a dynamically generated mass squared of magnitude

$$\Delta m^2 = \lambda T^2. \tag{3.16}$$

Example: The so-called ring or sunflower diagram, a circle with N bubbles attached to it (see Figure 3.3). Since it is a closed-line diagram, it contributes to total (free) energy. Its additional combinatorial factor can be calculated as follows. Mark one of the $-$ 'N' bubbles as the first one. The second is taken to be $N-1$, and so forth, so the combinatorial factor is

$$C_{\text{ring}} = \frac{(N-1)(N-2)\cdots}{N!} \frac{1}{=} \frac{1}{N}. \tag{3.17}$$

Figure 3.3. An example of the ring diagram of the eighth order, with $N = 8$ bubbles.

Thus their sum reads

$$Z_{\text{rings}} = T \sum_{n=\infty}^{\infty} \int \frac{d^3 k V}{(2\pi)^3} \sum_{N=2}^{\infty} \frac{1}{N} \left(\frac{3\lambda T^2}{k^2 + \omega_n^2} \right)^N. \tag{3.18}$$

Since one ($n = 0$) among the Matsubara frequencies is zero, we see that the individual diagrams have arbitrarily large powers of $1/k^2$, which implies that there are serious troubles in the infrared (small k) region of the momentum integral. Indeed, the r.h.s. of equation (3.18) contains arbitrarily large power $(1/k^2)^N$, and therefore most of the momentum integrals are power divergent at small k. This looks like serious trouble. Readers who have taken introductory QFT courses know that in this context, the divergent diagrams are usually thrown out as being unphysical.

However, it is not the case now, because the sum of all these diagrams is both finite and physical. To see this, let us do the sum over N first. The resulting expression is

$$\int \frac{d^3 k V}{(2\pi)^3} \left[-\log \left(1 - \frac{3\lambda T^2}{k^2} \right) - \frac{3\lambda T^2}{k^2} \right], \tag{3.19}$$

and we find that, in fact, this integral over k is nicely convergent at small k. This looks like a little miracle: *the sum of divergent ring diagrams is in fact IR finite!*

Furthermore, since in the integral the most relevant momenta scale as $k^2 \sim \lambda T^2$, the answer is a certain number (the dimensionless integral) times $(\lambda T^2)^{3/2}$. Note another little miracle: summing the series in integer powers of λ we find its noninteger power $3/2$! The lesson we learn from this example is that the IR divergences can be regulated by resummation of the diagrams. The appearance of divergent diagrams does *not* necessarily mean that the effect is large or unphysical: it may just signal that the result is not expandable into a good series with integer powers of the coupling, and yet it is still finite and even small for small coupling. In the case under consideration, we just have noninteger powers of the coupling λ, and it is obviously impossible to get such results from any particular diagram by itself.

Exercise *Evaluate the contribution to the free energy by "cactuses," which are ring diagrams with bubbles, which are also dressed by bubbles.*

3.4 Symmetry Breaking and the Landau Theory of Second-Order Phase Transitions

The major application for real scalar fields is in the context of the second-order phase transitions. We start by discussing these transitions in the framework of *Landau theory*. The physical origin of the field is not discussed; it is enough to state that it is some scalar field that can serve as the order parameter of the transition. This means that the field has zero expectation value $\langle \phi \rangle = 0$ in one phase and a nonzero vacuum expectation value (VEV) in another.

Landau proposed that the generic explanation for this phenomenon is that the potential for the field should be of the form

$$V = m^2 \phi^2 / 2 + \lambda \phi^4 + O(\phi^6). \tag{3.20}$$

Note that the equation includes only the even powers of the field: this is done to preserve the dynamical symmetry $\phi \rightarrow -\phi$. (This symmetry is called the Z_2 symmetry, by the finite group corresponding to it.) This effective potential has two parameters, the effective mass m and the coupling λ, which are assumed to be smooth functions of the temperature (or other variables of the problem).

When the parameter $m^2 > 0$, the minimum of the potential is at $\phi = 0$, But if at some temperature it can change sign and becomes $m^2 < 0$, then the potential possesses two minima, obtained from $dV/d\phi|_{\phi_*} = 0$, or

$$\phi_* = \pm\sqrt{\frac{-m^2}{4\lambda}}. \tag{3.21}$$

A system can randomly select one of these minima: this is called *spontaneous symmetry breaking*. If the system is very large, and the space dimension is not too small, then the transition between them is strongly suppressed, so once the choice is made, the whole system stays in it.

Expanding the potential near the new minimum, we find that the new effective mass is still positive, as it must be:

$$m_{\text{eff}}^2 = \frac{d^2 V}{d\phi^2}|_{\phi_*} = -2m^2 > 0. \tag{3.22}$$

So the transition point from symmetric to asymmetric phases is prescribed up to a point at which $m^2(T)$ changes sign:

$$m^2(T = T_c) = 0. \tag{3.23}$$

Let me stress once again that Landau's main assumption here is that the parameters of the effective action, $m^2(T)$ and $\lambda(T)$, should by themselves be smooth analytic functions without singularities at $T = T_c$. Under this assumption we can easily see what singularities should appear. For convenience, let us introduce a dimensionless temperature indicating closeness to the critical point:

$$t = \frac{T - T_c}{T_c}.$$ (3.24)

We have already determined that in the asymmetric phase

$$\phi_* \sim t^{1/2}.$$ (3.25)

This power of $1/2$ is the first Landau prediction for the *critical index* called β. The primary name α among the indices is, for historical reasons, reserved for that of the second derivative of the free energy:

$$\frac{\partial^2 F}{\partial T^2} \sim t^{-\alpha}.$$ (3.26)

It is easy to check that Landau theory predicts a finite jump in this quantity. The predictions of Landau's theory that we have worked out so far are

$$\alpha_L = 0, \quad \beta_L = 1/2$$

Two more important critical indices are related to the limits of the Green function. At $t \neq 0$ it is expected to fall off exponentially with some correlation length:

$$G(r) \sim e^{-\frac{r}{\xi(t)}}, \quad \xi(t) \sim |t|^{-\nu}.$$ (3.27)

In Landau theory, it is given by the inverse effective mass, and thus, as we already have established, Landau's prediction is $\nu_L = 1/2$.

At $t = 0$ there is no exponent, the correlation length is infinite ($\xi \to \infty$), and the correlator has to be power-like. This power is the definition of the last index η, we will discuss, namely,

$$G \sim \frac{1}{r^{d-2+\eta}},$$ (3.28)

where d is the space dimension. In the $d = 3$ case under consideration, Landau's theory lead to a Coulomb-like field correlator $G \sim 1/r$, so the predicted $\eta_L = 0$.

3.5 More on Critical Indices

Landau theory is based on the assumption of analyticity, which seems to be so natural that it was generally accepted for decades as nearly obvious. And yet over time, some clouds started to appear on the horizon, showing that the problem of the second-order phase transitions in two and three dimensions is nontrivial.

In 1944 Onsager had solved the $d = 2$ Ising model: his values for these indices disagreed with the Landau's prediction (see the values in Table 3.1)

In the 1950s and 1960s, experiments and eventually computer simulations for the $d = 3$ Ising model and other similar systems were performed, bringing (as usual) a mixture of

Table 3.1. Values of Major Critical Indices.

	Landau	2-d Ising	3-d Ising
α	0 (jump)	0 (log)	0.08
β	1/2	1/8	0.33
ν	1/2	1	0.629
η	0	1/4	0.04

Note: Index values as predicted by the Landau's theory, Onsager's solution to the two-dimensional Ising model, and empirical values for the three-dimensional Ising model, from numerical simulations.

good and bad news. The good news was an emerging *universality* conclusion: indices for all systems with the same number of order parameters (boiling liquids at their critical point, magnetics, restructuring crystals, etc.) converged to the same universal values.[2] The bad news was a persistent disagreement with Landau theory. The main assumption of analyticity of the effective action coefficients does not seem to be correct.

A huge amount of theoretical and experimental work has been done on this problem, especially in 1960s. The feasibility of a self-consistent solution with nontrivial powers of t was gradually established, and some important tools—like the operator product expansion and representations of the conformal group—were developed. Unfortunately we do not have time to discuss these developments here,[3] and I will instead just jump to the end of the process in 1970s.

General comment: In this book, we will systematically explore the main setting with few simple examples, do some straightforward calculations, and then, jumping over a lot of important work, arrive at some current understanding of the problem at hand. It is unfair to many individuals who have contributed to the research on relevant topics, but one cannot do otherwise in introductory books, like those in the *Nutshell* series.

3.6 Renormalization Group and Wilson's Epsilon-Expansion

The idea of a renormalization group (RG) appeared in the QFT context as a tool to do resummations of certain diagrams. We will discuss its continuous version (due to Murray Gell-Mann and Frances Low in the 1950s) later on in other applications. For simplicity of presentation, we now use its discrete version, in a form worked out by L. Kadanoff and then used in the famous works by K. Wilson.

It requires starting with the lattice representation of the field, in which space is discretized and points are arranged in a simple cubic lattice with step a. Any scalar field

---

[2] We mean here one-order parameter examples. Higher symmetry breaking leads to more order parameters, and thus other universality classes.

[3] I need to apologize for this omission, in particular to my personal friends A. Patashinsky, V. Pokrovsky, A. Polyakov and A. Migdal, who contributed significantly to this theory.

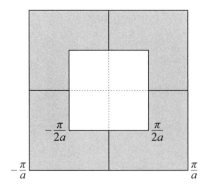

Figure 3.4. The shell in momentum space, that should be integrated out in one step of the renormalization group procedure.

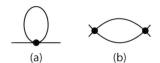

Figure 3.5. Diagrams for the mass-squared renormalization (a), the bubble and coupling renormalization (b), the fish.

configuration is represented by a set of discrete variables $\phi(x_i)$, the field values on the d-dimensional grid.[4] The RG "doubling step" is defined as the transition to the twice-as-coarse lattice, with step size doubled: $a \to 2a$. A sequence of such steps will eventually lead to a very large lattice, appropriate for studies of a large correlation length scale $\xi(t)$ at which the physics of the phase transition near the critical point is concentrated. To reach this goal, one may need many $(\log_2(\xi(t)/a))$ such doubling steps, during which the original atomic-scale specific action will flow into some universal effective action.

In momentum space the doubling procedure proceeds as follows. The original lattice has momenta located inside the square domain shown in Figure 3.4, and doubling of the lattice step a means that new momenta are located only inside the white square in the center, of the figure with half the size in each direction. The gray area around this square represents momenta no longer present there: so all effects associated with such momenta need to be accounted for. Their effect is not neglected but is preserved via appropriate renormalization of the effective action parameters.

Wilson suggested a rather crude but instructive way of performing this doubling transformation, which we now discuss. Let us include only two diagrams, to be referred to as the "bubble" and the "fish," shown in Figure 3.5. These are selected because they renormalize two parameters of the effective Lagrangian, the mass and the coupling. Here

---

[4] In the problem under consideration, time development is very slow and classical. This formally translates into keeping only the zeroth Matsubara frequencies in all sums, or ignoring the time variable completely. That is why time is not discretized or even mentioned here.

are the expressions for them and Wilson's approximate estimates of the results:

$$\text{bubble} = \lambda \int_{1/2<|p|<1} \frac{d^D p}{(2\pi)^D} \frac{1}{p^2 + m^2} \approx \frac{\lambda}{1+m^2} C \tag{3.29}$$

$$\text{fish} = \lambda^2 \int_{1/2<|p|<1} \frac{d^D p}{(2\pi)^D} \frac{1}{p^2 + m^2} \frac{1}{(\vec{p} - \vec{q})^2 + m^2} \approx \frac{\lambda^2}{(1+m^2)^2} C, \tag{3.30}$$

where the constant

$$C = \int_{1/2<|p|<1} \frac{d^D p}{(2\pi)^D}$$

represents the volume of the integrated (shaded in Figure 3.4) region. Note that in Wilson's units, the magnitudes of the momenta are normalized to $\pi/a = 1$ here. At the end of this chapter we will check the accuracy of his approximation, but for now we just accept it and see where it leads to.

The RG transformation for one doubling step

$$(m_n^2, \lambda_n) \rightarrow (m_{n+1}^2, \lambda_{n+1})$$

is complemented by stretching of the units, so that the new lattice appears to have the same size. The doubling relations take the following form:

$$(m)_{n+1}^2 = 2^2 \left[m_n^2 + 3\frac{\lambda_n C}{1+m_n^2} \right], \quad \lambda_{n+1} = 2^{4-D} \left[\lambda_n - 9\frac{\lambda_n^2 C}{(1+m_n^2)^2} \right].$$

The factors in front of the brackets are related to the dimension of the quantity: the mass squared has dimension 2 (thus the 2^2) and the coupling is dimensionless if $D = 4$ but in general its dimension is $4 - D$ (therefore the 2^{4-D} factor).

The central idea is that the sequence of such transformations converges to a *fixed point* (m_*, λ_*). Its value corresponds to a point in the $m - \lambda$ plain that is no longer changing, from n to $n+1$. At such fixed points the values of the parameters on the l.h.s. and the r.h.s. of the RG relation given above must coincide! This idea leads to the equations

$$m_*^2 = 4 \left[m_*^2 + 3\frac{\lambda_* C}{1+m_*^2} \right], \quad \lambda_* = 2^{4-D} \left[\lambda_* - 9\frac{\lambda_*^2 C}{(1+m_*^2)^2} \right].$$

From the second equation we get $\lambda_* = (2^{4-D} - 1)(1 + m_*^2)/(9C)$. Its form rather naturally brings us to Wilson's next brilliant idea. Let us consider the dimension of space to be a variable parameter, and consider

$$4 - D \equiv \epsilon \tag{3.31}$$

to be small. This leads to the fixed-point coupling

$$\lambda_* = \epsilon \times \log(2) \frac{1+m_*^2}{9C} \tag{3.32}$$

to be $O(\epsilon)$ and thus also small!

How small? Well, in the title of their early paper [12], the authors mentioned the space with $d = 3.99$ dimensions, or $\epsilon = 0.01$. And yet, as we will see below, eventually one can

```
Do[m = 0.4 * k;
 point[k] = {m,
    NIntegrate[1 / (2 * Pi) ^3 / (px^2 + py^2 + pz^2 + m^2),
     {px, 1 / 2, 1}, {py, 1 / 2, 1}, {pz, 1 / 2, 1}]}, {k, 10}]
ListPlot[Table[point[k], {k, 1, 10}]];
```

```
Show[%, Plot[.0004 / (1 + m^2), {m, 0, 4}]]
```

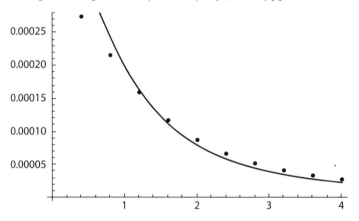

```
(* and now the fish at q=0, it has propagator squared *)
```

```
Do[m = 0.4 * k;
 point[k] = {m,
    NIntegrate[1 / (2 * Pi) ^3 / (px^2 + py^2 + pz^2 + m^2) ^2,
     {px, 1 / 2, 1}, {py, 1 / 2, 1}, {pz, 1 / 2, 1}]}, {k, 10}]
ListLogPlot[Table[point[k], {k, 1, 10}]];
```

```
Show[%, LogPlot[.0004 / (1 + m^2) ^2, {m, 0, 4}]]
```

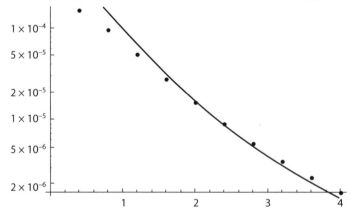

Figure 3.6. Results derived using mathematica 10, a Wolfram product. Dots are the exact result for the bubble and fish diagrams, integrated over the shell; lines are their parameterization à la Wilson used in the text.

use it at $\epsilon = 1$, and some people even used it for $\epsilon = 2$, with a surprising degree of success. The reason for this success can be seen by rewriting the first equations:

$$\lambda_* C = \epsilon \times \log(2)/9, \quad m_*^2 = -\epsilon \frac{4}{9}\log(2). \tag{3.33}$$

The small value of the coupling justifies the perturbative procedure. Furthermore, for $D = 3$ or $\epsilon = 1$, we find the l.h.s. to be just $\lambda_* C = 0.077 \ll 1$, still a really small value! Even for $D = 2$ or $\epsilon = 2$ it is not bad at all. Near the final fixed point the RG transformation can be linearized. There are two eigenvalues of the resulting matrix: one less than 1, which at large power leads to the approach to the fixed point, and one larger than 1, which leads to a departure from the fixed point. Playing with the RG transformation in the form given above or its simplified linear form, one finds that it is not easy but also not impossible to tune the initial values of the system's action, so that it will eventually approach the fixed point.

For a detailed pedagogical discussion of further calculations, the reader can consult the original papers, the review by Wilson and Kogut [13], or lectures by Parisi [14]. Let me illustrate the results by just one example, that of the index ν of the mass, which is the critical correlation length. Here are the subsequent terms of the epsilon expansion:

$$\nu = (1/2)(1 + .167\epsilon + 0.04\epsilon^2 - 0.016\epsilon^3 + \cdots). \tag{3.34}$$

Compare this to the values in Table 3.1, for $\epsilon = 1$ and even $\epsilon = 2$ (Onsager's $d = 2$ Ising model), and you will be surprised at how good the results are. Similar expressions for other indices were also derived. Since the epsilon expansion nicely explained the empirical values of indices, in 1982 Kenneth Wilson was awarded the Nobel Prize in physics. This story has been an inspiration to many theorists ever since. Indeed, with only two key diagrams, evaluated rather crudely, plus a very clever idea for implementing the RG transformation, he was able to go a very long way! The lesson is that as $t \to 0$ and for the infinite scale of the correlation length, the mass parameter does not go to zero, as Landau assumed, but to a specific nonzero value. It was exactly such as to be able to delicately balance the charge renormalization by a nonzero coupling constant.

Exercise *Calculate the integral over the shell in momentum space, shown in Figure 3.4, the expressions for the two diagrams (the bubble and the fish) and test the accuracy of Wilson's approximation (3.29) and (3.30). Figure 3.6 shows some parts of my Mathematica solution of this. The agreement is not bad.*

4 | Complex Scalar Fields

In this chapter we extend the theory a bit, from real to complex scalar fields. The latter of course can be considered as two real fields, in particular, the modulus and the phase. These two fields do not appear on an equal footing, and the result is an extended symmetry of the problem. The Lagrangian is basically insensitive to the constant phase; in other words, it is $U(1)$ symmetric. This symmetry leads to charge conservation, the concept of nonzero chemical potential, and eventually to important modifications of the Feynman rules (see Section 4.2). For pedagogic reasons, we discuss the settings both in relativistic and nonrelativistic versions. The former has a much more symmetric formulation, with second-order time derivatives in the field equations, while the latter, with only the first-order time derivative, is sufficient for most applications.

However, this basic symmetry—and the particle conservation that goes with it—can be spontaneously broken in the vacuum wave function, or thermal ensemble. As we discuss in Section 4.4, the phenomenon of Bose-Einstein condensation (familiar from the ideal Bose gas discussed in all statistical mechanics classes) breaks this symmetry, since the condensate should spontaneously select certain values for its phase. We extend the Bose-Einstein condensation (BEC) theory to a weakly coupled Bose gas, and following Belyaev, we resum certain diagrams and obtain the properties of "quasiparticles" in it in Section 4.5. A summary of the weakly coupled Bose gas is given in Section 4.6.

4.1 Phase Symmetry and Its Breaking

Let us promote the one-component real scalar field we have discussed so far to a complex-valued one. Of course, a complex field can be viewed as two real fields $\text{Re}(\phi)$ and $\text{Im}(\phi)$, as ϕ and its complex conjugate ϕ^*, or as modulus and phase. As convenient, we will freely move between these different representations.

The Lagrangian we will discuss looks nearly the same as before, but it now contains the quadratic mass term $m^2\phi\phi^*$ and a quartic $\lambda(\phi\phi^*)^2$ term. Note however that this new theory has higher symmetry then the old one: a reflection Z_2 symmetry of the real fields is now promoted to a rotation with an arbitrary global (that is, coordinate-independent) phase

$$\phi \to e^{i\alpha}\phi, \tag{4.1}$$

known as *the global U(1) symmetry*. Let us now discuss breaking of this new symmetry, still in the spirit of Landau theory from Chapter 3. When the coefficient of the quadratic $m^2\phi\phi^*$ term changes sign to negative, instead of a double-well potential with two degenerate minima, one finds the so-called Mexican hat potential, with a minimum at a circle. So there is a continuum of available degenerate vacua, with fixed $|\phi|$ and an arbitrary phase. The *spontaneous breaking* of the symmetry means that one random point on this circle is selected as the ground state.

In such asymmetric vacuums the excitations of the field are of two types. The *radial* ones (with fluctuations of the modulus) are massive, while the *tangential* ones (along the circle) remain massless. These massless modes are very special, and are called the *Nambu-Goldstone modes* (named for the people who first suggested that their properties possess some non trivial features). They can be described as space-dependent oscillations of the symmetry angle $\alpha(x)$. However, if this dependence is weak, at small momenta they become symmetry transformations and are therefore basically non interacting.

These phenomena are not abstract but occur in many real examples, as we will see on many occasions. We will see variations of it for Bose-Einstein condensates, superfluids, and multiple superconductors, as we proceed.

4.2 Conserved Charge and the Chemical Potential

Noether's famous theorem states that any continuous symmetry leads to a conservation law. In this case *particle number is conserved*, expressed as a conservation law of the current:[1]

$$j_\mu = (-i)(\phi^*\partial_\mu\phi - \phi\partial_\mu\phi^*) \sim \partial_\mu\alpha, \tag{4.2}$$

where $\alpha(x)$ is the phase of the field, $\phi(x) = |\phi(x)|e^{i\alpha(x)}$. If the phase is constant, there is no current, since the current is proportional to the field's gradient. The charge, as usual, is the integral of the current's fourth component, or the particle density:

$$Q = \int d^3x j_4.$$

Statistical mechanics courses usually start with microcanonical ensembles, with fixed charge or number of particles, and then teach us that it is more convenient to shift to a grand-canonical description, introducing the chemical potential conjugated to the

---

[1] Recall that in chapters devoted to relativistic theory, such as this one, we not only keep $\hbar = 1$ but also assume that the speed of light $c = 1$.

particle number. The corresponding generic expression for modified partition function reads

$$Z \equiv e^{-\Omega(T,\mu,V)/T} \equiv \text{Tr}[e^{-\frac{\hat{H}-\mu\hat{N}}{T}}]. \tag{4.3}$$

Note the main new element: the Hamiltonian $\hat{H}$ is shifted by a term proportional to the operator of particle number $\hat{N}$ (same as total charge Q defined above).

The *grand-canonical potential* $\Omega(T, \mu, V)$ thus defined is the thermodynamic potential in the three variables indicated, which means that all thermodynamical quantities can be obtained by its differentiation. Since V is the only extensive variable (i.e., depending on the system's size), Ω must be simply proportional to it. In fact $\Omega = -p(T,\mu)V$, where p is the pressure. So the derivative $\partial\Omega/\partial V = -p$ is trivial. Two other partial derivatives—over T and μ—lead to the entropy S and the particle number N, respectively.

Introduction of the chemical potential into the path integral with a relativistic complex field is our next task, which we do following [15]. (We will reduce this to the nonrelativistic case in the next section.)

Let us split the complex field into its real and imaginary parts adding a factor for future convenience:

$$\phi = \frac{\phi_1 + i\phi_2}{\sqrt{2}}. \tag{4.4}$$

From now on the reader is invited to follow all the i factors carefully, as this is important.

Another methodological change is the use of a bit more general path integral in phase space, which includes both the fields ϕ_i and conjugated momenta π_i.

For clarity, we start in Minkowskian real time. The Hamiltonian is

$$H = \frac{1}{2}[\pi_1^2 + \pi_2^2 + (\vec{\nabla}\phi_1)^2 + (\vec{\nabla}\phi_2)^2 + m^2\phi_1^2 + m^2\phi_2^2] + \lambda(\phi_1^2 + \phi_2^2)^2, \tag{4.5}$$

from which one calculates the action in a standard form. Let me just write its Euclidean time finite temperature version:

$$S_E = \int_0^\beta d\tau \int d^3x \left(i\pi_1\frac{\partial\phi_1}{\partial\tau} + i\pi_2\frac{\partial\phi_2}{\partial\tau} - H \right). \tag{4.6}$$

We will now add the chemical potential term. It adds a new term to the action:

$$\delta S = \mu \int_0^\beta d\tau \int d^3x j_4 = \mu \int_0^\beta d\tau \int d^3x(\phi_2\pi_1 - \phi_1\pi_2). \tag{4.7}$$

Note that this term couples the real and the imaginary parts of the field.

The *phase space* path integral we now need is

$$Z = \int D\pi_1\, D\pi_2\, D\phi_1\, D\phi_2 \exp(-S_E). \tag{4.8}$$

At this point the reader may wonder why we have not used it before in our quantum-mechanical examples. But it is a simple task to check that the Gaussian integral over momenta leads to the formulas we used in the previous chapters. Basically, for those examples we had no need to distinguish between the momenta and velocities, but we need it now.

While the integral over momenta is still Gaussian, linear terms proportional to μ are different and should be done properly. The reader is again invited to derive this Gaussian integral in detail, while I only present the answer:

$$Z = \int D\phi_1 D\phi_2 \exp \left\{ -\int_0^\beta d\tau \int d^3x \left[\frac{1}{2}\left(\frac{\partial\phi_1}{\partial\tau} - i\mu\phi_2\right)^2 + \frac{1}{2}\left(\frac{\partial\phi_2}{\partial\tau} + i\mu\phi_1\right)^2 \right. \right.$$
$$\left. \left. + (\vec\nabla\phi_1)^2 + (\vec\nabla\phi_2)^2 + m^2\phi_1^2 + m^2\phi_2^2 \right] + \lambda(\phi_1^2 + \phi_2^2)^2 \right\}, \tag{4.9}$$

in which the time derivatives are appended by certain $O(\mu)$ terms, with imaginary i. Note why this i has appeared: we have moved the time to imaginary time τ at the beginning of the formalism, but we keep μ real, as it was in the Minkowski world.

(At this point, let me provide an alternative derivation, which for some readers may be more intuitive. Since we have a charge, we can imagine it to be an electric charge, a source of an electromagnetic Abelian vector field. We know that in electrodynamics, to which we have generalized our theory for the moment, each derivative should be promoted to its "long" or covariant version, with a vector potential. The kinetic part of the Lagrangian of electrodynamics is

$$(-i\partial_\mu + eA_\mu)\phi^*(i\partial_\mu + eA_\mu)\phi. \tag{4.10}$$

The chemical potential can be seen as a part of the electric potential $\mu = eA_0$ of constant magnitude. This derivation leads to the same result as we obtained above.)

The next steps are very standard. Proceeding to Fourier transforms in both Euclidean time and space, the Lagrangian can be written in a compact matrix form

$$(\phi_1\phi_2)\begin{pmatrix} \omega_n^2 + \omega_k^2 - \mu^2 & -2\mu\omega_n \\ +2\mu\omega_n & \omega_n^2 + \omega_k^2 - \mu^2 \end{pmatrix}\begin{pmatrix} \phi_1 \\ \phi_2 \end{pmatrix},$$

with $\omega_n = 2\pi T n$ being the Matsubara frequencies, and $\omega_k = \sqrt{k^2 + m^2}$, as usual. The appearance in the finite-μ theory of the nondiagonal elements suggests that one should diagonalize this matrix before proceeding to a particle interpretation. Standard calculations (which the reader may perform) for a noninteracting ($\lambda = 0$) case result in the following answer:

$$-\log(Z) = \int \frac{d^3k\, V}{(2\pi)^3}\left[\beta\omega_k + \log\left(1 - e^{-\beta(\omega_k - \mu)}\right) + \log\left(1 - e^{-\beta(\omega_k + \mu)}\right)\right], \tag{4.11}$$

which has an obvious interpretation. The last two terms are the contributions from particles and antiparticles. The opposite signs of μ in these two terms simply stems from their opposite charges. The particle number (total charge) is obtained by the differentiation $\partial\log(Z)/\partial\mu$, which also leads to the obvious expression for the total charge of the ideal Bose gas

$$Q = \int \frac{d^3k\, V}{(2\pi)^3}\left(\frac{1}{e^{\beta(\omega_k - \mu)} - 1} - \frac{1}{e^{\beta(\omega_k + \mu)} - 1}\right) \tag{4.12}$$

with the opposite sign for particle and antiparticle densities. Recall that we are in full relativistic theory, so normalization of the chemical potential is relativistic as well. In fact the above expression for Q is finite only when $-m < \mu < m$: when μ approaches its limiting values, the *Bose-Einstein condensation* (BEC) takes place,[2] either of particles or antiparticles.

Finally, let us return to the Sommerfeld-Watson method for calculation of the Matsubara sums. Introduction of the nonzero chemical potential $\mu \neq 0$ results in a shift of the poles and the original line of the contour C by μ in the real direction. The points in Figure 3.2 are therefore shifted horizontally by the amount μ. The contour itself and the sequence of poles still go in the imaginary direction, as before. However, because the poles are shifted, the "right" and "left" parts are now redefined: it is now $\mathrm{Re}(z) - \mu$ that should be positive or negative. Thus the singularities of the function F in the sums separate into "particles" and "holes" and as a result have different thermal factors, since the energy is now counted relative to μ. Even the simple integral over imaginary axes, without thermal factors, is now different: because its rotation to real axes have different rules of avoiding singularities, the integral is now dependent on μ. These modifications will be especially relevant for fermionic systems, so we return to explanations of the Feynman rules at finite μ later in Chapter 7.

4.3 Nonrelativistic Approximation and Normalization

We will discuss matter made of relativistic particles—quark-gluon plasma—near the end of the book. For the rest of the book we deal with nonrelativistic particles, and so we may simplify the relativistic theory. If the energy scale mc^2 is many orders of magnitude larger than all relevant energies (such as the temperature), we can formally let $c \to \infty$ and exclude it from consideration.

The nonrelativistic normalization of the energy (as well as of the chemical potential) is defined without Einstein's mc^2 term: we return to the common idea that a particle at rest has *zero* energy:

$$\omega_k - mc^2 \equiv \omega_k^{\mathrm{nr}} \approx \frac{k^2}{2m}, \quad \mu - mc^2 \equiv \mu_{\mathrm{nr}} \ll mc^2, \tag{4.13}$$

and it is these combinations that are assumed to be $O(T)$. Thus the particle term in the expression above is $O(1)$ and is retained. The antiparticle term with $\omega_k + mc^2 \gg T$ is suppressed exponentially $(\exp(-mc^2/T))$ and is thus ignored in the nonrelativistic theory: for example, when discussing the electron gas, we will never mention positrons. In such cases, we return to the familiar textbook expression for the nonrelativistic ideal Bose gas:

$$N = \int \frac{d^3k\,V}{(2\pi)^3} \frac{1}{\exp\left[\frac{\beta k^2}{2m_{\mathrm{nr}}} - \beta\mu\right] - 1}. \tag{4.14}$$

---

[2] The history of the study of BEC will be discussed in Chapter 6 devoted to trapped ultracold atoms. The physics of BEC will be elucidated in many other chapters as well.

Note that in this definition the nonrelativistic chemical potential of the Bose gas must be negative, $\mu_{nr} < 0$. Otherwise, there would be a pole and the momentum integral would diverge.

To simplify the notation further, we would like to change the definition of the fields as well, introducing their nonrelativistic version ψ,

$$\psi \equiv \frac{\phi}{\sqrt{2mc^2}}, \tag{4.15}$$

so that the quadratic form in the action reduces to its nonrelativistic form, with only linear dependence on energy:

$$(\omega^2 - k^2 - m^2)\phi^*\phi \approx (2m\omega_{nr} - k^2)\frac{\psi^*\psi}{2m} = (\omega_{nr} - k^2/2m)\psi^*\psi. \tag{4.16}$$

The Euclidean action then has only a single time derivative:

$$S = \int_0^\beta d\tau \int d^3x \left[\psi^* \left(\frac{\partial}{\partial\tau} - \mu_{nr} \right) \psi + \frac{(\vec{\nabla}\psi)^2}{2m} \right]. \tag{4.17}$$

Note that if we were to derive the classical equations of motion from this action, it would look like the usual Schrödinger equation for the wave function ψ, but with the energy shifted by μ_{nr}.

The corresponding Green function—the correlator of two ψ_s—now has the simplified form

$$G = \frac{1}{i\omega_n - \mu_{nr} + k^2/(2m)} \tag{4.18}$$

with only a single pole in the denominator. Yet we will still use the Sommerfeld-Watson method of Matsubara summation, and it will remain important whether the pole is in the right or left semi-plane: depending on its position, the pole corresponds to a particle or a hole. This distinction will become crucial later, in the Chapter 7 and after, devoted to fermionic systems.

The bubble diagram in Figure 3.5, defined in terms of the local field correlator, now gives

$$\langle \psi^*(0)\psi(0) \rangle = T \sum_n \int \frac{d^3k}{(2\pi)^3} \frac{1}{i\omega_n - \mu + k^2/(2m)}$$

$$= \int \frac{d^3k}{(2\pi)^3} \left[\frac{1}{2} + \frac{1}{\exp(p^2/(2m - \mu_{nr}))} \right]. \tag{4.19}$$

Omitting the first divergent term, we find that the second is just the particle density, as one would indeed expect from the local value of the Schrödinger wave function squared.

4.4 Weakly Coupled Bose Gas and Bose-Einstein Condensation

These expressions are all well defined when $\mu < 0$, which is the case for $T > T_c$. As it was famously noticed by Einstein, at $T \to T_c$, one has $\mu \to 0$. (Once again, this in nonrelativistic notation; in relativistic ones $\mu \to m$, the particle mass.) In this limit the particle density goes to the finite limit

$$n_T = \int \frac{d^3 p}{(2\pi)^3} \frac{1}{e^{p^2/(2mT)} - 1}.$$ (4.20)

So, what happens if we set the density of particles larger than that? As Einstein explained, the excess particles all go into "the condensate," accumulating in the lowest energy state, $\vec{p} = 0$ in a large box. The condensate density n_0 is simply their difference

$$n_0 = n - n_T$$ (4.21)

and for all $T < T_c$ the value of the chemical potential remains equal to zero. All of this is of course part of standard statistical mechanics courses.

Exercise *Einstein's condition for condensation naturally assumes that the gas is nonrelativistic: $T \ll m$. Derive a similar relation between T and density n in the opposite case $T \gg m$, in which one needs to include antiparticles as well. Generalize your result to a generic mass in between, and plot $n_c(T, m)/T^3$ for an arbitrary m/T ratio.*

The problem of a (weakly) interacting Bose gas is among the central problems of the many-body theory in general and of this book in particular. The real hero of this field is C. N. Yang, who worked on it from classic papers [17] until very recent works he described at a Stony Brook seminar a few years ago.

Unfortunately, the Yang et al. papers are too complicated for this book. Fortunately, a much simpler way to derive the properties of quasiparticles in a weakly interacting Bose gas was discovered, by Belyaev in 1958 [18], whom we follow for the rest of the chapter.[3]

The main issue is the following: at $T < T_c$ one needs to include the component of the Bose gas that resides in the Bose-Einstein condensate. The required modification of the Feynman diagrams boils down to the introduction of a new contribution, recognizing that there exist extra wave functions of the condensate with the value

$$\psi_0 = \sqrt{n_0}.$$

While we denote propagators of the ordinary noncondensate particles by solid lines (with arrows, indicating the flow of the charge and thus separating ϕ from ϕ^*), the condensate particles will be shown by dashed lines in diagrams. Since the condensate is expected to

---

[3] Let me add here some personal and historical comments. Spartak Belyaev was my PhD advisor, during 1970–1974. Unfortunately, during this time he was already the rector (president) of the university, so he had little time for me and no time to teach many-body theory. So, I only learned about his papers decades later. I was stunned by their beautiful simplicity—hopefully the reader will share this opinion. Lev Gorkov applied the same idea, of anomalous Green functions, to superconductivity: we will see how this approach works in Chapter 9. Decades later, in 2004, Belyaev and Gorkov were co-recipients of the prestigious Eugene Feenberg Memorial Medal, the main award in the field of many-body physics, essentially for their early papers just mentioned. Spartak Belyaev that year came to United States to receive this prize, via Stony Brook. We then discussed many new ideas and papers, but had no time to devote to history of the anomalous Green functions. Unfortunately that was the last chance for me to ask questions about it. Spartak Belyaev passed away on January 5, 2017, in his ninety-fourth year, and Lev Gorkov died just few days before him.

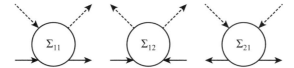

Figure 4.1. The diagrams representing the anomalous mass operator, including the contribution from the condensate (dashed lines).

be static and homogenous, these lines will carry zero frequency and zero momentum. The condensate field is real, but for clarity, we will still put arrows on the field lines in diagrams, indicating the charge flow.

The main consequence of this new contribution is that the particle number of the noncondensate particles is no longer conserved, since a particle can now "hide" inside the condensate. In particular, the mass operator Σ obtains some anomalous terms with disappearing or reappearing noncondensate particles, as shown in the Figure 4.1. (Note that the total number of particles is still of course conserved.)

Arrows, as usual, correspond to the particle flow, from incoming ϕ to outgoing ϕ^*. Note that since the dashed condensate particle lines carry zero frequency and momentum, those for the noncondensate particles remain unchanged along the solid lines. Only the direction of the arrow may change, as shown in Figure 4.1. Therefore, although these diagrams include four lines, they are not treated as 4-vertices but as 2-line mass operators, since only the solid lines are counted. The diagrams in which the direction of the arrow changes are called "anomalous" mass operators.

We have shown before that one can resum a sequence of diagram with alternating Green functions and mass operators to all orders by summing a geometric series, which puts the mass operator in the denominator:

$$G = G_0 + G_0 \Sigma G_0 + G_0 \Sigma G_0 \Sigma G_0 + \cdots = \frac{1}{G_0^{-1} - \Sigma}. \tag{4.22}$$

Note that this resummed G can also be obtained by solving the following recurrent relation equation for it:

$$G = G_0 + G_0 \Sigma G. \tag{4.23}$$

(Don't overlook the G without zero in the last term.)

All we need to do while considering the $T < T_c$ Bose gas with the condensate is to perform the same resummation, but now with the anomalous Green function and mass operators included. The same iterative equation gives this resummation:

$$\hat{G}(\omega_n, p) = \hat{G}_0(\omega_n, p) + \hat{G}_0(\omega_n, p)\hat{\Sigma}(\omega_n, p)\hat{G}(\omega_n, p) \tag{4.24}$$

where the hats indicate the natural change of notation: all objects are promoted to 2×2 matrices, which are multiplied in the order as written.

The meaning of the matrices and their indices are explained in Figure 4.1. Products of matrices, as usual, imply summations over the indices, which automatically takes care of summing all possible diagrams. Here $\hat{G}$ is the total resummed version of the zeroth-order Green function $\hat{G}_0$.

Simple manipulation with matrices leads to the following solution of this equation:

$$\hat{G} = \frac{1}{(\hat{G}_0)^{-1} - \hat{\Sigma}}, \tag{4.25}$$

where a matrix appearing in denominator means its matrix inversion.

The mass operator can be easily evaluated to the lowest order. The diagrams shown in Figure 4.1 are then just the 4-point vertex, or the coupling constant, times the condensate wavefunction squared, or the condensate density:

$$\Sigma = \begin{pmatrix} \Sigma_{11} & \Sigma_{12} \\ \Sigma_{21} & \Sigma_{22} \end{pmatrix} = \lambda n_0 \begin{pmatrix} 2 & 1 \\ 1 & 2 \end{pmatrix}. \tag{4.26}$$

Here the numbers in the r.h.s. matrix simply count the number of contributing diagrams: there is only one diagram of the type Σ_{12}, Σ_{21} but two of the type Σ_{11} and Σ_{22}, because the condensate arrow now can be reversed (see Figure 4.1).

Basically, this is all one needs to calculate. Consider for simplicity the zero-temperature case, in which, to zeroth order, all particles are in the condensate: $n = n_0$, $n_T = 0$. One should include the lowest-order shift in G_0, caused by the shift in chemical potential, which is $\mu = \lambda n_0$, and plug in the mass operator (4.26) into the expression for $\hat{G}$. The result for the inverse of the resummed Green function (the denominator of (4.25), rotated for clarity to Minkowski energy ω) takes the form

$$G^{-1} = \begin{pmatrix} \omega + \lambda n_0 + p^2/2m & \lambda n_0 \\ \lambda n_0 & -\omega + \lambda n_0 + p^2/2m \end{pmatrix}. \tag{4.27}$$

The last simple step is to find the inverse of this matrix, and then we get the final form of Belyaev's full propagator:

$$G = \frac{2}{-4\omega^2 m^2 + 4\lambda n_0 \times m \times p^2 + p^4} \begin{pmatrix} -2m^2\omega + 2\lambda n_0 m^2 + p^2 m & -2\lambda n_0 m^2 \\ -2\lambda n_0 m^2 & 2m^2\omega + 2\lambda n_0 m^2 + p^2 m \end{pmatrix}.$$

4.5 Renormalized Quasiparticles and Condensates

We have now obtained the resummed Green function (also known as "dressed propagator") in a weakly coupled low-T Bose gas and are thus in a position to digest the information that this expression contains.

Recall that the poles of the Green functions always provide the dispersion relations of the excitations of the system, which can be produced by a vanishingly small external source. Therefore, let us have a look at the zeros of the denominator we took out of the matrix in the expression for G just derived. We see that the poles are at the following locations:

$$\omega_{\pm}(p) = \pm \frac{\sqrt{4\lambda n_0 m p^2 + p^4}}{2m}. \tag{4.28}$$

At zero condensate ($n_0 \to 0$) or at large momenta ($p \to \infty$), these poles correspond to the usual particles of the gas, with $\omega(p) = \pm p^2/(2m)$. However, if the condensate is nonzero and at small $p \to 0$, we find a completely new dispersion law, namely, a linear dependence in momentum:

$$\omega_{\pm} = \pm p \sqrt{\lambda n_0/m}.$$

Thus at small p the excitations are not particles but waves, propagating with a fixed speed $v = \sqrt{\lambda n_0/m}$. In fact this is the speed of sound, and thus what we have found are quanta of sound waves, known as "phonons."

Now let me confess that these results were historically obtained long before the calculation we just followed: what we found are known as Bogolyubov quasiparticles, already derived in [16] in 1947. The phenomenon we have come across is a particular case of *mode mixing*: the excitation mode corresponding to scalar particles and to the sound are mixed together, into one common *quasiparticle*. This of course is only possible for $T < T_c$, because only in this case is particle number conservation broken by the condensate. At $T > T_c$ the particle mode and the sounds cannot mix, and they form two different excitations, because they have different quantum numbers.

Further insights come from the resummed Green function we found, rewritten in the form of two poles and their residuals:

$$G = \frac{u_p^2}{\omega - \omega_+ + i\delta} + \frac{v_p^2}{\omega - \omega_- - i\delta}. \tag{4.29}$$

The coefficients u_p, v_p can be simply read from the formula for G above. In fact we have already analytically continued the Green function in Minkowski world: how the poles should be handled is explained by the sign of the infinitesimal imaginary part δ. Note that the sign is different for the two poles: this follows from the rule of rotation of the contour from the Euclidean to the real frequency. The signs of the resulting imaginary part of the Green function fixes the interpretation of each term: the former term describes a creation of a quasiparticle, with the energy ω_+, while the latter term describe an absorption of a quasiparticle, with the same energy.

Now let us think a bit about what problem we are discussing and what this expression tells us. We are discussing a cold ($T = 0$), weakly coupled Bose gas. To zeroth order, the condensate is made of $\vec{p} = 0$ (that is, nonmoving) particles. The resummed Green function tells us something else.

One can only absorb something which is already there. So the second term in G implies the presence of moving ($p \neq 0$) particles in the condensate. A little thought indeed leads to the following *condensate renormalization* expression:

$$n = n_0 + \int \frac{d^3 p}{(2\pi)^3} v_p^2. \tag{4.30}$$

The explicit expression we now need is

$$v_p^2 = \frac{p^2/(2m) + \lambda n_0}{2\omega_+} - \frac{1}{2}. \tag{4.31}$$

Let us first check whether the momentum integral in (4.30) is convergent. At large p the leading terms cancel the $1/2$, and what is left is $O(1/p^4)$, so the integral over large p is convergent. For small p, $v_p^2 \sim 1/p$, but due to the $d^3 p$ term, the integral in this region is also convergent. Dimensional analysis of the integral suggests its magnitude to be $\sim (\lambda n_0 m)^{3/2}$.

For better physical interpretation of all these results, let us elucidate the definition of the coupling constant. It can be related to the quantum-mechanical scattering length a of the particles[4]

$$\lambda = \frac{4\pi\hbar^2}{m} a. \tag{4.32}$$

Since a is a length, we can form a dimensionless combination with the particle density and write the weak coupling condition as

$$a n^{1/3} = \frac{a}{R} \ll 1. \tag{4.33}$$

The scattering length is small relative to interparticle spacing. For liquid He it is of the order 1, and indeed most of the atoms at $T = 0$ are not in the $p = 0$ state. For trapped ultracold atomic gases (to which we will turn in Chapter 6), this combination is indeed small, typically less than 10^{-2}.

The result we obtained above, the *depletion of the condensate* induced by the interaction of its particles, can of course be written with all the numerical factors and the explicit dimensionless parameter

$$\frac{n - n_0}{n_0} = \frac{8}{3} \left(\frac{n_0 a^3}{\pi} \right)^{1/2}. \tag{4.34}$$

Note that while being parametrically small, it has a noninteger power $3/2$ of the small parameter. Since no particular diagram can produce this power, such an answer can only be obtained by resummation. (This is already the second example of its kind in this book; hence the reader should already be prepared for this conclusion.)

The most spectacular of Belyaev's results that come from these diagrams is the quasiparticle lifetime. At $T > T_c$ there is no condensate, particle number conservation is strict, and at a quartic vertex, a particle can only decay into three others. But for $T < T_c$ one of the lines can be a condensate line, which carries zero energy and momentum. So one can have a quasiparticle that can decay into two.

---

[4] The fact that one can consistently substitute scattering amplitude in place of the coupling is nontrivial, and is due to V. Galitsky. Instead of explaining this technical point, let me just comment on the dimensions of quantities included. The nonrelativistic field squared is density, thus its dimension is $[\psi^2] = L^{-3}$. The coupling dimension comes from the interaction term of the Lagrangian, $[\lambda] = L^2$, consistent with the formulas in which $[a] = L$ and $[1/m] = L$.

Without presenting the calculation itself, let me show some of the results. Let us concentrate on the small p part of it (i.e., the sounds). The $O(a^{3/2})$ corrections in the sound dispersion relation are

$$\omega_+ = p\sqrt{n_0 a}\left[1 + \frac{7}{6\pi^2}(n_0 a^3)^{1/2}\right] - i\frac{3}{640\pi}(n_0 a^3)^{1/2}p^5. \tag{4.35}$$

We will further discuss the last term, the sound decay, in Section 5.4.

4.6 Summary of Perturbation Theory for a Weakly Coupled Bose Gas

Recall that our small parameter was defined in (4.33), where a is the scattering length (basically, atomic size), and R is the interparticle separation. Simply speaking, the gas satisfying this condition is interacting but dilute.

The main phenomenon is BEC, the presence of the condensate. In the diagrams in Figure 4.1, it is accounted for by additional condensate lines (the dashed lines). Resummation of these lines is done by transforming the Green functions into 2×2 matrices, with normal terms on the diagonal and anomalous ones off the diagonal. The new quasiparticles are mixtures of collective excitations—phonons—at small p and particles at large p. Their real and imaginary parts at zero T and small p are given by (4.35).

Here is the summary for the total energy of the cold weakly interacting Bose gas in increasing powers of our small parameter:

$$E = \left(\frac{2\pi a n_0^2}{m}\right)\left(1 + \frac{128}{15\sqrt{\pi}}(n_0 a^3)^{1/2} + \left[\frac{8}{3}(4\pi - 3^{3/2})\log(n_0 a^3) + \kappa\right](n_0 a^3)\right). \tag{4.36}$$

The first term is from an early Bogolyubov paper [16] from 1947: it is contributed by direct 2-particle scattering, the open double-bubble lowest-order diagram. The second term is due to a very elaborate paper by T. D. Lee and C. N. Yang [17] from 1957.[5] It also follows from the resummation of the Green function we did above, following Belyaev's 1958 paper. The next term was found by another distinguished theorist, T. T. Wu (et al.) one year later [19]: it included resuming the so-called fish-in-fish or 3-into-3 scattering diagrams. The last (known to me) term, called κ here, without details, was calculated by Braaten and Nieto [20] in 1996 (three decades later!). In this last work not only the scattering length a but, for the first time, the scattering range[6] r_0 enters. Higher-order terms would include further new parameters, describing the scattering amplitude of particles, and so these terms would be not so universal.

The perturbative domain is quite limited (see the next exercise), and that is why these theoretical expressions had to wait, for about four decades, to be tested experimentally in the 1990s: we will return to them in Chapter 6 devoted to trapped ultracold gases.

---

[5] I found it remarkable that their Bose gas paper was written in the same year as their famous parity-violation paper, which was promptly confirmed experimentally, and gave them both the Nobel Prize.

[6] One can look up its definitions in a quantum mechanics textbook.

When these tests confirmed the theory under consideration, developed in 1950s, their authors received a variety of awards. The lesson of this story is, in my opinion, that any important theory must be developed, if possible, regardless of whether its direct experimental verification is at the moment possible.

Exercise   *Plot the four terms in (4.36), as a function of the small parameter defined in (4.33), and determine the region of applicability of the weak coupling expansion in which the series converges reasonably well.*

5 | Liquid ^{4}He

The quest to approach absolute zero temperature during the twentieth century fueled one of the frontiers of physics, led to a spectacular array of discoveries, and pushed the temperatures attainable from a few Kelvins at the beginning of the century to nano-Kelvins today. One of the systems that has fascinated physicists during this period is liquid helium, and so we start the chapter with a brief summary of its history.

The superfluidity of liquid helium and its spectacular experimental manifestations is discussed in Section 5.2. Elementary excitations in it are discussed in Section 5.3, and their decays in Section 5.4. Landau's criterion for superfluidity is derived in Section 5.5. The physics of the rotation of a glass of superfluid and the vortices that develop in it are discussed in Section 5.6.

5.1 History

At the beginning of the twentieth century the frontier of low-temperature physics was in Leiden, Holland. A lot of gases had been liquified there, and in 1908 the lightest noble gas, and of course the most difficult one, helium, had also surrendered to Heike Kamerlingh Onnes and his team. They proudly published its critical temperature, $T_c = 4.2$ K.

Yet the most important discovery was still ahead. Another critical point was found two decades later, in 1927. It is now known as the lambda point, at

$$T_c = 2.17 \text{ K}, \tag{5.1}$$

because the measurements of the specific heat had resulted in a curve reminiscent of the Greek letter lambda. This critical point was clearly a new phase transition, but to what? Helium was a liquid on both sides: no liquid-to-liquid phase transitions had ever been

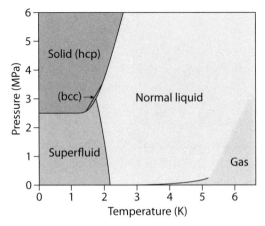

Figure 5.1. The phase diagram of ^{4}He. Adapted from the web page of LTL/Helsinki University of Technology, http://ltl.tkk.fi/research/theory/helium.html.

seen prior to that one. For lack of ideas about it[1] or a better name, the new phase was called He-II.

Exercise *Calculate the temperature of the Bose-Einstein condensation of liquid ^{4}He if it were an ideal gas (which it is not!). The mass density of liquid ^{4}He at small T is $\rho = 0.145 \, g/cm^3$.*

All attempts at standard atmospheric pressure to make helium solid by cooling it failed: it is the only substance that remains liquid even as $T \to 0$. Helium solidifies at higher pressure, of about 25 atmospheres; see the low-T part of the phase diagram shown in Figure 5.1.

In 1938 Kapitza[2] [26] and Allen and Misener [27] discovered the crucial difference between He-I and He-II liquids: the latter is a *superfluid*. We will discuss some of the spectacular experiments explaining what that means later in this chapter. Let me now only comment on the intense theoretical debates of that time. F. London related superfluidity to BEC, while Tisza and Landau argued that it cannot be a superfluid, since a Bose gas with condensate should not be a superfluid. We will see how these debates were later resolved.

This chapter covers only an introduction to this classical subject: for extensive coverage, the reader can consult the classic textbooks [24] and [25].

Before we discuss specific experiments, some general reminders of the properties of liquid He are in order.

---

[1] The next exercise suggests an estimate testing whether this new phase has a Bose-Einstein condensate. One may thus wonder if that idea came to mind to the Leiden group after their discovery. It might have happened— the Einstein paper on BEC was from 1925, two years prior to the helium II discovery—but in fact the estimate proposed is due to F. London, who made it in 1938.

[2] Forty years later P. Kapitza was awarded the 1978 Nobel Prize for this discovery.

Its mass density is $\rho = 145\,\text{kg/m}^3$. For comparison, water has mass density of $1{,}000\,\text{kg/m}^3$, but the He atom has 4 nucleons, while a water molecule H_2O has $2 + 16 = 18$ nucleons: thus the number density is in fact rather similar.

A simple measure of interatomic interaction is the speed of sound c_s. It is about $200\,\text{m/s}$ in helium and $1{,}400\,\text{m/s}$ in water. Even air, which is $1{,}000$ times more dilute, has a speed of sound higher than that of liquid helium, of about $300\,\text{m/s}$! So the interparticle interaction in liquid helium is indeed weak.

The interatomic interaction between two He atoms is very short range and has a typical weakly attractive part at large distances and a repulsive core at small ones. The potential well has a depth of only $29\,\text{K}$. On one hand, it is tiny compared to atomic scale of about $1\,\text{eV} \sim 10^4\,\text{K}$ typical for atomic energies. But on the another hand, we will consider low temperatures on the order of a few Kelvins. At such low temperatures, it is impossible to get far from the minimum of the potential well. As a result, any motion of He atoms should be well coordinated, so as not to change the distance between them too much from that minimum. (We will return to motion of atoms in liquid helium when we discuss numerical evaluation of the path integrals in Chapter 13.)

5.2 Superfluidity in Experiments

It is impossible to speak about superfluidity without recalling the definition of *viscosity*. In general, viscosity is proportional to the particle mean free path in matter, allowing neighboring hydrodynamic cells to exchange momentum and to create a friction in case the cells move with a bit different velocities: thus it enters the hydrodynamical equations of motion as a coefficient of terms containing an extra gradient of velocity. The most obvious effect of viscosity is one that everyone is familiar with: river flow is stronger in the middle of the stream than near its banks.

Standard measurement of liquid viscosity is done as follows. The fluid being measured is forced to flow through a thin tube of radius R and length Δl, under pressure difference Δp: the volume flow through the tube per second is given by

$$\frac{dV}{dt} = \frac{1}{\eta}\left(\frac{\pi}{8}R^4\right)\frac{\Delta p}{\Delta l}. \tag{5.2}$$

Solution of the Navier-Stokes equations for flow in a tube is known as Poiseuille flow. The quantity η is the shear viscosity, η times the velocity gradient is the shear force per area, so its standard unit is $N \times s/m^2$. For air it is 0.018, and for liquid He-I it is on the order of 10^{-6}. Roughly speaking, that means that the mean free path of the atom is four orders of magnitude smaller. This is not surprising, because three orders is due to the density difference. In short, liquid He-I is already a rather "good" fluid, meaning that its η is rather small. As T drops, the thermal motion of the atoms goes down, and it is expected that η decreases further.

What Kapitsa, Allen, and Misener actually discovered was that in the He-II phase, and for flow velocity below some critical value $v < v_c$, no viscosity was seen at all. Flow

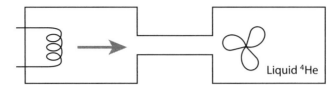

Figure 5.2. The flow between two hermetically closed vessels filled with He-II. The left vessel contains a heater, and the right one a propeller sensitive to fluid motion.

continues for a long time even when the force Δp is absent! Even more striking, one could bend a tube of the material into a closed torus, and generate flow that would go on *forever*! This discovery led to the term "superflow."

As a result, liquid He-II can flow though very thin tubes (or even microscopic pores of the tubes), or though tubes being plugged by a cotton wad. It is also able to propagate in thin films on the glass surface of an open container, climbing up to its edge and then down the outside of the vessel, thus escaping from its container! If an empty glass is put into a vessel with liquid He-II, the glass will soon be filled to the same level as the liquid surrounding it.[3]

Experiments with heating and flowing helium revealed that there seem to be *two distinct fluids* in helium. After flowing through thin tubes and surface films, the liquid He-II is not the same as it was before: in fact it became much cooler(!). One of many wonderful experiments is sketched in Figure 5.2. Two closed vessels, completely filled with liquid He-II, are connected by a tube. The left vessel has a small heater, and the right one has a small propeller that can rotate when there is fluid flow. When the heater is turned on, fluid flow through the tube, left to right, is detected. But how is that possible? Both vessels are hermetically closed, there is no extra fluid coming from anywhere, and the weight of both vessels remains the same.

Here is one explanation, put forth by the *two-fluid model* (e.g., see [29]): He-II consists of two fluids, one "normal" with mass density ρ_n and the other "super" with density ρ_s. They can move with different—even opposite—velocities, $\vec{v}_n \neq \vec{v}_s$. So the total flow current

$$\vec{j} = \rho_n \vec{v}_n + \rho_s \vec{v}_s \tag{5.3}$$

in this experiment is zero, yet the two components are nonzero: they just flow in opposite directions. The propeller indicates only one flow, namely, that of the normal component, since the superfluid fraction is frictionless. The heater is the source of the normal component.

Yet here is another—and much simpler—explanation: He-II has its ground state at rest. What is moving through the tube are certain elementary excitations, created by the heater. Eventually this point of view took over, because it turned out that normal component may indeed even not be a fluid. At very low T the excitations have such a large mean free

---

[3] Imagine yourself in a boat in a helium-II pond, learning that this fluid can easily get inside your boat just by climbing up and down the sides.

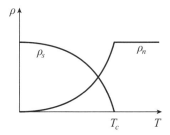

Figure 5.3. Sketch of the temperature dependence of both densities, super and normal, resulting from the Andronikashvili oscillating disk experiment.

path that they start moving ballistically, from the heater straight to the propeller. This was checked by inserting collimators, creating quasiparticle beams.

Many other experiments are nicely explained by the two-fluid model. One, which was a spectacular success, is the *Andronikashvili oscillating disk experiment* [31]. A stack of disks is suspended on a wire, normal to the stack, and the disks can oscillate on the wire. The distance between adjacent disks is small enough, so if the disks are immersed in a fluid, the fluid rotates together with the disks. Oscillation frequency depends on the oscillating mass, that of the discs plus that of the entrained fluid in between them. Experiments performed with He-II had shown that only the normal component of the fluid was entrained and oscillated together with the disks, which allowed experimentalists to measure its fraction of the normal component as a function of the temperature (see Figure 5.3).

5.3 Elementary Excitations

So what are these excitations in liquid He-II? Are they simply He atoms, which are not in the ground state of the fluid? Suppose they are. Then their energy is $\epsilon_p = p^2/(2m)$, and from expressions for the Bose gas, we deduce that their density $\sim (mT/\hbar^2)^{3/2}$ and the energy density $\sim m^{3/2} T^{5/2}/\hbar^3$. This can be compared to data on the thermodynamics of the fluid.

Another possibility is that at low T the experiments not the individual particles but collective excitations, such as sounds. The famous case of the specific heat of solids shows that the lowest excitations are indeed phonons, the quanta of sound. That is why the free energy of solids at small T is $\sim T^4$, as it is for the photons of blackbody radiation. All one needs to do to see this analogy is to substitute the speed of sound for of the speed of light and change the number of polarizations.[4]

And indeed, at sufficiently low T, this is the case: the free energy at small T is $\sim T^4$ for liquid He-II as well. Yet accurate experiments, done in particular by W. H. Keesom and H. P. Keesom [28], have found strong deviations from the phonon model at higher T. An explanation of those deviations was proposed by Landau in 1947 [30], who suggested

---

[4] This analogy was pointed out by none other than the young Einstein himself.

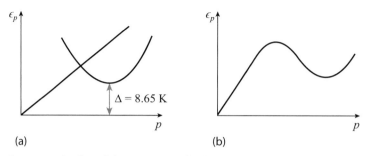

Figure 5.4. Sketches of the two options for the Landau dispersion curve: (a) two separate excitation modes or (b) just one.

a model with a qualitatively new shape for the quasiparticle dispersion curve (the energy versus the momentum, ϵ_p or $\omega(p)$) sketched in Figure 5.4a. The linear dispersion relation $\omega = c_s p$ for phonons at small p was supplemented by a turn downward, toward a new parabolic minimum at some nonzero p_{min}. The energy of the corresponding minimum, $\Delta = \omega(p_{min}) = 8.65$ K, is called the roton gap. The excitations around the minimum add to the thermodynamic quantities a contribution $\sim \exp(-\Delta/T)$, which described the data at low T rather well.

Of course, nobody at the time knew what the physics of this new dispersion curve was. The legend goes that I. E. Tamm, present at the Landau's talk introducing this idea, suggested the name "rotons," which stuck to it. But, it remained unclear whether the phenomenon implied some relation to rotations, and if so what form of rotation it was supposed to be. Later Feynman proposed a model ascribing rotons to certain vortex rings. While those indeed do exist in He-II and play an important role—to be discussed in Section 5.6—it eventually turned out that this idea was wrong, and the rotons are not related to the vortex rings at all.

Experimentally, the dispersion curve has been directly measured, in particular by Palevsky et al. in 1958 [32]. The idea of the experiment, proposed by Placzek and van Hove [34] and others, was to perform an inelastic neutron scattering on liquid He-II, in which one quasiparticle is produced. The dispersion curve can then be directly deduces from the energy and momentum transferred to the medium. Once the dispersion curve was measured, the roton minimum predicted by Landau was indeed found. It also resolved the question of whether there are two separate modes or a single continuous phonon-roton curve. The latter option, sketched in Figure 5.4b, turned out to be the case.

Eventually, many-body theory development also explained the physical nature of the rotons. They turned out to be just the longitudinal oscillations, the same as phonons. The particular deformation of their dispersion curve was explained by Feynman [33]: it is caused by the formfactor of a liquid, enhancing scattering in the liquid at momenta corresponding to the wavelength equal to the mean interparticle distance. Its nature is therefore related to angular peaks observed in scattering in crystals.

I already mentioned the value of the roton gap, $\Delta = 8.65$ K. Quite fortunately, it happens to be a bit larger than the binding energy of a single He atom to the liquid, $B = 7.15$ K. So a roton, propagating upward in a liquid He-II and striking its surface from below, can knock one atom upward, out of the liquid! It took a long time for experimentalists to be able to detect those events on an individual basis: they first succeeded in the 1980s. By that time, advances in electronics allowed experimentalists to follow the events in real time, from the moment when a pulsed heater was switched on and produced rotons, to the detected atoms jumping up out of the liquid. Relating the time of roton propagation, from the heater to the glass surface, to its energy allowed observers to study the dispersion curve with better accuracy.

5.4 Quasiparticle Decays in a Weakly Coupled Bose Gas and in He-II

The term "quasiparticle" implies that, while some of the excitation modes may retain the quantum numbers and certain properties of the original constituents from which, the material was built, many of these properties are unrelated to them. Many can only exist in a specific phases of matter.

Like real elementary particles studied nowadays in high-energy colliders, these quasi-particles can participate in various elastic and inelastic collision processes. One example just discussed in the preceding section is roton production in inelastic neutron scattering. Those processes of course determine multiple kinetic properties of matter. But since this is an introductory book, we will not go into this vast subject here.

The example we will discuss is the simplest possible process: their decays into two or more other quasiparticles. But before we do so, let us go over some terminological conventions. An excitation mode is called a "quasiparticle" provided it lives long enough. This may sound like a poorly defined notion, and it is: the time scale in question depends on the physical processes under consideration. To explain what is meant here, let me give an example from physics of the QCD vacuum and hadrons, to which we will turn by the end of the book. Quarks are elementary constituents of matter, and in the Standard Model of particle physics, they are neatly organized into three families of electroweak doublets: (u, d), (c, s), and (t, b). The former two are actually nearly massless, but certain phenomena—to be discussed in Section 17.4—make them into quark quasiparticles with masses on the order of $m_{\text{eff}} \approx 400$ MeV. While other phenomena—to be discussed in Section 16.6—still prevent them from moving freely and becoming simply particles, we know about this m_{eff} from the fact that typical masses of quark-antiquark mesons are $\approx 2m_{\text{eff}}$, of three-quark baryons are $\approx 3m_{\text{eff}}$, and so forth. The only exception that does not fit the idea of quark quasiparticles is the top quark t, for the reason that its lifetime is an order of magnitude too short to "get properly dressed." So, even though it is still quite well defined as a quantum state, its mass is about a factor of 100 larger than the width; since the top quark cannot form mesons or baryons, it has no relevance to hadronic physics.

Let us now return to low-temperature physics and phonons. Any quasiparticle can decay, provided it is allowed to do so by conservation of the relevant quantum numbers, the total energy, and momentum. To elucidate the requirements for phonons in particu-lar, let us start with collinear kinematics first, taking all three momenta to be in the same direction. Then there are just two conditions to be satisfied:

$$\epsilon(p) = \epsilon(p_1) + \epsilon(p_2), \quad p = p_1 + p_2. \tag{5.4}$$

Let us start with phonons at small momenta. In this case one can expand the dispersion relation to the cubic term

$$\epsilon(p) = c_s p + C p^3 + O(p^5) \tag{5.5}$$

and substitute it into the energy conservation equation. The linear terms on both sides trivially cancel because of momentum conservation. Indeed, strictly collinear decays of truly massless particles with linear dispersion curves are allowed. However, they would not proceed in practice, because there is zero phase space available for the reaction.

Let us see what happens if the corrections to the linear dispersion law are included, starting from the weakly coupled Bose gas we discussed in Chapter 4. The sign of the cubic coefficient C is crucial. Note that $p^3 = (p_1 + p_2)^3 > p_1^3 + p_2^3$, because all momenta are presumed to be positive. So if $C < 0$, the initial energy is insufficient for the final state, and thus no decays are possible. If it is positive, $C > 0$, then there is more energy than needed in the initial quasiparticle: now both equations can be satisfied by allowing some nonzero transverse momenta in the final state. However, those components will be parametrically small at low p, because the nonlinear term is itself small in this case. So, a phonon can split into two (or more) phonons, located in a small angular cone around its direction, provided the dispersion relation has positive $C > 0$.

The quasiparticle dispersion relation in a weakly coupled Bose gas we have already calculated in Chapter 4 (See (4.28)). Let us expand it at small momenta to the next term:

$$\epsilon_p = \left(\frac{1}{2m}\right)\sqrt{p^4 + 4\lambda n_0 m p^2} \approx p\sqrt{n_0\lambda/m} + p^3 \frac{\sqrt{n_0\lambda/m)}}{8mn_0\lambda} + O(p^5). \tag{5.6}$$

So the cubic coefficient is positive and thus the near-collinear decay of phonons is indeed allowed. Writing $p = \sqrt{p_l^2 + p_t^2} \approx p_l + p_t^2/(2p_l)$ for p_1, p_2 and plugging it into the (5.6) we find the following expression for energy conservation,

$$C p^3 = C(p_{l1}^3 + p_{l2}^3) + p_t^2 \left(\frac{1}{2p_{l1}} + \frac{1}{2p_{l2}}\right), \tag{5.7}$$

which yields a positive solution for p_t at given p_{l1}, p_{l2}. For example, for the symmetric point $p_{l1} = p_{l2} \approx p/2$, we have $\alpha p^4(1 - 1/8 - 1/8)/2 = p_t^2$. Quasiparticles at low p have transverse momentum $p_t \sim p^2 \ll p$. So the angle of the decay is small and so is the decay phase space. This is the reason small-momentum phonons have very long lifetimes.

In the particular case of a weakly coupled Bose gas, calculation of the precise lifetime of quasiparticles—known as the Belyaev lifetime—was one of the main successes of the approach based on Feynman diagram resummation. Without derivation, let me explain just one crucial feature of this system: the power p^5 in the resulting $1 \to 2$ phonon decay rate. It comes from

$$\gamma_B = \frac{9c_s}{2^7\pi^2 mn_0} \int d^3 p_1 |p| |p_1| |\vec{p} - \vec{p}_1| \delta\left[\epsilon(p) - \epsilon(p_1) - \epsilon(\vec{p} - \vec{p}_1)\right] \sim p^5. \tag{5.8}$$

Let us now count the powers of the momenta. The integration over transverse and longitudinal momenta can be separated: $d^3p = dp_l \times dp_t^2/2$. Recall that only small angles contribute, $p_t^2 \sim p^4$, so the differential is actually $O(p \times p^4)$. Three momenta add another p^3. The argument of the delta function is $O(p^3)$, which should be counted as the negative power -3. The total power is then $p^{1+4+3-3} = p^5$.

Exercise *The phonons are the so-called Goldstone modes related to spontaneously broken Lorentz symmetry (matter is at rest in some particular frame, not the others). Therefore, a general Goldstone theorem requires that any Goldstone mode with momentum going to zero does not interact, as it becomes a symmetry transformation. Since matrix elements are analytic in momenta, and the amplitude vanishes for each vanishing momentum, the lowest-order option is just their product. Show that the decay matrix element squared is proportional to the product of all three momenta moduli.*

Quasiparticle decays in liquid ^{4}He are, however, entirely different. The phonon dispersion relation is curved in the opposite direction, as was proposed by Landau and is seen in Figure 5.4b. The coefficient of the cubic term is negative, $C < 0$, and therefore phonon decay processes into two is forbidden.

This conclusion holds at small momenta. Let us now discuss whether the $1 \to 2$ decays may in fact occur near the roton minimum. The roton dispersion curve near minimum can be approximated by

$$\epsilon_p = \Delta + (p - p_0)^2/(2m^*),$$

With some new parameter m*. One needs at least energy $> 2\Delta$ to create two rotons. Can it actually happen with both momenta of the final rotons being at the minimum, with the modulus of the momentum being p_0, while still producing the total momentum $\vec{p} = \vec{p}_1 + \vec{p}_2$ appropriate for the double energy, $\epsilon_p = 2\Delta$? Yes, by selecting a proper angle between the two secondary phonons, it is in fact possible. Therefore, rotons with momenta satisfying $(p - p_0)^2/(2M^*) > \Delta$ are indeed able to decay into two rotons.

Interestingly, above this momentum threshold, the roton dispersion curve measured by neutron scattering becomes so wide that it effectively loses its meaning as a dispersion curve. The verdict is that when the rotons are too short-lived, they no longer deserve the name "quasiparticles"; the roton curve thus effectively ends. For more details on quasiparticle decays, see the classic book by Khalatnikov [25].

5.5 Landau's Criterion for Superfluidity

Suppose a body of mass M moves in He-II. What are the conditions required for it to emit an elementary excitation? Energy and momentum conservations require

$$\frac{Mv_i^2}{2} = \frac{Mv_f^2}{2} + \epsilon_p$$
$$M\vec{v}_i = M\vec{v}_f + \vec{p}.$$

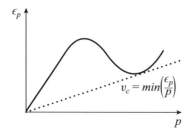

Figure 5.5. Sketch of the phonon-roton excitation dispersion relation ω_p. The dotted line indicates the r.h.s. of the Landau superfluidity condition.

Eliminating the final velocity v_f by solving for it in the second equation and substituting the result into the first, we get

$$\epsilon_p - (\vec{p}\vec{v}_i) + \frac{p^2}{2M} = 0.$$

If M is very large, the last term can be dropped, and since the cosine of any angle is less than 1, the following condition should be satisfied:

$$v_i > v_c = \min\left(\frac{\epsilon_p}{p}\right). \tag{5.9}$$

This argument by Landau, very characteristically, relies only on general conservation laws: so it must be correct! However, let us test whether it works in practice. The critical velocity is thus given by a line tangent to the dispersion curve, as shown by the dotted line in Figure 5.5. Using the data for the phonon-roton curve, however, one finds it to be

$$v_c \approx 58\,\text{m/s}. \tag{5.10}$$

While generally being of the same order as the critical velocity measured in experiments, it is not in agreement with them, being systematically larger. We will address this puzzle later in the chapter. As a hint, let me now just say that the criterion is correct but phonons/rotons are not the only excitations in the system.

Landau emphasized that this criterion, if applied to an ideal Bose gas with excitations $\epsilon_p = p^2/(2m)$, leads to zero critical velocity v_c. So an ideal Bose gas, while possessing a condensate, is not a superfluid! More generally, Landau argued that relating He-II phase to Bose-Einstein condensation is misleading. As we will see, he was not correct; superflow and the condensate are in fact directly related.

5.6 Rotation and Vortices

Let us now forget about the rotons for a while and discuss true rotation of a superfluid. What happens if one takes a glass filled with superfluid He-II and starts rotating it?

From what we know so far, one would expect the so-called "Landau state" (actually predicted by him): the normal component ρ_n of the fluid would begin to rotate, while the super component ρ_s would not and remains at rest. Indeed, isn't that what happens in the Andronikashvili experiments we already discussed in Section 5.2?

To test different suggestions made on the issue, the experimentalists conducted such experiments and studied the depth of the meniscus[5] from the top of the glass. If part of the fluid does not rotate, one would expect this depth to be reduced. Quite embarrassingly, however, in experiments with He-II, its magnitude turned out to be the same as for the ordinary fluid, rotating as whole with the same frequency. This puzzle will be explained at the end of this section.

To understand rotation of liquid He-II, one needs to recognize the existence of some collective "condensate" of atoms in the ground state of the liquid, possessing some collective wave function in which macroscopic numbers of atoms participate coherently. All of them together define a collective complex wavefunction ψ, normalized to the density of the participating particles. One can imagine a simple effective action for it, and thus all general expressions about complex fields (wavefunctions) will be valid. In particular, the supercurrent should be related to the gradient of its phase α:

$$j_s = \hbar |\psi|^2 \vec{\nabla}\alpha.$$

In a toroidal geometry (as for a tube closed into a ring with the circular superflow we discussed in Section 5.1) the phase of the wavefunction should be well defined. This implies that variation of the phase over the ring cannot be arbitrary but periodic: therefore there is quantization of the flow,

$$\Delta\alpha = 2\pi n,$$

with integer n. This is exactly the same phenomenon as quantization of angular momentum in quantum mechanics to some integer times $\hbar$. So the velocity circulation of the superflow over a closed line C must be quantized as

$$\kappa = \int_C \vec{v}_s \vec{dl} = \frac{\hbar}{m_4} 2\pi n = n \left(0.998 \times 10^{-7} \frac{m^2}{s} \right), \tag{5.11}$$

where m_4 is the mass of a helium-4 atom.

It does not take a lot of thinking to realize that this quantization is in direct contradiction to an idea that a superfluid can be rotated as a whole. Indeed, if this were possible, the velocity at radius r would be proportional to r and circulation be $\sim r^2$, not a constant, as required. In fact the only allowed $v(r)$ dependence is the inverse one, $v(r) \sim 1/r$. Thus, the superfluid can only be rotated as a vortex!

Such *quantized vortices* in liquid helium do in fact exist, and they lead to many interesting phenomena. However, to see them is not easy, because the quantum of circulation is rather small. No matter how slowly the glass of liquid rotates and how small a sample is taken, the number of vortices is expected to be huge (millions).

---

[5] The *meniscus* is the central depression of liquid surface, displaying an equipotential surface, well known to anyone who has rotated tea in a cup with a spoon.

In a vortex the velocity $v(r) \sim 1/r$ becomes infinite at the center. Is this a problem? Not if the energy density $\rho_s(r)v^2(r)/2$ remains finite, which means $\rho_s(r) \sim r^2$, or the modulus of the condensate wavefunction vanishes at the center. Thus any vortex can be viewed as a toroidal motion around its center line, a "hole" in condensate that is excluded from the quantization condition. Where there is zero modulus, there can be any phase!

What happens at large distances from a vortex? Since the velocity circulation remains fixed and the velocity component in cylindrical coordinates is $v_\phi(r) \sim 1/r$, the vortex energy

$$E = \int d^3x \rho v^2 / 2 \sim \int dr\, r(1/r)^2 \sim \log\left(\frac{r_{max}}{r_{min}}\right) \tag{5.12}$$

depends logarithmically both on the upper (size of the glass) r_{max} and the lower (microscopic inner radius) r_{min} sizes of it.

A rotating glass contains many vortices, and their long-range repulsive interactions make up some kind of two-dimensional plasma state, with parallel ones repelling each other. At low T that rotating superfluid develops—as its ground state—a two-dimensional crystal of such vortices. Perhaps at this point it is useful to recall that by shifting to a rotating frame (like on rotating Earth), the Hamiltonian changes to

$$\hat{H} \rightarrow \hat{H} - \vec{\Omega}\hat{\vec{L}},$$

where Ω is the rotation frequency, and $\hat{\vec{L}}$ is the operator for orbital momentum. So vortices that make negative additions to the energy are present in the ground state, which then looks like zillions of mini-tornadoes, existing forever since there is no friction.

Let us cut right to the answer. The glass of rotating He-II is not rotating as a whole at all: in fact it is full of vortices, each with $n = 1$, in a two-dimensional crystalline lattice, in a plane normal to rotation axes $\vec{\Omega}$. To see them directly was not easy, but eventually it became possible.

Among other puzzles, the model describing the glass of rotating He-II by a plasma of vortices solved the puzzle about the meniscus as well. The density of the vortex plasma state was calculated, and it happens to be exactly such as to produce the value of the meniscus as if there were rotation as a whole, on average. Only by using a microscope can one see that in fact the surface is not smooth but covered by millions of little holes, the vortex lines.[6]

Let us now proceed to closed vortices, known also as vortex rings (see Figure 5.6). From the flow velocity direction it is clear that if the velocity at large distances goes to zero, the vortex should move as a whole, in the direction of the vertical arrow.

Before I can explain experiments with vortex rings, we need to understand how to produce them. For that one needs to embed into liquid helium certain objects that can be externally manipulated. Various positive or negative ions can be used for that purpose. In

[6] Does the vortex state contradict to the Andronikashvili experiments with the oscillating discs? No, it does not. Andronikashvili oscillations have time scales too short to create the vortex crystal, which can only be formed rather slowly in a persistent rotation.

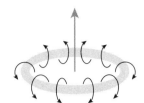

Figure 5.6. Sketch of a vortex ring.

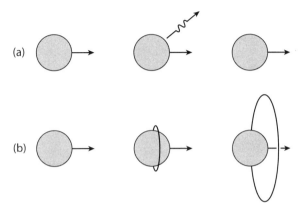

Figure 5.7. The moving ion can either (a) emit elementary phonon-roton excitation or (b) create a vortex ring.

fact the best positive ion is a bare ^{4}He nucleus, without electrons. Conveniently, these are abundantly produced by heavy isotopes undergoing α decays (e.g., ^{210}Po).

The bare ^{4}He nuclei interact with neutral atoms attractively: this can be seen as a result of their creating an increased density or small additional pressure around them. Recall that the liquid is not far from solidification, so a charge of $+2$ is enough to create small "snowflakes" of solid He around the bare nucleus! Its size is not even small, as seen from the fact that its typical mass is about 40 masses of He!

The most elementary negative ion is just an electron, also available to experimentalists from nuclear beta decays or electron guns. Extra electrons repel other electrons in He atoms: so they, in contrast, creates small voids around themselves.

Moving in a fluid, a snowflake or a void experiences a certain amount of friction, by means of two mechanisms, sketched in figure 5.7. If a vortex ring is produced, the ion will move with it. By supplying some energy to the ion (by applying a certain voltage across some electrodes imbedded in the helium), one feeds the ring and increases its excitation energy. This makes the ring larger, but a bit paradoxically, causes the ring to decrease its velocity: the velocity of the ring is $v \sim 1/R$. Therefore, in time-sensitive experiments, there are ions with and without vortex rings: the excitations simply separate into two groups, moving with completely different speeds.

What is curious is that these rings can be grown to become truly macroscopic beasts, with a radius reaching up to 10^4 angstroms, and their total energy can be as large as 100 eV, a million times the ambient temperature. Unlike rotons, with $\Delta \sim 8$ K gaps, large vortex rings cannot be excited thermally by a heater and instead need some special source, like a moving ion. When the ring gets too large, the ion may decouple and move by itself, again accelerating and producing another—initially smaller—ring.

Vortex rings are rather exotic objects but are still very important ones. They have radii comparable to radii of the tiniest pores discussed when I introduced superfluidity in Section 5.1. And it is the production of these vortex rings that eventually explained the observed critical velocity of the superflow! (To my limited knowledge, the generation rate of the vortex rings remains a bit of a mystery: its T dependence suggests some roton assistance near T_c.)

6 Bose-Einstein Condensation of Trapped Ultracold Atoms

Although we have already discussed the BEC phenomenon itself in ideal and weakly coupled Bose gases, this chapter starts with a digression on its convoluted history. We then focus on specific applications of the weakly coupled theory, such as condensate depletion (Section 6.2). The nonlinear equation for the condensate, known as the Gross-Pitaevsky equation, is introduced for repulsive gas interactions in Section 6.3. In the case of attractive interactions there are interesting instabilities of the condensate, mini-supernovas, which we also discuss in that section.

Rotation in BEC systems should be of the same type as in rotating liquid helium—in the form of an ensemble of quantized vortices. The experiments revealing that this is indeed the case are discussed in Section 6.4.

6.1 History of BEC

In 1924 Satyendra Nath Bose, unknown in the West and originally from Calcutta[1] but at that time a young physics professor in Dacca (now in Bangladesh), wrote a paper about the proper statistical calculation for Plank's quantum hypothesis. He could not publish it and wrote a letter to Einstein, which is preserved and can be found on the Internet, in English. Einstein understood the importance of Bose's paper, translated the paper into German and submitted it, as a member of the German (Prussian) Academy, to its journal. The main point of Bose was that all photons are indistinguishable. Appropriately, not only

---

[1] Bose had a long life in Calcutta, and an Indian friend of mine, Bikash Sinha, was in fact mentored by him as a boy: so I am, so to speak, one handshake removed from Bose. Einstein was much closer, in Princeton, New Jersey, and he even rented a house at Stony Brook for few summers, not so far from mine. And yet, there was no Stony Brook University then, and all the Princeton people I know are too young to have known Einstein. My "one handshake" with Einstein is due to my teacher at Novosibirsk, Yuri Rumer, who got a fellowship administered by Einstein in 1920s and was even invited for a dinner in Einstein's house in Berlin.

photons but also all other particles having integer spin are what we now call "bosons"[2] (the name is due to Dirac).

This is a good time to mention the fundamental spin and statistics theorem, stating that all particles (atoms, molecules, nuclei, hadrons) with integer j are *bosons*, and those with semi-integer j are *fermions*.[3]

The year after Bose's paper, 1925, Einstein wrote another paper, this one by himself: it was not about photons, but about monoatomic (noble) gases. Not only had Einstein enormously generalized applications of the Bose statistics by applying it to atoms, but he found a completely new phenomenon: at high enough densities, a finite fraction of particles must belong to a single condensate state, with zero momentum, since the Bose expression at some T and $\mu = 0$ can only describe a limited number of particles (what we called n_T in Chapter 4). Thus BEC is basically the wrong name: one should rather call it "Einstein's condensation," a consequence of Bose's statistics. Yet one cannot of course argue with the accepted terminology.

It took about 70 years between Einstein's paper and the actual observation of BECs in trapped atoms. This long time was required to develop the appropriate refrigeration methods to reach sufficiently low temperatures. There were three competing groups:

1. JILA group (Anderson et al.) using Rb
2. MIT group (Devis et al.) using Na
3. Rice University group (Bradly et al.) using ^{7}Li

The reader who wants to know much more about the subject than will be found in this chapter should look at two excellent books by Pethick and Smith [35] and Pitaevskii and Stringari [36].

Note that all the atoms used by these experimentalists are alkali metals, so they have a single electron above the last closed shell, and thus they all have semi-integer atomic angular momentum. To make them bosons, one should only use isotopes for which the nuclei also have a semi-integer j, so that the total momentum of the atom+nucleus is integer.

I already mentioned that the major technical goal/achievement of these experimentalists was of course reaching sufficiently low temperatures. While in the 1920s the low-temperature frontier, in Leiden, reached T on the order of a Kelvin (a factor of 300 below room temperature), BEC in dilute gases had to wait until the attainable T range moved down by another 6 (now 8) orders of magnitude. To emphasize this great achievement, let me compare it to the progress in high-energy physics. During the twentieth century that field moved from producing an atomic energy of 1 eV (of say, X-rays) to the LHC

---

[2] Bose, once asked by a journalist what he felt about the fact that he never got a Nobel Prize, proudly answered that he had already gotten all the recognition he deserves. And indeed, having half of all fundamental particles bearing your name, like the latest "Higgs boson," should indeed be enough.

[3] Amusing historical remark: the fundamental bosons—W, Z, H—were all discovered at CERN, while the fundamental fermions—positron, muon, tau-lepton, and all types of quarks and neutrinos—were all discovered in the United States. A popular joke has it that Europeans like to be in a collective, while Americans prefer individualism.

collider, with energies on the order of $10\,\text{TeV} = 10^{13}\,\text{eV}$, covering 13 orders of magnitude. The experiments in low-temperature physics moved down during the same time by more than 10 orders of magnitude. Unfortunately, I am even less of an expert on cooling than on accelerators, so I am not able to describe how exactly such cooling has been achieved.

Let me instead describe the setting itself. A certain number N of atoms are trapped in a potential that is approximately harmonic:

$$V(x, y, z) = (m/2)(\omega_x^2 x^2 + \omega_y^2 y^2 + \omega_z^2 z^2). \tag{6.1}$$

There is no thermal contact with any macroscopic surfaces or bodies, of course: the atoms are suspended in a vacuum, trapped by lasers and magnetic fields. Their number is incomparably small compared to the Avogadro number of atoms in macroscopic bodies: a typical N is 10^6 or fewer.

At high T, when one can use classical statistics, the probability distribution of atoms over coordinates in a harmonic trap is Gaussian, $\sim \exp(-V(x, y, z)/T)$, in shape. When T is comparable to quantum level spacings $\hbar\omega_i$, one should recognize their discreteness, and perform the appropriate statistical sums. At vanishingly low $T \ll \hbar\omega_i$, all Bose particles are expected to occupy only one lowest state. The distribution in this case is defined by the ground-state wavefunction $P \sim |\psi_0|^2$, where quantum mechanics tells us that

$$\psi_0 = \Pi_{i=1,2,3} \left(\frac{m\omega_i}{\pi\hbar}\right)^{1/2} \exp(-m\omega_i x_i^2/\hbar). \tag{6.2}$$

Note that its shape is also Gaussian,[4] but the width is completely different and is much smaller.

The problem we are addressing now is of course very different from what we examined in the Chapter 1. We have not one but many atoms. Also, what can be measured are not the coordinates but the momenta of atoms. It is done in a very simple way: by switching off the trap and letting the particles flyout. Since particle-particle interaction is negligible, each particle conserves its instantaneous momentum. The larger their momenta are, the farther the particles fly from the origin. At some time the cloud is illuminated by a laser pulse, and a photo is taken of the cloud. Since the clouds are extremely dilute and cold, they are destroyed by the laser pulse, so only a single photograph of each cloud can be made. As cooling progresses, it was seen that the momentum distribution develops a peak at small p, the condensate (see Figure 6.1). Two Nobel prizes were given to six people: one prize in 1997 for cooling methods; the other was awarded in 2001 for BEC observation to Eric Cornell, Wolfgang Ketterle, and Carl Wieman.

It sounds simple enough. But recall that we are speaking about atoms, not light quanta. Their number in the trap, and subsequently in the condensate, may be very much smaller than Avogadro's number but still very large typically $N = 10^6$. To force them into a single coherent quantum state, in which they are described by one collective wave function, is indeed a spectacular achievement. To appreciate it, let me give some comparison of the

---

[4] In fact we already derived from the path integral that a particle in a harmonic trap has Gaussian spatial distribution at any temperature, with a T-dependent widths; see equation (1.17).

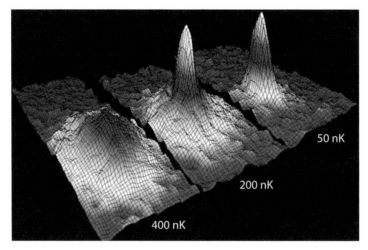

Figure 6.1. Momentum distribution of the trapped Bose gas, from JILA experiments, 1995. The peak at small momenta at low T (the right plot) is the Bose-Einstein condensate. This image is from JILA, University of Colorado, Boulder, and NIST, credited to Mike Matthews, from hyperphysics. Phy-astr. gsu.edu/hBose/quantum/rubbec.html.

thermal velocities of atoms and the corresponding wavelengths, at room temperature and at very cold ones:

$$T = 300\,\text{K}: \ mv^2/2 = T \rightarrow v \sim 1\,\text{km/s}; \quad \lambda = \hbar/mv \sim 10^{-8}\,\text{cm}$$

$$T = 10\,\text{nK}: \quad v \sim 1\,\text{mm/s}; \quad \lambda \sim 10^{-3}\,\text{cm}.$$

To make a condensate, one needs the particle waves to interfere. Qualitatively, one needs the de Broglie wavelength to reach the interparticle distance $R = n^{-1/3}$. The numbers just given thus explain why such low temperatures are required.

6.2 Depletion of the Condensate

Diluteness of trapped Bose atoms makes them a perfect object for application of the diagrammatic theory of weakly coupled Bose gases discussed in Chapter 4. As an interesting and quite nontrivial example, let me pick up an expression for depletion of the condensate due to interparticle interaction:

$$n = n_0 + \int \frac{d^3 p}{(2\pi)^3} v_\text{p}^2, \tag{6.3}$$

where v_p is the residual of the quasiparticle pole in the propagator

$$v_\text{p}^2 = \frac{p^2/(2m) + \lambda n_0}{2\omega_+} - \frac{1}{2}. \tag{6.4}$$

The effect, after the integration, is on the order of $\sim (\lambda n_0 m)^{3/2} \sim (a R)^{3/2}$.

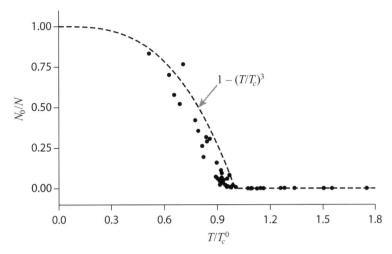

Figure 6.2. The fraction of atoms in a condensate, as a function of the normalized temperature T/T_c. The dashed line corresponds to $1 - (T/T_c)^3$ behavior expected for an ideal noninteracting gas.

In Figure 6.2 the measured amplitude of the lowest momentum peak is compared to the ideal gas expression (the dashed line). The power of T in the density in this case is not 3/2, as in Einstein's textbook expression, but 3: this is because the gas is in a harmonic trap, not in a box. The phase space volume element is $d^3 p d^3 x$, adding another 3/2. The depletion is quite visible in the figure. The figure is from the experiment that observed the effect for the first time: since then the comparison has been made more exact.

In general, the diagrammatic theory (with resummations) of weakly coupled Bose gases has been tested in detail. Let me add that, using the so-called Feshbach resonance in a magnetic field, the experimentalists learned to change the scattering length a over a wide range, nearly from minus infinity to infinity, producing a perfect laboratory for many-body theory. The perturbative and nonperturbative formulas were tested from weak to very strong coupling, especially for Fermi gases, as we discuss below.

6.3 Gross-Pitaevsky Equation and Repulsive/Attractive Bose Gas

At very low T one can ignore the out-of-condensate atoms and focus on the condensate alone. The point is, one can write down a Schrödinger-like equation for the condensate's collective wave function.

In the ideal gas, in which atoms do not interact, the condensate is formed in the lowest *single particle state* of the trap, to which we already alluded above. That wavefunction is a solution of the ordinary Schrödinger equation, that for a single atom. Adapted from J. R. Ensher, D. S. Jin, M. R. Matthews, C. E. Wieman, and E. A. Cornell, "Bose-Einstein Condensation in a Dilute Gas: Measurement of Energy and Ground-State Occupation," *Physical Review Letters* 77, 4984 (1996).

However, if the particles do interact, the collective wavefunction of the condensate satisfies a modified *nonlinear* Schrödinger-like equation, which follows from variation of the effective action. Known as the Gross-Pitaevsky equation, it has the form

$$i\hbar \frac{\partial \psi}{\partial t} = \left(-\frac{\hbar^2}{2m}\vec{\nabla}^2 + V_{\text{trap}} + g|\psi|^2 \right)\psi, \tag{6.5}$$

where the last self-interaction term has coupling $g = 4\lambda$ in terms of the notation we used in Chapter 4.

Since the equation is nonlinear, there is no superposition principle, and the normalization issue needs to be explained. The time dependence is given by an input μ, $\psi \sim e^{-i\mu t}$, so the l.h.s. is just $\mu\psi$. Then the nonlinear equation needs to be solved and the solution plugged into the expression for the particle number:

$$N(\mu) = \int d^3x |\psi(x)|^2. \tag{6.6}$$

The equation and solution depend on some input value of μ, which should be determined from the normalization of (6.6) to the actual number of atoms.

Let me compare scales of the kinetic energy (the first term on the r.h.s. of equation (6.5)) with the interaction term (the last term):

$$E_{\text{kin}} \sim \frac{N\hbar^2}{mR^2}, \quad E_{\text{int}} \sim \frac{N^2 a\hbar^2}{mR^3}, \tag{6.7}$$

where R is the spatial size of the wave function. The scattering length is of atomic scale, much smaller than R, but N can be as large as a million or so. Thus the last term can be dominant in some cases.

Suppose this is the case, and one can also ignore for now the gradient term (I will explain this neglect shortly). Then the Gross-Pitaevsky equation becomes algebraic, and since the ground-state wavefunction has no sign changes ($\psi > 0$) it can be dropped. A simple solution arises:

$$\Psi^2(r) = n(r) = \frac{2\mu - kr^2}{(8\pi Na/R_{\text{osc}})}R_{\text{osc}}^{-3}, \quad r < \sqrt{\frac{2\mu}{k}}, \tag{6.8}$$

where I have substituted in the parabolic potential $kr^2/2$ of the trap. Note also the weak coupling times the large particle number in the denominator.

The lesson is then as follows: the shape of the condensate is not Gaussian but is in fact parabolic! A look at the experimental results in Figure 6.3 confirms that this is indeed the case. This comparison nicely demonstrates the very meaning of chemical potential μ: it is indeed basically like the water level in a coffee pot: the density is nonzero only below the level.

The integral over this shape provides the relation of the particle number to the chemical potential:

$$\mu = \frac{\hbar\omega_{\text{trap}}}{2}\left(\frac{15Na}{R}\right)^{2/5}. \tag{6.9}$$

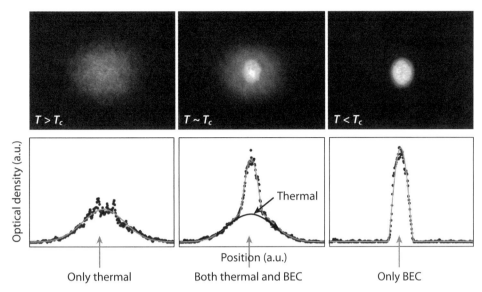

Figure 6.3. The spatial density distribution of the trapped Bose gas at three different temperatures. Credited to University of Michigan BEC group, from http://cold-atoms.physics.lsa.umich.edu/.

Note that here a small parameter a/R is multiplied by a large particle number N, which makes the product not small at all.

(A very skeptical reader may wonder whether we are justified in omitting the gradient term. It is only important at the edges, at distances on the order of the so-called healing length

$$\xi = \sqrt{8\pi na},\qquad(6.10)$$

which is small, since it contains the small a/R parameter.)

Without mentioning it, it has been assumed above that $a > 0$; that is, the atomic interaction is repulsive. What happens in the opposite case of an attractive interaction? Formally, nearly all expressions above become apparently meaningless if $a < 0$, since they contain noninteger powers of it, leading to imaginary parts, perhaps indicating some instabilities in the system.

Here comes my recollection and my own (5-cent) role in the development of this theory. R. Hulett from Rice University was giving a physics colloquium at Stony Brook in 1995, describing the Rice group's exciting new data on BEC formation with ^{7}Li atoms. At the end, during a discussion session, he said: There is only one problem we don't understand. The measurements of the scattering length gave a negative value, namely,

$$a_{7\text{Li}} = -1.45\,\text{nm},\qquad(6.11)$$

which means our atoms are attracted to one another. And yet, all classic papers on interacting BEC—such as the Lee-Yang paper of 1957, he said—insisted that BECs can only be formed if the interaction is repulsive. C. N. Yang, sitting of course in the first row just in front of the speaker, nodded affirmatively. "And yet, we still do see BEC in our experiments!" replied Hulett.

After the colloquium I went to my office and soon figured out what was happening. All classic papers, like Lee-Yang et al., considered an infinite homogeneous system with macroscopically large $N \to \infty$ and effective action with interaction $\sim a\psi^4$. Of course, in the case with $a < 0$, this term leads to unlimited negative energy and growth of ψ.

In a trap, N is large but finite, and it is possible that the condensed state is metastable even for $a < 0$. Indeed, in the Hamiltonian

$$H = \int d^3x \left[\frac{\hbar^2}{2m} |\vec{\nabla}\psi|^2 + V|\psi|^2 + \frac{N2\pi a\hbar^2}{m}|\psi|^4 \right] \tag{6.12}$$

with the normalization condition (6.6), the three terms scale with the size of the cloud R as $1/R^2$, R^2, and $1/R^3$, respectively. The last one dominates at small R, and if it has a negative coefficient, as is the case for ^{7}Li (but not others atoms used for BECs!), the potential goes to $-\infty$ at small R, which means that the system collapses. However, the other two terms may well create a metastable minimum, separated by a barrier from the small-R region. By substituting the actual N used in these experiments, I found that this is where the experimental BEC cloud was sitting. I predicted that with the further increase in N, one would eventually observe what I called a "mini-supernova," an internal collapse of the condensate, with atoms leaving the trap. I told Yang about my results, and he agreed that they made sense. I then wrote a paper [38]. I learned later that the idea itself happened to be known in general terms, but I was perhaps the first to point out that the metastability conditions were in fact fulfilled for the experimental setup of the Rice experiment. In the paper I not only estimated the critical number of atoms to be reached for metastability to be lifted and classical collapse to occur, but also evaluated a quantum tunneling rate through the barrier for N close to critical value.

Exercise *For ^{7}Li atoms with attractive interaction, derive the expression for the critical number of particles above which the metastable minimum cannot exist.*

Being on sabbatical at MIT in 1997, at lunch I told Kerson Huang about this paper. He became interested in the problem and he and his group did excellent studies of the time-dependent process of collapse by solving the time-dependent Gross-Pitaevsky equation [39] numerically. Their paper predicted periodic events of the appearance and collapse of the condensate. The name they used for the phenomenon, a *bose-nova*, stuck to it. The phenomenon was finally observed experimentally few years later, again by Hulett's group [40].

6.4 Rotation

After BEC for trapped atoms was convincingly demonstrated, the next intriguing question was whether this system can display superfluidity. Great experiments, mostly by the MIT group, managed to produce a rotating system of trapped atoms, observe a superfluid transition with some nonzero critical velocity [37], and eventually produced visible lattices

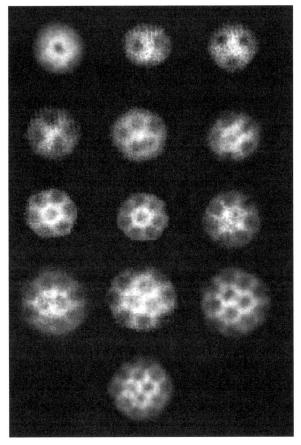

Figure 6.4. Vortices in the Bose-Einstein condensate, observed by the MIT group. Photo credit: Andre Schirotzek (MIT), www.rll.mit.edu/cua-pub/ketterle-group/experimental-setup/BEC/image gallery.html.

of vortices. These pictures, shown in Figure 6.4, show that the experimentalists mastered the art of excitation of the cloud to practically any desired number of vortices. These experiments put to rest any doubts about the superfluidity of the BEC.

Should one be surprised? I think these experiments are as close to an experimental miracle as one can possibly get. Indeed, such an experiment puts about a million atoms into a single coherent state described by a single-argument wave function. Then it is put into a rotational state, which settles down into a crystal of, say, a dozen vortices. The system has only 100×100 atoms in the plane normal to the rotation axes, so each vortex— is made of a restricted number of atoms. And still each is involved in the same complex collective dance of the atoms, reproduced experimentally event by event.

Currently people who study trapped Bose gases have become even more ambitious. The atoms can be split into some one-, two- or three-dimensional arrays of traps, and experimentalists are attempting to use their states as the basis for a quantum computer. This fascinating subject, however, goes well beyond the scope of this book (and my expertise), so we stop at this hint.

7 | The Electron Gas

So far in this book we have discussed scalar fields only. Their quanta have spin zero and, therefore, they are bosons. However, particles that obey Fermi-Dirac statistics (fermions, for short) play a more important role in Nature, in many-body theory, and, consequently, in this book. The electron gas we discuss in this chapter is the first classic example. There are many more examples, and therefore we start this chapter by enumerating which fermion systems we will be discussing in the rest of this book.

In Section 7.2 we generalize our discussion of the Matsubara frequency sums, performed via the Sommerfeld-Watson trick, to fermionic systems. In Section 7.3 we then discuss the lowest-order diagrams, those for a weakly coupled Fermi gas. The results are then applied to the cold electron gas (neutralized by positively charged "jellium") in Section 7.4.

Renormalization of the Coulombic forces using the polarization diagram is discussed in Section 7.5. The derived expression, via the Lindhard function, explains screening of the charge as well as the generation of a new quasiparticle, known as a *plasmon*. In Section 7.6 we switch to a discussion of an electron system apparently forming a strongly coupled liquid.

7.1 Fermi Systems Discussed in This Book

Electron gas in metals was the classical system of interest, for obvious reasons. It is of course the basis of quantum condensed-matter theory. We will not discuss specific metals but limit our study to the so-called jellium model, with homogeneous positive charges compensating the negative charges of the electrons. The model was historically important, and it is in this context that the resummation methods had first been applied. The Fermi energy or μ is in this case on the order of the atomic energy:

$\epsilon_f \sim me^4/(2\hbar^2) \sim 13$ eV. Whether the interaction parameter is weak or strong depends on the details.

Electron gas in white dwarfs is mentioned just for completeness. In this case gravitational compression is such that the corresponding density increases by many orders of magnitude, so that the Fermi energy reaches $\epsilon_f \sim 0.3$ MeV, which is in this case comparable to the electron mass, and the electrons become semi-relativistic. We will not discuss this case in detail.

Liquid ^{3}He, unlike its bosonic cousin ^{4}He, this substance does not show BEC-like transition at temperatures of a few kelvins. The interesting phase transition temperature is in this case about factor of 1,000 lower: 10^{-3} K. Here it does form two very different and very interesting superfluid phases, which we will discuss in Chapter 10.

Nuclear matter, both in nuclei and neutron stars, is an object of many-body theory on which historically perhaps maximal amount of theoretical human-years has been spent. It is a quantum fluid, in many respects similar to the electron gas, except that the basic interaction is much more complicated. I will use it as an example of very powerful applications of the renormalization group method.

Cold quark matter has not been directly experimentally observed, but it is perhaps present at the cores of neutron stars. It is, however, a methodologically interesting example of an ultrarelativistic Fermi gas, with the chemical potential μ much larger than the quark masses, $\mu \gg m_q$. Cold quark matter also is believed to develop the so-called *color superconductivity* phenomena, with several interesting phases. We will study them near the end of the book, in Chapter 20.

Hot quark-gluon plasma is produced experimentally in heavy ion collisions and is relatively well studied. It is made of interacting relativistic quarks and gluons. It is theoretically interesting because the interaction shifts from weak to strong as a function of temperature, resulting in certain phase transitions to hadronic matter. Experimentally accessible temperatures are up to the scale of 0.5 GeV, see Chapter 16.

Trapped/cooled fermionic atoms are a new toy that physicists have developed, starting at the beginning of this millennium. It is an extremely versatile tool, since the scattering length a can be changed by the Feshbach resonance over a very wide range. The parameters of the trap and the cloud are comparable to that of bosonic atoms we just studied in Chapter 6. The temperature can be either large or small, compared to the Fermi energy. These systems also transition from the usual superconductivity, with loosely bound Cooper pairs, to BECs of certain deeply bound binary molecules. In short, one can hardly think of a better device to study quantum many-body theory in experiment! See chapter 12.

For all of these systems, we will follow similar paths, starting from the general setting and noninteracting cases, then including the interaction perturbatively and calculating the first few diagrams and trying to do certain resummations of higher-order ones. The renormalization group (RG) flow will lead us to certain effective theories (e.g., Landau Fermi liquid theory), and then we will consider the pairing phenomena and eventually various forms of superconductivity.

7.2 Path Integrals and Matsubara Sums for Fermions

With all this fascinating physics ahead of us, we should still start, so to say, at square one, with the requisite formalism, ensuring that we have a solid foundation and the right tools to understand the phenomena of interest.

The first of these tools is the definition of the path integrals for anticommuting Grassmann variables θ_i (describing fermions with some internal quantum number i). It was developed by F. Berezin (summarized in [41]; see also [42]) and can be found in any modern QFT course, such as [1].

This book is not the place for the formal exposition of theory. So, without much ado, here is the complete table of integrals for a single Grassmann variable:

$$\int 1 d\theta = 0, \quad \int \theta d\theta = 1. \tag{7.1}$$

There are no integrals with higher powers of θ listed, because self-anticommuting $\{\theta, \theta\} = 0$ implies that for the square, and thus any power above one $n > 1$, the combination is zero: $\theta^n = 0$.

Based on that, the generic formula for the multifermionic Gaussian integral with two Grassmanian vectors θ_i, η_i and a matrix A^{ij} connecting them into some bi-linear form, is

$$\int \exp(\theta^T A \eta) d\theta d\eta = \det(A). \tag{7.2}$$

This result can be proved by direct Taylor expansion of the exponent to first order.[1]

For fermions, one needs to append the periodicity rules to the Euclidean time circle, leading to sums over the Matsubara frequencies different from those we derived for bosons in Section 3.2. Recall that bosonic fields have periodic paths on the circle $\tau \in [0, \beta]$, thus the frequencies are $\omega_n = 2\pi n/\beta$ for integer n.

Now, rotating a vector (or spin-1 field) by an angle $\Delta\phi = 2\pi$ (i.e., making a circle) brings it to the same direction as it originally had. More generally, a rotation of any field is performed by the rotation operator $\exp(i\Delta\phi \hat{J})$, where $\hat{J}$ is the angular momentum operator. Therefore, spin-1/2 fields get only half of the the rotation angle, and for $\Delta\phi = 2\pi$, the spinor rotates only by π. This means it changes sign: $e^{i\pi} = -1$. The same *antiperiodic* boundary conditions need to be postulated for the spinor field on the Matsubara circle, namely,

$$\psi(\tau + \beta) = -\psi(\tau).$$

One consequence of this is an extra factor $(-1)^{\text{fermionic loops}}$ that has to be included in all Feynman diagrams. Another is that for fermionic fields, the Matsubara frequencies take

---

[1] Recall that for the analogous bosonic integral, one also gets a determinant, but to a different power: $-1/2$. Since the fermionic determinant is in the numerator and the bosonic in the denominator, theories with supersymmetry allow for complete or partial cancellation between both determinants. Due in part to such simplifications, supersymmetric theories are much easier to study, and they became a large industry in the 1980s and 1990s. Supersymmetry is also crucial for string theory.

on semi-integer values:

$$\omega_n = \frac{2\pi}{\beta}\left(n + \frac{1}{2}\right).$$

Let us see what the result of this is in the summation method à la Sommerfeld-Watson. To use it, one first needs to invent a new function, with poles at the semi-integer points along the imaginary axis of z. The solution is $f_F = 1/(e^{2\pi z} + 1)$ (rather than $f_B = 1/(e^{2\pi z} - 1)$ for bosons). All other steps are exactly the same: there appear three terms, residuals of poles of the function in the right semi-plane, in the left semi-plane, plus an integral over the imaginary axes. The only difference is that Bose-Einstein distribution for on-shell particles is replaced by the Fermi-Dirac one, as you might expect.

Example: Let us take the simplest nonrelativistic Green function for a fermion. In the frequency-momentum representation, it is the inverse of the action operator, and thus has the same form as for a boson,

$$G = \frac{1}{i\omega_n - \mu + p^2/(2m)}, \tag{7.3}$$

except the values that the Matsubara frequencies have are now different. The bubble diagram, in which a fermion returns to the same point, is the in frequency-momentum representation a sum-integral:

$$T\sum_n \int \frac{d^3 p\, V}{(2\pi)^3} \frac{1}{i\omega_n - \mu + p^2/(2m)}.$$

The pole of the denominator is at z_* such that $2\pi T z_* = p^2/(2m) - \mu$. If the r.h.s. is positive, then the corresponding term from the right semi-plane contributes

$$\int \frac{d^3 p\, V}{(2\pi)^3} \frac{1}{e^{(p^2/(2m)-\mu)/T} + 1},$$

which is nothing other than the familiar expression[2] for the fermionic density at temperature T and chemical potential μ. Note that the Fermi-Dirac distribution function naturally follows from the antiperiodicity condition over the time circle.

As for bosonic fields, a nonzero μ shifts the poles in the propagators. To see the effect of this shift, let us take $T \to 0$, so that no thermal contributions survive. The only difference, shown in Figure 7.1, is that the vertical contour C_μ rotates[3] into a thick solid contour of integration at nonzero μ. Since it is different from that at $\mu = 0$ (called C_0, with the integration line indicated in the figure by the dashed lines), the contribution of poles

---

[2] In the relativistic case there are two poles, and thus two contributions appear, from particles and antiparticles, as usual. Note that the bubble diagram is not the density: recall that in this case the current has a derivative; thus one needs to include an extra frequency to get the density.

[3] Rotation is clockwise, and no poles on the real axes are allowed to be crossed, to keep the same value of the integral.

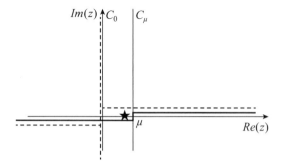

Figure 7.1. Integration contours on the complex energy plane.

such as that at $p^2/(2m) < \mu$ (shown by the star) needs to be included as

$$\int_{p^2/(2m)<\mu} \frac{d^3 p \, V}{(2\pi)^3}.$$

This is the total density of fermions at $T = 0$ occupying the Fermi sphere.

7.3 Weakly Interacting Fermi Gas

We will be considering cold or *degenerate*[4] Fermi gases, in which the temperature is much smaller than the Fermi energy: $T \ll \mu = \epsilon_F$.

As explained in all quantum and statistical mechanics courses, at $T = 0$ the fermions occupy all available states, one by one as the Pauli principle requires, that are inside the Fermi sphere. Here are familiar expressions for the particle number and total energy of the ideal Fermi gas:

$$N = 2_{\text{spin}} \int_{p<p_F} \frac{d^3 p \, V}{(2\pi)^3} = \frac{p_F^3 V}{3\pi^2} \tag{7.4}$$

$$E = 2_{\text{spin}} \int_{p<p_F} \frac{p^2}{2m} \frac{d^3 p \, V}{(2\pi)^3} = \frac{3}{5} \frac{p_F^2}{2m} N. \tag{7.5}$$

Let us now proceed to the ensemble of *weakly coupled* fermions. The naive meaning of the word "weakly" perhaps implies that the potential of particle interaction is small compared to their energies: $V \ll \epsilon_F = p_F^2/(2m)$. For trapped gases, which are very dilute and cold, the energies are, say, $10\,\text{nK} = 10^{-8}\,\text{K}$, while atomic potentials are of the order $V \sim 1\,\text{eV} \sim 10^4\,\text{K}$. So one may think there is a discrepancy of 12 orders of magnitude, between weak potentials and those one has to confront in real life! Yet it is not the actual potentials that are relevant, but the scattering amplitudes. At low collision energies, quantum scattering amplitudes can be characterized by just a number, the scattering

---

[4] For ordinary metals, $\mu \sim 10^4\,\text{K}$, and so room temperature (about 300 K), is "cold." Touching a metal, we do feel colder contact than, say, when touching an insulator, but this is because of the higher heat transfer rate, not different T values.

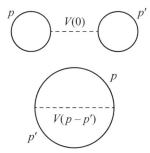

Figure 7.2. Two types of diagrams for the Fermi gas energy: direct and exchange diagrams.

length a, and the actual small parameter is (as for Bose gases) the ratio of the scattering length to interparticle distance:

$$an^{-1/3} \ll 1$$

We will calculate corrections to the ideal Fermi gas formulas in the form of some series in this *diluteness parameter*.

The corrections to the partition function of the system to first order are given by the two diagrams shown in Figure 7.2.

The interaction between fermions, shown by the dashed lines in the figure, is assumed to be some instantaneous potential, depending on the distance between the particles:

$$\frac{1}{2} \int d^3x n(x) \tilde{V}(x-y) n(y) d^3y. \tag{7.6}$$

We put a tilde on the potential and no tilde on its Fourier transform: so $V(0)$ in top diagram has zero momentum transfer, not zero distance. The expression for the lowest-order interaction (at $T = 0$) is a double integral over the Fermi sphere:

$$\frac{E}{V} = \frac{3}{5} \frac{p_F^2}{2m} \frac{N}{V} + \left(\frac{1}{2}\right) 2_{\text{spin}} \int \frac{d^3 p}{(2\pi)^3} \frac{d^3 p'}{(2\pi)^3} [2_{\text{spin}} V(0) - V(\vec{p}-\vec{p}')], \tag{7.7}$$

where $V(q)$ is the Fourier transform of the potential $\tilde{V}(r)$. Two terms in (7.7) are known as the *direct and exchange* energy terms. Note that the first is the interaction of two different particles, while the second is the interaction of a particle with itself: that is why the number of spin components is squared in the former term but linear in the latter. Note the relative negative sign of the second term: it is a consequence of a general Feynman rule, demanding that each fermion loop have an extra minus sign.

If all momenta are much smaller than the atomic one, $V(\vec{p}-\vec{p}') \approx V(0)$, and from quantum mechanics textbooks (the part related to the Born approximation for low-energy collisions), we also find

$$V(0) = -\frac{4\pi\hbar^2}{m} f(k,k) \rightarrow \frac{4\pi\hbar^2}{m} a. \tag{7.8}$$

So the integrals decouple from each other and are just the volumes of the Fermi sphere. Therefore we can complete evaluation of the first-order correction to the energy of weakly coupled Fermi gas:

$$\frac{E}{V} = \frac{p_F^2}{2m} \left[\frac{3}{5} + \frac{2}{3\pi} p_F a + O(a^2) \right].$$

(7.9)

Much more work is needed to get the next two terms, so let me just give the answer:

$$\frac{E}{V} = \frac{p_F^2}{2m} \left[\frac{3}{5} + \frac{2}{3\pi} p_F a + \frac{4}{35\pi^2} (11 - 2\log(2)) (p_F a)^2 + .23 (p_F a)^3 + \cdots \right].$$

(7.10)

The a^2 term is due to K. Huang and C. N. Yang [43], and the last one to C. DeDominicis and P. C. Martin [44], and V. N. Efimov and M. Ya. Amusya [45].

Let me mention one more result on quasiparticles by Galitskii [46] for the lifetime near the Fermi surface due to particle-hole excitation:

$$\Gamma_p = \frac{p_F^2}{2m} \frac{2}{\pi} (p_F a)^2 \left(\frac{p_F - p}{p_F} \right)^2.$$

(7.11)

The last factor shows that at the Fermi surface, the quasiparticle lives forever, because the available decay phase space goes to zero.

7.4 Cold Electron Gas

The problem of an electron gas combined with a homogeneous compensating positive charge is known as the "jellium model." It is a classic many-body problem. We do not have the time to follow many aspects of it—such as the so-called Wigner crystallization at small densities [47]—but will narrowly follow a diagrammatic development, more or less following Pines [48].

The dimensionless parameter in what follows is in a way inverse to the diluteness parameter we used above: it characterizes the density measured in units of the Bohr radius $a_0 = \hbar^2/(me^2)$,

$$\frac{4\pi}{3} r_s^3 = \frac{1}{na_0^3}.$$

(7.12)

Small r_s thus corresponds to the "dense regime," while the large r_s case is the "dilute" one. Here are its values for alkali metals:

$$r_s(\text{Li}) = 3.24, \quad r_s(\text{Na}) = 3.96, \quad r_s(\text{K}) = 4.96, \quad r_s(\text{Rb}) = 5.23, \quad r_s(\text{Cs}) = 5.63,$$

(7.13)

suggesting that, in those metals, the valence electron gas is indeed relatively dilute.

For completeness, let us start with the ideal gas and present a couple of expression in terms of it for the Fermi momentum and energy per particle

$$p_F a_0 = \frac{1.92}{r_s}, \quad \frac{E}{N} = \frac{2.21}{r_s^2}.$$

(7.14)

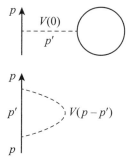

Figure 7.3. Two diagrams for the mass operator, the direct (top) and the exchange (bottom) diagrams.

The diagrams for direct and exchange energy are the same as in the previous section. However, I now prefer to plot them as the mass operator shown in Figure 7.3. The direct (or Hartree) term has a loop that gives the total density of the electrons. However, another diagram (not shown) in which the dashed line goes to the density of positively charged jellium gives the same contribution with the opposite sign and so cancels it. Thus the only one that we need to calculate is the exchange (called the Fock) diagram, the second in Figure 7.3.

To calculate it, let us first evaluate the Fourier transform of the Coulomb potential,

$$V(q) = \int d^3 e^{i\vec{q}\vec{x}} \frac{e^2}{r} = 2\pi \int d\cos(\theta) r^2 dr e^{iqr\cos(\theta)} \frac{e^2}{r} = \frac{4\pi e^2}{q^2}, \tag{7.15}$$

and then do the integral over $\vec{p}'$ in the loop:

$$\Sigma(p) = -\int \frac{d^3 p'}{(2\pi)^3} \frac{4\pi e^2}{(\vec{p}-\vec{p}')^2}|_{p' < p_F} = -\frac{e^2}{\pi} \int^{p_F} p'^2 dp' \int_{-1}^{1} \frac{d\cos(\theta)}{p^2 + p'^2 - 2pp'\cos(\theta)}$$

$$= -\frac{e^2 p_F}{\pi} \left[1 + \frac{1 - (p/p_F)^2}{(p/p_F)} \log \left|\frac{1 + (p/p_F)}{1 - (p/p_F)}\right|\right]. \tag{7.16}$$

The function in the square brackets is a bit complicated, so in Figure 7.4 I show what it looks like. Note that it is continuous but has a singularity of a derivative at the Fermi surface. If one calculates the so-called effective mass m^* defined by standard relation

$$\frac{m}{m^*} = 1 + \frac{\partial \Sigma}{\partial \epsilon_p}|_{p=p_F}, \tag{7.17}$$

one finds a (logarithmic) singularity.[5]

Integrating this mass operator once again (completing the "oyster" diagram of figure 7.2) over p, we find the exchange energy:

$$\frac{E}{N} = -\frac{3}{4} \frac{e^2 p_F}{\pi} = -\frac{0.916}{r_s}. \tag{7.18}$$

Note the sign. The electron interacts with itself: normally it is a repulsive Coulomb force, but the extra minus in the exchange makes it attractive. Together with the kinetic energy

---

[5] This equation may not look too terrible: but it is the first cloud in the sky, indicating that the Landau-Fermi liquid assumption of smooth analytic behavior near the Fermi surface may in fact be wrong, even in the lowest order of the perturbation theory.

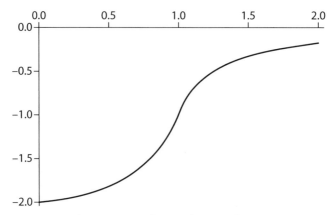

Figure 7.4. $\Sigma\pi/e^2 p_F$ as a function of p/p_F.

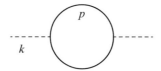

Figure 7.5. The polarization diagram for Π.

(7.14) written at the beginning of the section, it makes a nice minimum at a certain $r_s \sim 4.8$. This is not bad as a numerical prediction for the alkali metals density.

We will not calculate the so-called correlation energy by Gell-Mann and Bruckner [50]: we will discuss its analog in the quark-gluon plasma in Chapter 15. It results in the addition of the following two terms:

$$E_{\text{corr}} = -0.094 + 0.0622 \log(r_s) \tag{7.19}$$

to the kinetic and exchange energy calculated above.

7.5 The Polarization Diagram and the Lindhard Function

Renormalization occurs not only for particles but also for their interaction potential V. This happens, in the lowest order, due to the *polarization function* [49] given by the loop diagram with two particle lines, as shown in Figure 7.5 and corresponding to the expression

$$\Pi(\nu_m, \vec{k}) = 2_{\text{spin}} T \sum_n \int \frac{d^3 p}{(2\pi)^3} \left(\frac{1}{i\omega_n - \epsilon_p + \mu} \right) \left(\frac{1}{i\nu_n + i\omega_n - \epsilon_{\vec{p}+\vec{k}} + \mu} \right)$$

$$= 2_{\text{spin}} T \sum_n \int \frac{d^3 p}{(2\pi)^3} \frac{1}{\epsilon_p - \epsilon_{\vec{p}+\vec{k}} + i\nu_n} \left[\frac{1}{i\omega_n - \epsilon_p + \mu} - \frac{1}{i\nu_n + i\omega_n - \epsilon_{\vec{p}+\vec{k}} + \mu} \right]$$

$$= 2_{\text{spin}} \int \frac{d^3 p}{(2\pi)^3} \frac{n_F(\epsilon_p) - n_F(\epsilon_{\vec{p}+\vec{k}})}{\epsilon_p - \epsilon_{\vec{p}+\vec{k}} + i\nu_n}. \tag{7.20}$$

Let me repeat in words what has been done in equation (7.20). First, a product of two propagators with two poles has been rewritten as a sum of two terms with a single pole each. Second, the Matsubara sum has been calculated, using the Sommerfeld-Watson trick we discussed in Chapter 3 and the beginning of Chapter 7: it leads to the appearance of Fermi distribution functions. The arguments of those functions correspond to the poles, the zeros of the denominators in the intermediate expression. The terminology used for this step is "cutting the line," or "putting the particle on-shell and picking it from the heat bath." Note that the final expression has two terms, corresponding to "cutting" of either one or the other line in a loop diagram. The other, "uncut" line, remains virtual or "off-shell."

Proceeding in general with the remaining momentum integration is a bit hard, and we focus first on some special cases. Let us assume that the temperature T is small and evaluate first the static case with zero external frequency $\nu_n = 0$ at small $k \ll p_F$. The ratio in the last expression can then be approximated by $\partial n_F(\epsilon)/\partial\epsilon \approx \delta(\epsilon - \epsilon_F)$, and the integral is readily done, yielding

$$\Pi(\nu = 0, \vec{k} \to 0) = -mp_F/\pi^2. \tag{7.21}$$

This result allows us to derive a "dressed static Coulomb potential,"

$$V_{\text{dressed by }\Pi} = \frac{4\pi e^2}{q^2 - \Pi(\nu = 0, \vec{k} \to 0)} = \frac{4\pi e^2}{q^2 + mp_F/\pi^2}, \tag{7.22}$$

whose Fourier transform—the coordinate-space potential—is the *Debye screened* potential:

$$\tilde{V}_{\text{dressed by }\Pi} = \frac{e^2}{r} \exp(-\sqrt{mp_F/\pi^2} r). \tag{7.23}$$

In general, the integration to be done has two variables, the angle θ between $\vec{k}$ and $\vec{p}$, and the magnitude p. Since the three-dimensional solid angle contains $d\cos(\theta)$ and the denominator is linear in $\cos(\theta)$, there appear some logs, as you will see below. The general case needs more work, and without going into detail let me present its final expression (Lindhard [49]).

The real part of the integral is

$$\text{Re}\Pi = \frac{mp_F}{2\pi^2} \left\{ -1 + \frac{1}{2k}\left[1 - \left(\frac{\nu}{k} - \frac{k}{2}\right)^2\right] \ln\left|\frac{1 + \nu/k - k/2}{1 - \nu/k + k/2}\right| - \frac{1}{2k}\left[1 - \left(\frac{\nu}{k} + \frac{k}{2}\right)^2\right] \ln\left|\frac{1 + \nu/k + k/2}{1 - \nu/k - k/2}\right|\right\},$$

where the external energy and momentum have been rescaled to proper units, $\nu =$ frequency $\times m/p_F^2$, $k =$ momentum$/p_F$ and thus become dimensionless.

Its high frequency (or "antistatic") limit $1 \gg \nu \gg k$ gives $\text{Re}\Pi = \frac{mp_F}{3\pi^2}\frac{k^2}{\nu^2}$, and thus the dressed potential propagator looks like

$$\frac{1}{-\omega^2 + k^2 - \Pi} = \frac{1}{-\omega^2 + k^2 + \frac{mp_F}{3\pi^2}\frac{k^2}{\omega^2}}. \tag{7.24}$$

The pole in ω is shifting to a new position, giving us a modified dispersion relation $\omega(k)$ of a new collective mode present in the Coulomb systems in place of the phonons, the so-called *plasmons*.

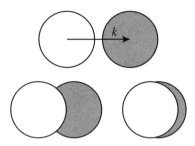

Figure 7.6. Three cases for ImΠ.

The Lindhard integral also has an imaginary part. Formally, it corresponds to cases when the argument of one of the four logarithms becomes negative, so the logarithm has an imaginary part: $\log(-|a|) = i\pi + \log|a|$. Note that formally it corresponds to "cutting both propagators" of the loop diagram and thus putting not one but *both* intermediate particles on shell. Of course this is how it should be, since physical decay places two particles on shell.

The diagram is then completely split into two (conjugate) parts, and thus describes the modulus squared of certain amplitude $|M|^2$, describing a process in which the external energy-momentum leads to the to creation of two excitations. What we see here, with the example of the loop diagram, is the simplest case of the *optical theorem* or the *unitarity relations* at work. Recall that the quantum-mechanical optical theorem relates the imaginary part of the forward scattering amplitude to the total cross section, which is the probability and thus includes certain amplitudes squared.[6]

This digression may sound very complicated, so I need to explain what is happening in this case in straightforward language. The external energy/momentum applied to a cold Fermi gas may kick a particle from inside the Fermi sphere to the outside, creating a particle-hole pair. There are three cases: large $k > 2$, intermediate $k + k^2/2 < v < k - k^2/2$, and small $v < k - k^2/2$. The overlaps of the initial and final (shifted by k) Fermi sphere are different for these three cases (see Figure 7.6). The imaginary part of the forward scattering amplitude explains excitations of the metals and nuclei.

The so-called correlation energy in the electron gas was obtained by resummation of bubble diagrams in a classic paper by Gell-Mann and Bruckner [50]. We will not describe it now, but postpone the discussion of the method until we come to the quark-gluon plasma in Section 15.4, for which a similar resummation was made.

7.6 The Strongly Coupled Electron Fluid in Graphene

The electric current in metals is a coherent flow of electrons, but it is very different from, say, the flow of water in a river or in a tube. In the latter cases large-scale behavior is

---

[6] A large industry has developed around calculating multi-loop perturbative QFT diagrams by first evaluating various "cuts" these diagrams have and then reconstructing the whole answer from these pieces. Unfortunately, we do not have time for that.

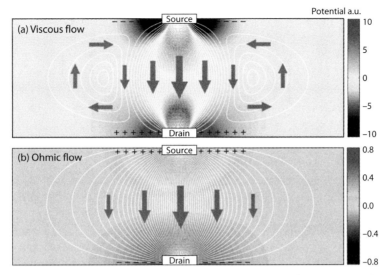

Figure 7.7. The potential map and current streamlines for viscous and Ohmic flows. From Levitov and Falkovich [22]. Reprinted by permission from Springer Nature.

driven by hydrodynamics in general and viscosity in particular, imposing a certain profile on the velocity of the flow, with near-zero velocity values at the boundaries. The usual Ohmic flow of electrons is different: the main friction comes from the interaction with the underlying lattice of atoms—collisions with impurities and phonons—interrupting the electron acceleration by the electric field. The current $\vec{j}$ is related to the electric field $\vec{E}$ locally. The interaction between electrons is ignored, since the total momentum of the electrons is then conserved.

Recent developments, however, have shown a regime in which the flow of electrons is in fact very much like the hydrodynamical flow of a liquid.

The setting is graphene, a two-dimensional form of carbon. For our present purposes it is enough to say that electron properties can be approximated by a relativistic-like linear dispersion curve $\omega = v_F k$ with the Fermi surface at zero energy, $\epsilon_F = 0$. The effective Coulomb coupling is large,

$$\alpha = \frac{e^2}{\epsilon \hbar v_F} \sim O(1), \tag{7.25}$$

so the electron system is strongly coupled. (Here ϵ is the dielectric constant.)

What is also important for the argument is that the material is strong and has few impurities. At low T, scattering on phonons can also be neglected. So, there is a chance to see viscous flow of the electrons. Paradoxically, if this happens, the total resistance to flow in the hydrodynamic regime should be *smaller* than in the usual Ohmic (ballistic) regime, as already was pointed out in 1960s [21].

Here we will follow Levitov and Falkovich [22], where the interested reader can also find references to the historical development of these ideas. The pictures of viscous and Ohmic flows are compared in Figure 7.7. In the former case one finds two vortices formed, one on each side of the flow. The fluid friction with the middle of the flow causes the peripheral flow to have the direction opposite that of the electric field. So the essence of the proposal was to look for a negative flow at the edges, relative to the main voltage.

The experiment has been done very recently [23]. It has indeed been observed that the resistance decreases with T, and negative flow occurs at the edges. The whole phenomenon is analogous to the transition from the Knudsen to Poiseuille regimes, well known in fluid dynamics, but now seen for the electron gas for the first time.

8 | Nuclear Matter

Unlike the electron gas, for which the zeroth-order interaction is just the electric Coulomb potential, the interactions of nucleons are not simple. Therefore in the next section we present an extremely simplified version of the nuclear forces, known as the Walecka model. After that, we are ready to discuss the lowest-order diagrams (Section 8.2). The effect of the nuclear "core"—strong repulsive potential at small distances—is discussed in Section 8.2.

Then, jumping over decades of hard work by many people who tried to include an ever-growing number of diagrams, we proceed in Section 8.3 to the formulation of the effective nuclear forces derived with the help of the RG. It is then used in Section 8.4 to describe the properties of nuclear matter.

The chapter ends with Section 8.5, in which I briefly describe the so-called relativistic mean-field model for the optical potential, which naturally follows from the Walecka model. This approach is highly successful for the description of polarized nucleons scattering off nuclei.

8.1 Nuclear Forces

Readers familiar with nuclear forces, deuterons, and similar issues discussed in detail in nuclear physics courses can skip this elementary introduction. Special books with in-depth discussions include Brown [51] and the relatively recent one by Walecka [52]. Disclamer: Nuclear forces and their effects are by no means fully decribed by the simplified version we discuss in this introductory chapter: I simplify for pedagogic reasons.

For a general orientation, let me emphasize the differences between nuclear matter and the electron gas problem we just studied in Chapter 7. Both are, in the zeroth approximation, nonrelativistic Fermi gases. Nuclear matter has protons and neutrons, which we sometimes will be writing as ψ_i, $i = 1, 2$, the isospin doublet of nucleons. Recall that the $SU(2)$ isospin symmetry exists because the masses of u, d quarks, while not the same, are both too small to be important. It is an approximate "flavor" symmetry of QCD, extended to the $SU(3)$ symmetry with the strange quark in 1960s, and then to all six quark species now known.

The electrons in metals have velocities comparable to their velocity in atoms:

$$\frac{v}{c} \sim \frac{e^2}{\hbar c} \approx \frac{1}{137} \ll 1. \tag{8.1}$$

As we will see shortly, the velocities of nucleons are also small, but not this small: $v/c \sim 1/3$. Thus magnetic and other relativistic corrections $O(v^2)$ are more important, while still not dominant. Furthermore, since the mesons carrying the interaction between the nucleons include scalar σ and vector ω (and others), and their fields transform differently under Lorentz transformation, it will be convenient to keep the theory fully relativistic from the start. (This will also prepare us to discuss quark matter in subsequent chapters.)

The input interaction in the electron gas problem is very simple: it is the repulsive Coulombic force. In coordinate and momentum representation it is

$$\tilde{V}(r) = \frac{e^2}{r}, \quad V(q) = \frac{4\pi e^2}{\vec{q}^2}. \tag{8.2}$$

Its long-range nature, $r \to \infty, q \to 0$, seems to be a problem, but we learned that the polarization of the electron gas "dresses" it to a screened version

$$\tilde{V}_{\text{screened}}(r) = \frac{e^2}{r} e^{-\kappa r}, \quad V_{\text{screened}}(q) = \frac{4\pi e^2}{\vec{q}^2 + \kappa^2}, \tag{8.3}$$

where $\kappa^2 = -\Pi(v = 0, q \to 0) > 0$.

In spite of my earlier comment on always including relativistic effects, let us start with instantaneous potentials. (Later we can develop the nucleon-meson effective relativistic field theory, including full meson propagators and time-delay effects.)

The nuclear forces are finite range from the beginning, since none of the mesons are massless. At first we will include only the isoscalar scalar σ and vector ω ones (the so-called Walecka model) and use the following notations for the couplings:

$$\tilde{V}(r) = -\frac{g_\sigma^2}{4\pi r} e^{-m_\sigma r} + \frac{g_\omega^2}{4\pi r} e^{-m_\omega r}. \tag{8.4}$$

Note that 4π is absent in the coordinate version of the Coulomb's law but is present in its Fourier transform. In the notation used here, it is the other way around. This notation is the one used historically, because in QFT, units for the fields and couplings used are defined differently from those in electrodynamics. So $\alpha_{\text{QED}} = e^2$, $\alpha_s(\text{QCD}) = g^2/4\pi$, and so forth.

With appropriate masses and couplings, this potential is plotted in Figure 8.1.

Comment 1: The vector potential is repulsive, like a Coulomb potential for two electrons. It should change sign for nucleon-antinucleon interactions. The scalar exchange is always attractive.

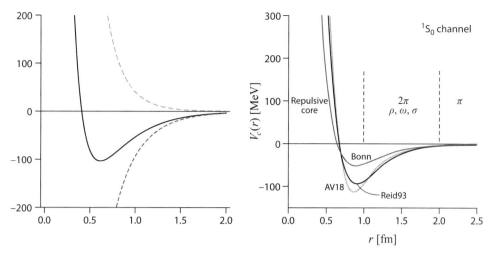

Figure 8.1. (Left) The potential (MeV) versus distance (fm) in the Walecka model. The solid curve is the sum of its two components (dashed). (Right) Several phenomenological nucleon-nucleon potentials, for the central or $^1S_0$ channel.

Comment 2: The masses of those mesons are $m_\sigma \approx 500$ MeV, $m_\omega \approx 770$ MeV. Thus the attraction wins at large distances. To translate the energy units to distance, recall that in the physicist's units, $\hbar = c = 1$, one has 197 MeV $\approx 1/$fm, where 1 fm = 1 femtometer = 10^{-15} m.

Comment 3: The coupling values selected by Serot and Walecka [52] are

$$g_\sigma^2 = 267.1 \left(\frac{m_\sigma^2}{m_N^2} \right), \quad g_\omega^2 = 195.9 \left(\frac{m_\omega^2}{m_N^2} \right). \tag{8.5}$$

Note that the ω coupling is stronger, thus dominating at small distances. Note further that these two terms nearly cancel each other, leaving us with a relatively shallow potential, $V < 100$ MeV $\sim m_N/10$. Further cancellation happens between the potential and kinetic energies. As a result, nuclear physics has its own energy scale, which is different from that of hadronic physics.

Comment 4: These couplings are not based on a fit to nucleon scattering data and deuteron binding, as are other phenomenological forces also displayed for comparison, but on nuclear matter properties in the mean-field approximation. We will use these properties, as we are focusing on nuclear matter in this chapter.

Comment 5: The forces we so far have not discussed include isospin-1 mesons, pseudoscalar π and vector ρ in particular. Isospin matrices $\tau_{ij}^a \tau_{i'j'}^a$ are included in the interaction points, where the dummy $a = 1, 2, 3$ counts $SU(2)$ generators, and ij and $i'j' = 1, 2$ are proton-neutron isostates. The pseudoscalarity of π leads also to two spin matrices called σ with spinor greek indices, so the potential looks rather complicated:

$$V_\pi \sim (\tau_{ij}^a \tau_{i'j'}^a)(\vec{\sigma}_{\alpha,\beta} \vec{q})(\vec{\sigma}_{\alpha',\beta'} \vec{q}) \frac{1}{q^2 + m_\pi^2} \tag{8.6}$$

The next level would include even more complex spin-orbit and tensor forces of relativistic origin in the potential. I will not go into the history of multiple empirical potentials with a dozen parameters, fitted to deuteron binding and scattering phase shifts. Three of them are shown in Figure 8.1 (bottom panel). We will discuss them in Section 8.3, when we discuss the renormalization group.

8.2 Nuclear Matter in the Relativistic Mean-Field Model

Heavy nuclei are basically spheres of homogeneous density made up of protons and neutrons, with a radius $R = r_0 A^{1/3}$, where $A = Z + N$ is the nucleon number, and $r_0 = 1.07$ fm. The nuclear matter density

$$n = \frac{A}{4\pi R^3/3} = \frac{3}{4\pi r_0^3} \approx .16 \, \text{fm}^{-3} \tag{8.7}$$

is its main parameter. In reality, heavy nuclei are distorted by the electric field of the protons, but for simplicity we will switch QED off, keeping strong nuclear forces only. The Fermi momentum follows from it,

$$n = 2_{\text{spin}} 2_{\text{isospin}} \int_{p < p_F} \frac{d^3 p}{(2\pi)^3}, \tag{8.8}$$

producing $p_F \approx 1.42 \, \text{fm}^{-1}$. Empirically we know that in nuclear matter, the binding per nucleon is $E/N \approx -16$ MeV.

The interaction, to first nonrelativistic order, is a sum of the direct and the exchange diagrams. The first diagram contains $V(q = 0)$, which, unlike in the electron gas, does not cancel out, but it is easy to calculate:

$$\frac{E}{N} = \frac{3}{5} \frac{p_F^2}{2m} + \frac{n}{2} \left(-\frac{g_\sigma^2}{m_\sigma^2} + \frac{g_\omega^2}{m_\omega^2} \right) + (\text{exchange}). \tag{8.9}$$

The first kinetic energy term is numerically equal to 25 MeV. The exchange term includes a double integral over momenta and $V(\vec{p} - \vec{p}')$: I leave it to be evalvated as an exercise, as it closely follows what was done for the electron gas.

Exercise *Calculate the exchange (Fock) energy of nuclear matter, using the ω, σ exchange forces with parameters defined above. Compare this contribution to the kinetic energy and the direct term (Hartree).*

The next level of discussion would logically be based on the weakly coupled Fermi gas, which we discussed before. In this case, however, a problem arises.

There are two binary channels, with different isospin: one is the deuteron (p-n) iso-singlet case, with small binding (≈ 2 MeV) and another is iso-triplet (nn,pp,np) in which there is a virtual state at energy 70 keV. As a result, both scattering lengths are huge. It clearly makes no sense to use them in expressions we derived before for weakly coupled Fermi gases.

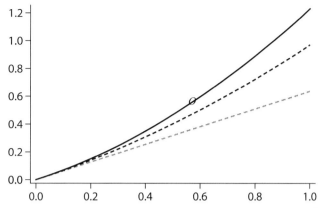

Figure 8.2. Nuclear matter energy due to a repulsive core, plotted as $p_F^2/(2m)$ versus $p_F r_{core}$. The three lines are three progressively more accurate approximations (bottom to top) in (8.10). The Letter O indicates the value at nuclear matter density: but the curve can be used for larger densities and values of p_F as well.

One may however think of including the radius of the nuclear repulsive core, $r_{core} \approx$ 0.4 fm, as a scattering length. In other words, consider a simple potential $V(r < r_{core}) = \infty$, zero otherwise, which leads to $a = r_{core}$. Let us substitute this into the corresponding expression (corrected slightly from expressions for a weakly coupled gas, to account for the fact that there are now two species, n and p):

$$\frac{E/A}{p_F^2/(2m)} = \frac{2}{\pi}[(p_F r_{core}) + \frac{6}{35\pi}(11 - 2\log 2)(p_F r_{core})^2] + 0.26(p_F r_{core})^3. \qquad (8.10)$$

Figure 8.2 shows this curve. Substituting in the values of p_F and r_{core}, we get a repulsive contribution of about $0.6\,p_F^2/(2m)$, very similar to the original kinetic energy of the Fermi gas of $(3/5)\,p_F^2/(2m)$. At higher densities—up to 8 times higher in the center of neutron stars—the p_F is twice as large, and the repulsion due to the core becomes dominant. We can evaluate how repulsive the equation of state becomes from the largest mass observed for a neutron star, currently about twice the mass of the Sun.[1]

Now let us return to the fully relativistic formulation and work out the mean-field approximation, following Serot and Walecka [52]. Its Lagrangian density is

$$L = \frac{1}{2}\left(\partial_\mu\phi\partial^\mu\phi - m_s^2\phi^2\right) - \frac{1}{4}F_{\mu\nu}F^{\mu\nu} + \frac{1}{2}m_v^2 V_\mu V^\mu + \bar\psi\left[\gamma_\mu(i\partial^\mu - g_v V^\mu) - (M - g_s\phi)\right]\psi, \qquad (8.11)$$

where the subscripts S, V mean scalar and vector respectively and, the Abelian field strength of the vector field $F_{\mu\nu} \equiv \partial_\mu V_\nu - \partial_\nu V_\mu$ is the same as in electrodynamics. There

---

[1] The first observation of two neutron stars merging was made in August 2017. The observations—in gravitational radiation (by LIGO/VIRGO), gamma rays, X-rays, and optical all the way to radio emission—have only set upper limits on the tidal deformability and the radii of the two neutron stars. Unfortunately, the most interesting part of the rotational curve, near touch-down, falls in a higher frequency range that is outside the current LIGO/VIRGO sensitivity window. So we must wait until these detectors are improved before we can get a better upper limit on the repulsive term in the equation of state.

are thus three fields—the Dirac nucleons ψ, vector omega-mesons V_μ, and scalar sigma-mesons ϕ—interacting with each other in a relativistically invariant way. Their masses are considered to be an input.

The relevant equations of motion are

$$\left(\partial_\mu \phi \partial^\mu + m_s^2\right)\phi = g_s \bar{\psi}\psi \tag{8.12}$$

$$\partial_\mu F^{\mu\nu} + m_v^2 V^\nu = g_v \bar{\psi}\gamma_\mu \psi \tag{8.13}$$

$$\left[\gamma_\mu(i\partial^\mu - g_v V^\mu) - (M - g_s\phi)\right]\psi = 0. \tag{8.14}$$

Mean field theory keeps only the classical (nonfluctuating) part of the bosonic fields ϕ_0 and V_0. In homogeneous nuclear matter all derivatives are thus zero, and the first two equations become simply

$$m_s^2 \phi_0 = g_s \bar{\psi}\psi, \quad m_v^2 V_0^\nu = g_v \bar{\psi}\gamma_\mu \psi,$$

defining the magnitude of these fields in terms of the scalar and vector nucleon densities. However, note that they will play very different roles in what is to follow. The vector current is conserved, and the vector density (or just density) $n_B = \bar{\psi}\gamma_0\psi$ will be an input to the calculation. The scalar density $n_s = \bar{\psi}\psi$ is a dynamical quantity to be calculated.

The last (Dirac) equation tells us about the dispersion relation for the nucleon quasiparticles with momentum k:

$$E_k = k + g_v V_0 + \sqrt{k^2 + M_*^2}, \quad M_* = M - g_s\phi_0. \tag{8.15}$$

Note that if one expands the square root, the leading term of the mean potential $g_v V_0 - g_s\phi_0$ will be the same as that obtained from the usual nonrelativistic theory, but the kinetic energy term would be $k^2/(2M_*)$ rather than $k^2/(2M)$, and, as we will see, that difference is significant.

The total energy of the gas is

$$E_{mfa} = \frac{g_v^2}{2m_v^2}n_B^2 + \frac{m_s^2}{2g_s}(M - M_*)^2 + \frac{\gamma}{(2\pi)^2}\int_0^{k_F} d^3k \sqrt{k^2 + M_*^2}, \tag{8.16}$$

where the statistical weight γ equals 4 for symmetric nuclear matter and 2 for neutron matter in the neutron stars. Let me add that the two densities can now be written as integrals over the Fermi sphere:

$$n_B = \frac{\gamma}{(2\pi)^2}\int_0^{k_F} d^3k, \quad n_s = \frac{\gamma}{(2\pi)^2}\int_0^{k_F} d^3k \frac{M_*}{\sqrt{k^2 + M_*^2}}.$$

Note that the latter has scalar mass M_* in the numerator and the energy in the denominator, which is needed because the Lorentz-invariant integration measure is d^3p/E_p.

At this stage all is fixed except the scalar mean field (or alternatively, M_*): this is a parameter of our homogeneous field trial function, and as for any variational parameter,

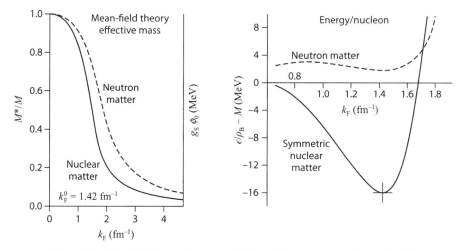

Figure 8.3. The effective nucleon mass (left) and the energy per baryon (right) versus the Fermi momentum, in symmetric nuclear matter (solid lines) and neutron matter (dashed lines). The cross indicates the empirical position of the nuclear matter binding. Adapted from Serot and Walecka [52].

it should be found from minimization of the ground-state energy. This leads to the following equation

$$M_* = M - \frac{g_s^2}{m_s^2} - \frac{\gamma}{(2\pi)^2} \int^{k_F} d^3k \frac{M_*}{\sqrt{k^2 + M_*^2}} \tag{8.17}$$

for M_*, to be solved numerically (e.g., by plotting the l.h.s. and r.h.s. and looking for intersections of the curves).

Exercise *Study equation (8.17) in the two limits, of small and large n_B, and prove that in the former case the energy per nucleon decreases, and in the latter it increases. This shows a minimum must exist in between the two limits.*

The solution for the mass and the energy are plotted in Figure 8.3, from [52]. The agreement with the empirical position of the nuclear matter binding is not a prediction; it is due to appropriate tuning of the couplings. What is a prediction is that $g_s \phi_0 \approx 330$ MeV, implying that M_* is significantly different from the nucleon mass, and thus the kinetic term is increased. However, the mean field is of the same order as in nonrelativistic theory, namely, $g_s \phi_0 - g_v V_0 \approx 70$ MeV.

8.3 Nuclear Forces and RG

We will jump over decades of the development of the theory of nuclear matter to relatively recent works based on the RG approach, following a review by Bogner, Furnstahl, and Schwenk [53]. The reader can find details in the original papers cited there. Here I illustrate only the main results.

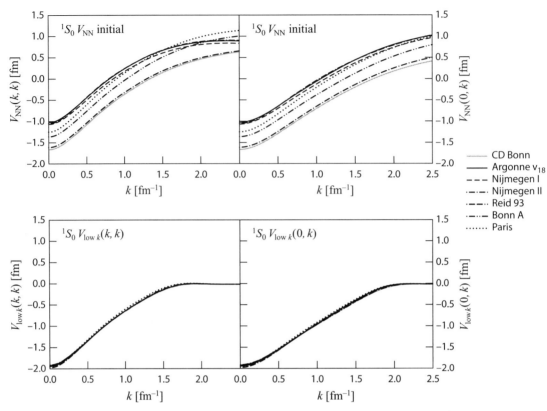

Figure 8.4. Diagonal (left) and off-diagonal (right) momentum-space matrix elements for various phenomenological nucleon-nucleon potentials initially (upper figures) and after RG evolution to low-momentum interactions V_{low_k} [5, 6] (lower figures) for a smooth regulator with $\Lambda = 2.0\,\text{fm}^{-1}$. Adapted from Bogner, Furnstahl, and Schwenk [53].

Different empirical potentials differ from one another, although they are fitted to basically the same set of data. This fact by itself tells us that some of the information they contain is irrelevant for nuclear physics. We can then ask: What is the smoothest potential that will do the same job as the rest of them? We need the smoothest because one hopes that in this case the perturbation theory will work better.

The RG program is to systematically integrate out momenta in the loops above a certain cutoff Λ. The reader can find out technically how it is done by consulting the literature. The result of the program is shown in Figure 8.4. The top part of the figure shows the Fourier transform of the empirical potentials. To a first approximation the difference between the different estimates seems to be just a vertical shift by a constant. What is a constant in momentum space is $\delta(r)$ in coordinate space: so the main difference in the estimates comes from the small r region. In other words, they are mostly different in the core region, in which the wave function is negligible anyway.

After RG application (the lower plots in Figure 8.4), two important changes take place. First, the potential for $k > \Lambda$ is now gone—it is zero. This is again basically achieved by shifting by a constant downward. The potential is larger at small k, so it does become

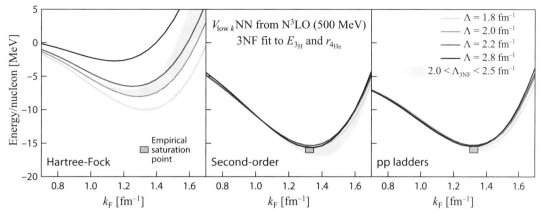

Figure 8.5. Nuclear-matter energy per particle as a function of Fermi momentum k_F at the Hartree-Fock level (left) and including second-order (middle) and particle-particle-ladder contributions (right), based on evolved N^3LO NN potentials and 3N forces fit to E(tritium) and r(^{4}He). Theoretical uncertainties are estimated by the NN (lines) and 3N (band) cutoff variations. Adapted from Bogner, Furnstahl, and Schwenk [53].

smoother. Note that this smoothened potential—called $V_{low\,k}$—does depend on the RG parameter Λ.

The second change is universality: after the RG filter has been applied, all phenomenological potentials collapse to the same(!) curve. This result confirms the suspicion that, after certain inessential information is removed, the remaining information is indeed the same in all of the models.

8.4 Nuclear Matter and RG

In Figure 8.5 the energy per nucleon is plotted, using the same $V_{low\,k}$. The Hartree-Fock approximation includes the direct + exchange diagrams that we have already encountered. The gray boxes in the figure indicate an empirical minimum for nuclear matter: this is the answer we want. In the approximations used so far we however have only gotten the position (density) correctly; the depth of the potential is too small. Detailed studies of few-body systems have shown that *three-particle forces* need to be included. The two-body forces, which are fitted to scattering experiments and deuterium, slightly miss the bindings of tritium (pnn), ^{3}He, and ^{4}He. These bindings can be parameterized and included. The results agree with nuclear matter as well (see Figure 8.5). The last panel in the figure shows that one can vary Λ in a certain window without any effect on the nuclear-matter properties.

8.5 Relativistic Mean Field and the Optical Potential

The fact that some collective fields of the nuclei (or nuclear matter) originate from scalar (σ, π) and some from vector (ω, ρ) meson exchanges have the following consequences.

We learned that at rest scalar and vector mean fields of the nucleons in nuclei are respectively attractive and repulsive, and to a significant degree, they cancel each other. What is left is their sum, which, averaged over a distance, gives a slight preference to the scalar mean field. For a passing nucleon, a nuclei, with an overall attractive potential well, acts as a convergent lens. (The mean potential is indeed called the "optical potential," due to this analogy.)

Suppose we consider a proton-nucleus scattering experiment with variable proton velocity. In the usual laboratory frame a nucleon moves with substantial velocity v, while the nuclei, a target, is at rest. Now let us think instead of what the fields generated by heavy nuclei look like in the *anti-laboratory* frame, in which the proton is at rest and the nucleus is moving. Transforming the nuclear fields, we naturally assume that the scalars are not subject to Lorentz transformation, while the vector fields should be modified as Lorentz vectors (e.g., the electromagnetic potentials). So the vectors are basically increased by the gamma factor of the Lorentz transformation, $\gamma = 1/\sqrt{1 - v^2}$.

Thus the vector component gets an extra Lorentz factor $\gamma > 1$ in the anti-laboratory frame; it gets enhanced. As a result, the attraction-repulsion balance is changed. At a certain collision velocity, the collective nuclear potential turns, through zero, from an attractive to a repulsive potential. Therefore, above that energy, the nuclei act like a divergent lenses! This is indeed what is observed.

Much more detailed predictions follow when the incoming nucleon beam is polarized. Again, the motions of the nucleon spin in the vector and scalar fields are different. The data on the polarized scattering have been well explained by these relativistic mean-field models with scalar and vector components.

The theory of nuclear forces can in general be elevated from the instantaneous potentials discussed above to effective QFT with Lagrangians that include dynamical mesonic fields. However, the status of such effective theories is a vast subject, and we do not have time to go into it in this book.

9 Cooper Pairing and the BCS Theory of Superconductivity

This is one of the central chapters of this book, devoted to the Cooper pairing phenomenon and the Bardeen-Cooper-Schrieffer (BCS) theory of superconductivity. The theory's historical development is briefly described in the next section, which is devoted to superconducting metals.

The pairing phenomenon itself is discussed in Section 9.2. I supplement this discussion by a digression, in Section 9.3, of more general applications of the RG in QFT settings. It is shown in Section 9.4 how renormalization of the local interaction between fermions leads to strengthening of the coupling as one approaches the Fermi surface.

The ultimate solution for the Green functions used to model superconductors is presented in Section 9.5. This solution inludes the "abnormal" Gorkov function describing the propagators that violate particle number conservation (because of the condensate of Cooper pairs). This formalism provides new dispersion relations for fermionic quasiparticles, using the "gapped" Fermi surface.

9.1 Superconducting Metals

Superconductivity was discovered by Dutch physicist Heike Kamerlingh Onnes on April 8, 1911, in Leiden. I have already mentioned him and his lab in Section 5.1 the experimental setting was more or less the same as that used in the discovery of liquid ^{4}He (which was promptly used as a coolant). Superconductivity was discovered and explained much earlier than was superfluidity.[1]

Returning to the experiments in Leiden: while measuring the conductivity of solid mercury, it was observed that all resistance to electric current abruptly disappeared at

---

[1] Today, a century later, the largest physics experimental devices, such as CERN's Large Hadron Collider, consist of accelerator rings made of superconducting magnets on a truly industrial scale (20 km or so.) Even so, a large amount of energy is dissipated by superfluid He-II.

the critical temperature of $T_c = 4.2$ K. This observation started decades of efforts by many physicists to understand the phenomenon, until the theoretical breakthrough of 1957, when Bardeen, Cooper, and Schrieffer [58] explained it. The BCS theory—in a more modern version, related to diagrams and the renormalization group—will be our subject in this and many subsequent chapters.

The following history of superconductors and of the quest for high-T_c superconductivity is a very interesting story, but it hardly belongs in this book. After all, this book is in the *Nutshell* series, which focuses on developments that have led to significant new ideas. Let me just say that after trying various alloys and producing industrial magnets, researchers in the field unexpectedly veered into studies of cuprates and related substances, which are not metals but a kind of ceramic. The theory remains very complicated, and the main lesson is that Cooper pairing may happen even in the $l = 2$ or d-wave channel, in spite of a huge repulsive centrifugal potential.

Recently (in 2014) it was discovered [59] that at high pressure, the sulfur-hydrogen compound H_3S becomes a superconductor at a record-breaking[2] $T_c = 203$ K! What is also nice is that it is a BCS superconductor, with attraction produced by phonons (as the isotope mass dependence tells us). As with most discoveries, it was quite unexpected at the time, but in retrospect not at all surprising. Indeed, H_3S (which, by the way, has a relatively simple cubic crystal structure) can be viewed as some approximation to *metallic hydrogen*—the famous "holy grail" of condensed matter physics, not yet created—with its smallest possible nuclear mass and thus the highest possible phonon frequencies.

Let us start with the main question: Where might an attractive interaction between electrons come from? Indeed, the two electrons in a Cooper pair have charge +2e and so must repel each other. The RG flow requires long-distance attraction to start growing. How these two requirements can be reconciled remained a puzzle until the discovery of the *isotope effect*: T_c for mercury isotopes was found to be dependent on the nuclear mass number. This discovery promptly lead to a resolution of the paradox: phonon exchange produced the needed attraction. It is rather weak and operates only at distances large enough for the repulsive Coulomb forces to be Debye screened; see more in [67].

The picture of the quasiparticle states, given by the dispersion curve

$$\epsilon_p = \pm\sqrt{\Delta^2 + v_F^2(p - p_F)^2} \tag{9.1}$$

is sketched in Figure 9.1. The finite gap Δ between the particle and hole states is its main feature. Direct measurements of the gap are possible if one places a contact between normal and superconducting metals that are separated by insulating barrier and measures the current as a function of voltage applied to the pair. When levels occupied in the metal become level with open levels in the superconductor, the current flows. Typical voltages (or Δ in volts) are on the order of 10^{-3} V, so the gap is easily measured.

---

[2] The previous highest BCS critical temperature was 40 K, so the jump is indeed without precedent. For perspective, 203 K is several degrees above the sublimation temperature of dry ice, which is routinely used to keep boxes of ice cream cold.

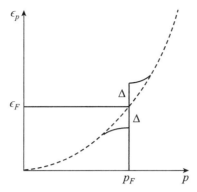

Figure 9.1. Sketch of the modified dispersion relation for particles and holes, showing the superconductor gaps.

Superconductivity is in fact the superfluidity of a charged fluid. Persistent currents can support magnetic fields in this state. I will not spend time on the Meissner effect and levitation as these phenomena widely known. Our first issue to discuss will be the application of the Landau critical velocity argument to superflow. The argument, discussed in Chapter 5 on liquid ^{4}He, is that for

$$v < \min(\epsilon_p/p) \tag{9.2}$$

there cannot be any friction, because excitations of quasiparticles are not possible. The critical velocity predicted from this argument is roughly $v_c \sim \Delta/p_F$ (more accurately, one needs to make a tangent plot to find the minimum value). Critical velocity times the condensate density times 2 (the charge of a Cooper pair) gives the critical electric current.

Before getting into the development of the theory, it is useful to look at an example and introduce some phenomenological numbers and their interrelations. As such an example, let us take a pure metal superconductor Sn (tin). It has one free electron per atom, and since the density is known, we can easily obtain the Fermi momentum: $p_F/\hbar \sim 1.2 \times 10^{10}$ m^{-1}. The BCS theory tells us that $\Delta(T=0) = 1.76\, T_c$, so from $T_c = 3.72$ K we get the value of the $\Delta(T=0) = 6.5$ K $= 8.3 \times 10^{-23}$ J. Therefore the Landau criterion gives $v_c \approx 70$ m/s.[3] The critical current $J_c = n(2e)v_c$ can be put in to the Maxwell equation $\vec{\text{curl}}\, H = \mu_0 \vec{J}$, and so we can estimate the critical magnetic field that a superconductor will tolerate. The estimate turns out to be $H_c \sim .1\, T$, close enough to measurements. The conclusion is that the Landau criterion for critical velocity works. It is in fact true not only for tin but also for all type-I superconductors (the distinction between two types is explained in Chapter 11.)

9.2 Cooper Pairing

A famous instability of the Fermi surface happens in the *particle-particle* and *hole-hole* channels, which leads to so-called Cooper pairing and to the superconductivity phenomenon. We will discuss pairing phenomena in a variety of settings.

---

[3] Compare this to the value of the critical velocity in ^{4}He, about 50 m/s, from the same criterion.

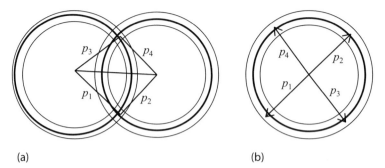

Figure 9.2. Panel (a) of the figure shows the kinematics of scattering $\vec{p}_1 + \vec{p}_2 = \vec{p}_3 + \vec{p}_4$. When all four momenta are forced to be close to the surface of the Fermi sphere (between the two thin rings flanking the heavy circle), the final momentum has to be inside the small diamond-shaped overlap regions. The exception is when the total momentum is zero: this case is shown in Panel (b). In this case the whole region near the Fermi surface is open for final momenta.

The scattering kinematics for two particles pp ($|p| > p_F$), or holes hh ($|p| < p_F$) with the initial momenta $\vec{p}_1$, $\vec{p}_2$ are displayed in Figure 9.2.

Two spheres (shown by heavy solid lines in the figure) have the radius of Fermi momentum p_F, and a couple of concentric spheres (shown by thin lines sandwiching the heavy) indicate the range occupied by excited quasiparticles (p or h) with in a certain energy range, say, the thermal excitations at some $T \ll \epsilon_F$.

Let us first imagine that our system is two-dimensional, and the Fermi spheres are the circles depicted in Figure 9.2. When the total momentum

$$\vec{p} = \vec{p}_1 + \vec{p}_2 = \vec{p}_3 + \vec{p}_4 \tag{9.3}$$

is nonzero, as in Figure 9.2a, all possible phase space for the final momenta $\vec{p}_3$, $\vec{p}_4$ consists of two small rhombs on the intersection of a pp ring with that from circles shifted by $\vec{p}$. In three dimensions, we have instead a thin rhombic ring connecting these two intersections and obtained by rotation of those intersections around the direction of the total momentum. As $T \to 0$, the volume of this space available for scattering vanishes as $\sim T^2 \epsilon_F$.

The situation is quite different if the total momentum is zero, $\vec{p} = \vec{p}_1 + \vec{p}_2 = 0$, as shown in Figure 2b. In this case the whole spherical region near the Fermi sphere is available, and as $T \to 0$, its volume goes to zero as $\sim T \epsilon_F^2$. For conditions in which ordinary superconductors are used, the temperature-to-Fermi-energy ratio is very small, $T/\epsilon_F \sim 10^{-4}$, and so this difference is very important: Cooper pairing happens at $\vec{p} = 0$.

Another point I want to make is that in the latter situation, the direction of $\vec{p}_1$ is irrelevant (as soon as $\vec{p}_2 = -\vec{p}_1$), and in low-energy s-wave scattering, the area of the Fermi sphere just enters as a common factor. It does not really matter whether we consider the two- or three-dimensional problem: only the one-dimensional component of the momentum (the radial one), really matters and enters the equations. Therefore all the equations for superconductivity (such as the famous BCS gap equation) are basically the same for any dimension.

Now recall a fact we all learned in quantum mechanics courses: in one dimension, unlike higher dimensions, any attractive potential creates at least one bound state. In the language of field theory, for renormalization of quartic coupling one should calculate a "fish" diagram. The main observation is that it is in fact logarithmically divergent near the Fermi surface. Indeed, when $\vec{p} = 0$ the momenta (but not energies) in both propagators in the loop are the same and we have (at small $T \to 0$)

$$F(\omega, 0) = \int \frac{d\omega}{2\pi} \frac{d^d p'}{(2\pi)^d} \frac{1}{i\omega' + \delta\epsilon_{p'}} \frac{1}{i\omega - i\omega' + \delta\epsilon_{p'}}. \tag{9.4}$$

In one dimension, $d = 1$, the integral of the denominators makes a divergent contribution as $\log(\omega)$ at small ω. In two and three dimensions, the momentum is distributed over the entire sphere, and integration over the solid angle leads to the surface area times the same log divergence $F \sim \log(\omega)$. The fish diagram thus needs to be regulated at some scale.

The resummation

$$\lambda + \lambda^2 F + \cdots = \frac{\lambda}{1 - \lambda F}, \tag{9.5}$$

which is similar to what we did before, tells us that even for very small $\lambda \ll 1$, a large log can compensate, and the gap equation

$$1 = \lambda F(\omega) \tag{9.6}$$

has a nontrivial solution, provided $\omega \sim \exp(-C/\lambda)$ with C the constant in front of the log.

9.3 A Continuous RG

The version of RG we discussed in Section 2.4, following the K. Wilson treatment of the second-order phase transitions, used a discrete block-spin transformation sequence. Recall that in this case a system defined on the lattice was changed: $a \to 2a$. Momenta belonging to certain momentum shells were integrated out, but the effect of that was not lost but included in the modified parameters of the effective action.

A systematic approach using continuous RG, developed originally by Gell-Mann and Low in 1954 [54], describes the renormalization of the charge. These authors introduced the so-called beta function, defined as the charge derivative over the scale at which it is defined:

$$\frac{\partial g}{\partial \log \mu} \equiv \beta(g). \tag{9.7}$$

If the beta function is known, equation (9.7) can be integrated:

$$\log(\frac{\mu}{\mu_0}) = \int_{g_0}^{g} \frac{dg'}{\beta(g')}. \tag{9.8}$$

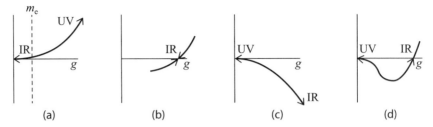

Figure 9.3. Schematic behavior of the beta functions in the four examples considered: (a) QED, (b) second-order phase transition, (c) QCD, and (d) QCD with a sufficiently large number of light flavors. The lines marked IR and UV with arrows indicate the direction of the charge motion at small and large momenta, respectively.

For small coupling values the beta function can be determined perturbatively, and its standard definition includes the first and second beta function coefficients,

$$\beta(g) = -b\frac{g^3}{16\pi^2} - b'\frac{g^5}{(16\pi^2)^2} + \cdots, \tag{9.9}$$

which are computed from one- and two-loop diagrams. Let us consider some examples with different beta functions (Figure 9.3).

Example 0: A very special class of QFTs are *conformal* (CFT). In these theories $\beta(g) = 0$ and the coupling is independent of the scale. The famous example in four dimensions is $\mathcal{N} = 4$ supersymmetric gluodynamics. It has four kinds of gluinos and six scalars, and their contributions to the beta function cancel that of the gluons, order by order. This theory was studied in 1990s and was historically central for a discovery of AdS/CFT duality.

Example 1: Quantum electrodynamics (QED) was the first QFT studied since the 1950s. In one loop the so-called vacuum polarization diagram leads to

$$\beta(e) = \frac{e^3}{12\pi^2}. \tag{9.10}$$

The solution for the charge as a function of scale is therefore

$$e^2(\mu) = \frac{e^2(\mu_0)}{1 - (e^2(\mu_0)/6\pi^2)\log(\mu/\mu_0)}, \tag{9.11}$$

and when μ grows (the charge is measured deeper inside the electron, at distances $\sim 1/\mu$), the charge becomes larger. Eventually one finds a zero of the denominator, the Landau pole, where the formula tells us the charge is infinite. Before taking this answer at its face value, recall that the formula has limited applicability, and when the coupling becomes of order 1, we cannot trust the perturbative results. What actually happens in QED at such small distances remains unclear. QED might be the first QFT studied, but by no means is it a well-defined theory: it lacks any nonperturbative definition. Therefore, we cannot use it to answer—or even in principle formulate how to answer—many questions.

Example 2: This example concerns second-order phase transitions, which we already discussed in Section 2.4. In this case the beta function changes sign and has a zero at g^*, the fixed point, so when the coupling reaches this value, its variation stops. Assuming that close to its zero it can be approximated by a linear function $\beta(g) = a(g - g^*)$, we find a power-like dependence near the fixed point:

$$g(\mu) - g^* = (g(\mu_0) - g^*)(\mu/\mu_0)^a. \tag{9.12}$$

Example 3: QCD-like gauge theories have the so-called ultraviolet fixed point, near which the coupling is small:

$$\beta(g) = -b\frac{g^3}{16\pi^2} - b'\frac{g^5}{(16\pi^2)^2} + \cdots \tag{9.13}$$

with positive

$$b = (11/3)\,N_c - (2/3)\,N_f. \tag{9.14}$$

where N_c and N_f are the number of colors and quark flavors, respectively; both can be taken to be 3 in QCD. This behavior is also known as "asymptotic freedom." In 1973 Politzer, Gross, and Wilczek not only discovered it but also related this feature to the experimentally observed weakly interacting point-like quarks, suggesting that the fundamental theory of strong interactions is QCD. We will derive it and briefly discuss other historic moments in Section 14.3.1, devoted to hot QCD.

Integrating the beta function, we obtain the explicit asymptotic freedom formula

$$g^2(\mu) = \frac{g_{\mu 0}^2}{1 + b(g_{\mu 0}^2/(8\pi^2))\log(\mu/\mu_0)} \tag{9.15}$$

and at large μ (small distances) the charge goes to zero. There is less and less color charge deep inside the quarks.

The expression can be written as

$$g^2(L) = \frac{8\pi^2}{[(11/3)\,N_c - (2/3)\,N_f]\log[1/(L\Lambda_{QCD})]}, \tag{9.16}$$

where the constant $g_{\mu 0}$ is traded for a dimensional one, Λ_{QCD}, giving to the strong interaction theory its natural scale.

Example 4: The QCD-like theories close to the $b = 0$ line have both infrared and ultraviolet fixed points (Banks and Zaks [55]). The expression for the second (two-loop) coefficient is

$$b' = (34/3)\,N_c^2 - (13/3)\,N_c\,N_f + (N_f/N_c), \tag{9.17}$$

and so when $b = 0$ (for example, in a theory with $N_c = 3$ colors and $N_f = 33/2$ flavors), b' is negative. Substituting expressions for b, b' into the definition of the beta function and ignoring higher order terms, one finds that it possesses a zero (fixed point) at

$$\frac{g_*^2}{16\pi^2} = |b/b'|. \tag{9.18}$$

Note that it is small if b/b' is small; in this case the charge is always small. Therefore hadrons cannot exist, and the correlation functions has a power-like, rather than an exponential, decrease with the distance.

Studies of this "conformal regime," at zero and nonzero temperatures, require handling many quark species, and therefore such treatments only recently became possible for lattice practitioners, due to advances in computing. There is evidence that QCD with $N_c = 3$ colors and $N_f = 12$ flavors is already in this regime. We will not discuss this interesting subject further, since it is too close to the cutting edge of supercomputer ability at the time of this writing.

9.4 RG and Cold Fermi Systems

After this digression into various QFTs, let us return to the problem at hand, which is a cold Fermi system. As we will see, a continuous version of RG for *running BCS coupling in a superconductor* is in complete analogy with other running couplings, in particular with the asymptotic freedom in QCD in example 3 of Section 9.3. The presentation of the setting will be strongly simplified (for a more comprehensive review of effective theories in cold Fermi system, see, e.g., [65]).

In view of the singular behavior near the Fermi surface, we can start by defining a regulated version of the theory, in which we first exclude the vicinity of the Fermi sphere by the width Δ. The exclusion is done both from outside and inside, meaning that only holes with energies $p_F v_F < \mu - \Delta$ and particles with $p_F v_F > \mu + \Delta$ are considered. After such a regulated theory is defined, we can consider changes in its parameters when the excluded region (namely, Δ) changes a bit. Solving the RG equation allows us to determine the effective coupling in the limit of approaching the Fermi surface itself.

As noted above, the most relevant interaction between the fermions is the scattering of two particles (or holes) with opposite momenta. Let us introduce the scattering amplitude from a pair with momenta $(\mathbf{p}, -\mathbf{p})$ to another pair with momenta $(\mathbf{k}, -\mathbf{k})$:

$$f(\theta) \equiv f(\mathbf{p}, \mathbf{k}) = T(\mathbf{p}, -\mathbf{p} \to \mathbf{k}, -\mathbf{k}).$$

As observed by Landau, near the Fermi surface the scattering amplitude depends only on the angle θ between $\mathbf{p}$ and $\mathbf{k}$. A positive f corresponds to a repulsive interaction, and a negative f means attraction.

In the spirit of the renormalization group, let us now integrate out all states with $e^{-1}\delta < |\epsilon_\mathbf{p}| < \delta$. According to quantum mechanics, the scattering through virtual states in this region gives a correction to the scattering amplitude. To account for these virtual processes, we need to correct the scattering amplitude:

$$f(\mathbf{p}, \mathbf{k}) \to f(\mathbf{p}, \mathbf{k}) - \sum_i \frac{T(\mathbf{p}, -\mathbf{p} \to i) T(i \to \mathbf{k}, -\mathbf{k})}{E_i - 2\epsilon_\mathbf{p}}, \tag{9.19}$$

where the sum is over all intermediate states i that should be integrated out. We assume that the initial and final particles are almost exactly located at the Fermi surface, so $\epsilon_\mathbf{p} = \epsilon_\mathbf{k} = 0$. The intermediate state energy E_i is not restricted by energy conservation and is integrated out.

The scattering through an intermediate state can be of two types:

1. The pair $(\mathbf{p}, -\mathbf{p})$ can scatter to an intermediate pair $(\mathbf{p}', -\mathbf{p}')$, which then goes to $(\mathbf{k}, -\mathbf{k})$. In this case, the intermediate state i is that with two particle excitations with momenta $\pm\mathbf{p}'$. The Pauli principle requires that $\mathbf{p}'$ is located above the Fermi surface. This state has $E_i = 2\epsilon_{\mathbf{p}'}$ and $T(\mathbf{p}, -\mathbf{p} \to i) = f(\mathbf{p}, \mathbf{p}')$, $T(i \to \mathbf{k}, -\mathbf{k}) = f(\mathbf{p}', \mathbf{k})$.

2. Alternatively, first a pair of particles inside the Fermi sea with momenta $(\mathbf{p}', -\mathbf{p}')$ can scatter to make the final pair $(\mathbf{k}, -\mathbf{k})$, and then the initial pair $(\mathbf{p}, -\mathbf{p})$ scatters to fill the holes vacated by the pair $(\mathbf{p}', -\mathbf{p}')$ below the Fermi surface. In this case, the intermediate state i consists of six elementary excitations: four particles with momenta $\pm\mathbf{p}$ and $\pm\mathbf{k}$, and two holes with momenta $\pm\mathbf{p}'$ located below the Fermi surface, $p' < p_F$. In this case, $E_i = -2\epsilon_{\mathbf{p}'}$, $T(\mathbf{p}, -\mathbf{p} \to i) = f(\mathbf{p}', \mathbf{k})$, $T(i \to \mathbf{k}, -\mathbf{k}) = f(\mathbf{p}, \mathbf{p}')$.

Now equation (9.19) becomes

$$f(\mathbf{p}, \mathbf{k}) \to f(\mathbf{p}, \mathbf{k}) - \int_{\mathbf{p}'} \frac{f(\mathbf{p}, \mathbf{p}')f(\mathbf{p}', \mathbf{k})}{2|\epsilon_{\mathbf{p}'}|}, \tag{9.20}$$

where the integration is over all $\mathbf{p}'$ belonging to a shell

$$\delta_{in} < |p - p_F| < \delta_{out}.$$

The integral over $|\mathbf{p}'|$ can be taken, and Eq. (9.20) then becomes

$$f(\mathbf{p}, \mathbf{k}) \to f(\mathbf{p}, \mathbf{k}) - \log\left(\frac{\delta_{out}}{\delta_{in}}\right) \frac{p_F^2}{2\pi^2} \int \frac{d\hat{\mathbf{p}}'}{4\pi} f(\mathbf{p}, \mathbf{p}')f(\mathbf{p}', \mathbf{k}),$$

where the remaining integration is over the solid angle of the directions of the unit vector $\hat{\mathbf{p}}'$.

We have thus derived the RG equation,

$$\frac{df(\mathbf{p}, \mathbf{k})}{dt} = -\frac{p_F^2}{2\pi^2} \int \frac{d\hat{\mathbf{p}}'}{4\pi} f(\mathbf{p}, \mathbf{p}')f(\mathbf{p}', \mathbf{k}), \tag{9.21}$$

where $t = -\log\delta$ goes to $+\infty$ at the Fermi surface. Often one refers to t as the "time" of the RG flow. The RG evolution of the scattering amplitude toward the Fermi surface we derived shows that the interaction is logarithmically growing. This behavior is similar to example 3 in Section 9.3, with the asymptotic freedom of the QCD coupling growing toward small momenta (the IR).

More specifically, following Landau's fermi-liquid theory [63], it is convenient to expand the scattering amplitude into partial waves,

$$f(\theta) = \sum_{l=0}^{\infty} (2l + 1) f_l P_l(\cos\theta),$$

or, inversely, $f_l = \frac{1}{2} \int_0^\pi d\theta \, \sin\theta \, P_l(\cos\theta) f(\theta)$. These partial-wave amplitudes f_l have independent RG flows defined by the equation

$$\frac{df_l}{dt} = -\frac{p_F^2}{2\pi^2} f_l^2 \tag{9.22}$$

leading to the solution

$$f_l(t) = \frac{f_l(0)}{1 + \frac{p_F^2}{2\pi^2} f_l(0)t}.$$

Note that if at $t = 0$, corresponding to some initial scale, all $f_l > 0$, and the interaction is repulsive in all channels, then the four-fermion interaction vanishes at the Fermi surface as $t \to \infty$. However, if one or a few initial values $f_l(0)$ are negative, it will grow and develop a singularity (analogous to the Landau pole in QED RG but in the IR limit) at $t = -\frac{2\pi^2}{p_F^2 f(0)}$. At this scale the interaction goes into a strong-coupling regime, which, as we will see shortly, results in the Cooper pairing of the fermions. In fact, different channels are competing, and the pole is reached first by the channel having the largest negative $f_l(0)$, that is, the strongest attraction at the start.

In Summary, the RG equation shows that any attractive channel will eventually lead to a strong coupling regime at the Fermi surface. This behavior is known as the *BCS instability* of the weakly coupled Fermi systems. The energy scale at which the Landau pole is reached, which is

$$\Delta \sim \exp\left(-\frac{2\pi^2}{p_F^2 f(0)}\right), \tag{9.23}$$

defines the superconductor parameters, such as the BCS gap and the critical temperature.

9.5 BCS Pairing and Gorkov's Anomalous Green Functions

To regulate the instability induced by the logarithmic rise of the interaction strength near the Fermi sphere, we need to recognize the presence of a *nonzero Bose condensate of Cooper pairs*. They are treated by the so-called Gorkov formalism [60], which is analogous to the Belyaev's formalism for bosonic BEC already discussed. Once again, we can define the propagator as a 2×2 matrix to include various possibilities, with pair condensates.

For simplicity, let us start by assuming that the condensed BCS channel corresponds to the spin-zero fermion pair. As in the case of Belyaev's matrix propagator, we define the propagator by

$$\hat{G}_{BCS} = \begin{pmatrix} \langle c_{k\uparrow} c_{k\uparrow}^\dagger \rangle & \langle c_{k\uparrow} c_{-k\downarrow} \rangle \\ \langle c_{-k\downarrow}^\dagger c_{k\uparrow}^\dagger \rangle & \langle c_{-k\downarrow}^\dagger c_{-k\downarrow} \rangle \end{pmatrix} \equiv \begin{pmatrix} G(k_0, \vec{k}, \Delta) & \bar{F}(k_0, \vec{k}, \Delta) \\ F(k_0, \vec{k}, \Delta) & \bar{G}(k_0, -\vec{k}, \Delta) \end{pmatrix}, \tag{9.24}$$

where small arrows in the l.h.s. indicate the direction of the spin, and c, $c^\dagger$ are annihilation and creation operators, respectively, familiar from the matrix form of the quantum

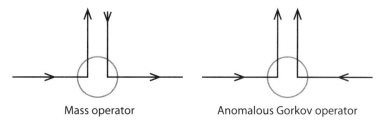

Figure 9.4. Two mass operators; the usual and the anomalous one, in Gorkov formalism.

harmonic oscillator problem. Δ is a parameter of the gap, to be determined later from (9.29). Note that the diagonal part, G, is the usual density-like operator, while the nondiagonal F is "anomalous," annihilating or creating a pair of particles. Once again, such creation/annihilation is only possible if another pair goes into the condensate, so above the critical temperature, when there is no condensate $F = 0$.

Gorkov's equations take the following self-explanatory form:

$$\left(i \frac{\partial}{\partial t} + \frac{\nabla^2}{2m}\right) G(x - x') - i\lambda F(0) \bar{F}(x - x') = \delta(x - x')$$

$$(9.25)$$

$$\left(i \frac{\partial}{\partial t} - \frac{\nabla^2}{2m} - 2\mu\right) F(x - x') - i\lambda \bar{F}(0) G(x - x') = \delta(x - x').$$

Their solution, resumming all normal and anomalous bubble diagrams, has the form

$$\hat{G}_{BCS} = \frac{1}{G_0^{-1} \bar{G}_0^{-1} - \Delta \bar{\Delta}} \begin{pmatrix} \bar{G}_0^{-1} & -\Delta \\ -\bar{\Delta} & G_0^{-1} \end{pmatrix},$$

$$(9.26)$$

where

$$G_0 = \frac{1}{k_0 - \epsilon_k + i\delta_{\epsilon_k}}, \qquad \bar{G}_0 = \frac{1}{k_0 + \epsilon_k + i\delta_{\epsilon_k}}$$

$$(9.27)$$

are the free-particle propagator and its conjugate at finite chemical potential. Here, $\epsilon_k = \omega_k - \mu$. Infinitesimal $\delta_{\epsilon_k} = \text{sgn}(\epsilon_k)\delta$ appear because Gorkov's equations as written are in real (Minkowskian) time. As before, their sign determines the pole position below or above the integration contour, and they directly follow from appropriate analytic continuation from the Euclidean version of these equations.

From these equations we can read on- and off-diagonal components of the Gorkov propagator (Figure 9.4). The off-diagonal (anomalous) propagator is

$$F(k_0, \vec{k}, \Delta) = \frac{-\Delta}{(k_0 - \epsilon_k + i\delta_{\epsilon_k})(k_0 + \epsilon_k + i\delta_{\epsilon_k}) - \Delta^2}.$$

$$(9.28)$$

The integrated loop over the normal Green function is equal to the density. The loop over anomalous Green function is equal to the gap self-energy Δ. Consistency of the obtained F and the gap leads to the so-called gap equation:

$$\Delta = -\lambda_{pp} \int \frac{d^4 p}{(2\pi)^4} F(p_0, \vec{p}, \Delta).$$

$$(9.29)$$

Here λ_{pp} is the effective coupling in the particle-particle spin-zero channel. Note that Δ enters on both sides: if it is zero, the equation is satisfied because $F \sim \Delta$ (9.28). If it is nonzero, we can cancel it from both sides and get a simpler equation

$$1 = \lambda_{pp} \int \frac{d^4 k}{(2\pi)^4} \frac{1}{(k_0 - \epsilon_k + i\delta_{\epsilon_k})(k_0 + \epsilon_k + i\delta_{\epsilon_k}) - \Delta^2},$$

(9.30)

in which the gap Δ appears only in the denominator of the r.h.s.

Note that in fact the expression for the r.h.s. looks very similar to the familiar fish diagram we discussed before (see equation (9.4)). There is only one—crucial—modification: it now has the Δ^2 term in the denominator, which regulates the divergence, producing $\log(\Delta)$. Think of the structure of the equation: an arbitrarily small initial coupling times an arbitrarily large logarithm of the gap equals 1!

This equation defines the magnitude of the gap, explaining why it must be exponentially small. As usual, the dispersion relation for the quasiparticles is calculated from the poles of G, which are at

$$\omega_{\pm} = \pm\sqrt{\epsilon_k^2 + \Delta^2}.$$

(9.31)

Note that between the particle branch and the hole branch there now appears a gap, of width 2Δ, in which there are no states at all (see Figure 9.1). The dangerous Fermi sphere surface is now "gapped." This is the origin of all the remarkable properties of superconductors!

10 | Pairing in Liquid ^{3}He and in Nuclear Matter

In this chapter we extend BCS theory to other fermionic systems, such as liquid ^{3}He at very low temperatures and nuclear matter.

However, we start with a nonpaired ("normal") liquid ^{3}He and consider Landau's Fermi-liquid theory in Section 10.1. This effective theory is very successful, even in the case where corrections to the original quasiparticles are no longer small. The particular prediction of this theory we discuss is the so-called zero-sound excitation mode, which substitutes for the usual sound mode at very low temperatures.

In Section 10.2 we will see how this mode can be found using the potential renormalization via the Lindhard function considered previously. One more instability of the Fermi surface—known as Peierls instability and leading to development of density and spin waves (and eventually to crystallization)—is also obtained in this way.

The superfluid phases of liquid ^{3}He are discussed in Section 10.3. Pairing in this system is then supplemented by the discussion, in Section 10.4, of pairing channels in nuclear matter.

10.1 Liquid ^{3}He and Landau's Fermi Liquids

Let me start this section with a small digression about ^{3}He: it is one of the most promising fuels to be used for energy production, via the nuclear reaction

$$^3\text{He} + d \rightarrow \ ^4\text{He} + p.$$

Unfortunately, it is very rare on Earth, because at Earth's temperature, it can evaporate into space due to its small mass. Yet it has been found in large quantities frozen in shady areas on the Moon's surface: perhaps one day collecting ^{3}He there will become an industry.

He atoms form a noble gas (one with no valence electrons or holes). Therefore they undergo only rather small and very short-range interactions, often parameterized by the potential

$$V(r) = 4\epsilon \left[\left(\frac{\sigma}{r} \right)^{12} - \left(\frac{\sigma}{r} \right)^{6} \right] \tag{10.1}$$

with

$$\sigma = 0.2648 \, \text{nm} \tag{10.2}$$

$$\epsilon = 1.48410^{-15} \, \text{erg} = 1.484 \times 10^{-15} \, \text{erg} \times (7.24 \times 10^{15} \, \text{K/erg}) \approx 10 \, \text{K}.$$

Note that while the Fourier transform of this potential does not exist, the corresponding scattering amplitude can of course be defined and is in fact well behaved. And as already mentioned numerous times, it is only the scattering length (amplitude) that actually matters, which then can be used as an effective coupling.

The density of the liquid, $n = 1.64 \times 10^{22} \, \text{cm}^{-3}$, in the Fermi gas approximation can immediately be converted to values of the Fermi momentum and then the Fermi energy

$$\epsilon_F = 4.97 \, \text{K},$$

which is, not surprisingly, comparable to the BEC temperature of liquid ^{4}He. Recall that the energy per atom is $(3/5)\epsilon_F$. We can also calculate the pressure for an ideal Fermi gas and then obtain the (squared) speed of (the usual density) sound

$$c_{\text{sound}}^2 = (2/3)\epsilon_F/m \approx \left(95 \, \frac{\text{m}}{\text{s}} \right)^2. \tag{10.3}$$

Note that it is even slower than in air and in liquid ^{4}He, indicating a weaker interaction. The weakness stems partly from a "Fermi blocking" of atomic scattering.

Before we consider very low temperatures and pairing, let us discuss the *normal* phase of liquid ^{3}He. It was historically important for the general development of a many-body theory of Fermi systems. In particular, it was used as the main by Landau in his 1956 paper [63] on Fermi liquids, which we will discuss next.

So far we have looked at Fermi systems using perturbative tools, the Feynman diagrams. These diagrams can be justified in the case of dilute trapped gases, but not really for the electron gas, liquid ^{3}He or nuclear matter. The Fermi-liquid theory by Landau is an effective theory that uses empirical scattering parameters and describes a vast range of phenomena happening at the Fermi surface.

The main essential assumption of this theory is that, while the fermion lines may be "dressed" by interactions in a complicated way, their quantum numbers remain the same. Most importantly, it is assumed that the Fermi surface—a sharp boundary between the occupied and unoccupied states—is still preserved. For an extensive review of applications of Landau's Fermi-liquid theory, see [64].

The energy of excitations near the Fermi surface expanded linearly:[1]

$$\xi_p = \epsilon_p - \epsilon_{p_F} \approx v_F(p - p_F). \tag{10.4}$$

The coefficient, called the Fermi velocity, may be hard to calculate by diagram resumma-tion, but it still can be used as an empirical parameter determined by experiment.

Landau argued that quasiparticles must be long-lived near the Fermi surface. Indeed, a particle decaying into itself plus a particle-hole pair has a small phase space. Parametri-cally, the imaginary part $\text{Im}(\xi_p) = \gamma_p$ of the quasiparticle energy is proportional to a small kinematical factor

$$\gamma_p \sim (p - p_F)^2. \tag{10.5}$$

Again, the coefficient in such an expression can be hard to calculate, but it can be treated as an empirical coefficient.

Now imagine a generic out-of-equilibrium distribution of quasiparticles δn_p, which we expect to be small enough to be treated as a dilute gas. The number of particles and holes should be balanced:

$$\int d^3 p \delta n_p = 0. \tag{10.6}$$

The energy can be rewritten, to first order in δn_p, as

$$\xi_{p,\sigma} = \xi^0_{p,\sigma} + \sum_{\sigma'} \int \frac{d^3 p'}{(2\pi)^3} f(\vec{p}, \sigma, \vec{p}\,', \sigma') \delta n_{p',\sigma'}, \tag{10.7}$$

where we explicitly indicate also spin degrees of freedom. The function f is the forward scattering amplitude of two particles, which can be split into the spin-independent and the spin-dependent parts:

$$f(\vec{p}, \sigma, \vec{p}\,', \sigma') = \phi(\vec{p}, \vec{p}\,') + (\vec{\sigma}_1 \vec{\sigma}_2) \zeta(\vec{p}, \vec{p}\,'), \tag{10.8}$$

where $\vec{\sigma}$ are spin matrices.

The next point by Landau is that *one only needs these functions at the surface of the Fermi sphere*. Therefore, they are actually functions of only one variable, the angle $\theta_{p,p'}$ between two momenta $\vec{p}$, $\vec{p}\,'$. This dependence can be expanded in standard polynomials of $z = \cos(\theta_{p,p'})$: the coefficients are known as the Landau Fermi-liquid parameters. Once these parameters are known, say, from a set of observables, one can use them to describe a large number of thermodynamical and kinetic properties. These assumptions were expected to hold at low temperatures $T \ll \epsilon_F$.

It is impossible not to mention one spectacular prediction by Landau, *zero sound*. Suppose that at some T there is a normal hydrodynamical sound, related to collisions: ideal gas thermodynamics leads to its velocity squared $c_s^2 = v_F^2/3$. In a Fermi system,

---

[1] Note also that Landau assumed that the dispersion relation is the same for particles and holes, or that one can simply Taylor expand ϵ_p at the surface. Thus he assumed analyticity of them at the Fermi surface, a very nontrivial assumption. Running ahead of the exposition, let me warn the reader that it will turn out to be true only for T not too low.

lowering T leads to collision times increasing as T^{-2}, and at some T quasiparticles start moving ballistically (without scattering). In this case the sound, as a collective mode, ceases to exist.

Yet, as one cools liquid ^{3}He to still lower temperature, another mode—zero sound—appears. Without derivation, let me give Landau's expression for its velocity (squared):

$$c^2_{\text{zero sound}} = \frac{p_F^2}{3m^2} + \frac{1}{6m} \sum_{\sigma,\sigma'} \int d\Omega_{p,p'} f(\theta_{p,p'})(1 - \cos(\theta_{p,p'})). \tag{10.9}$$

The r.h.s. thus includes two angular harmonics, with $l = 0,1$, of the scattering amplitude, or two Landau parameters. Zero sound exists because of some elasticity of the Fermi sphere and quasiparticle ensemble, without them colliding, so it is also known as *collisionless sound*. Similarly, some spin waves may exist and be parameterized as well.

Simplifying equation (10.9), we finds the following expression for the ratio of its speed to that of ordinary sound:

$$\frac{c_{\text{zero sound}}}{c_{\text{sound}}} = \left(\frac{1 + F_0^s}{1 + F_1^s/3} \right)^{1/2}. \tag{10.10}$$

The numerical values of the s-wave Landau parameters in channels with two-particle spin 0 and 1 (the lower index) are

$$F_0^s \approx 10, \quad F_1^s \approx 6. \tag{10.11}$$

We give such details to emphasize that corrections to the ideal Fermi gas picture are not at all small: the scattering F parameters are in fact significantly different from being small in (10.11). So, perturbation theory is not applicable; the fermionic quasiparticles are renormalized strongly and are rather far from the original particles. And, nevertheless, Landau's Fermi-liquid theory—as an effective parameterization at the Fermi surface—works remarkably well. In particular, the experimental value for the zero-sound velocity $c_{\text{zero sound}} \approx 182$ m/s and many other properties were even predicted before the experiments!

The treatment based on RG elevated the status of the Landau Fermi liquid theory from a nice model to a systematically derived approximation. One can start following the effective description along the RG path. Unfortunately, as the strength of the interaction in certain channels grows, one enters a regime where matter-induced renormalization becomes substantial, and the perturbation theory looses its applicability. Still, we understand its applicability domain: at scales where this interaction is not yet strong enough to generate superfluid Cooper pairing.

10.2 Zero Sound and the Peierls Instability from the Resummed Lindhard Diagram

Instead of following Landau's arguments, I point out that we can trace zero sound using something we already calculated while discussing the electron gas. The sequence of

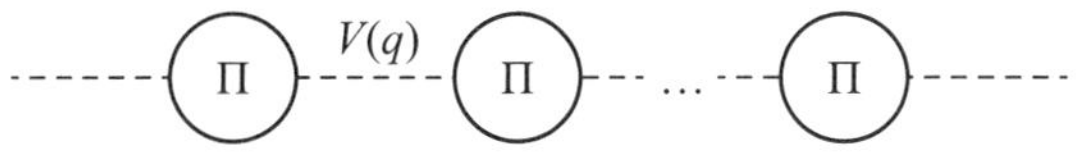

Figure 10.1. The sequence of diagrams for potential renormalization, with insertions by any number of the polarization operators Π.

diagrams shown in Figure 10.1 sums into a dressed form

$$V(q) + V(q)\Pi(\omega, q)V(q) + \cdots = \frac{V(q)}{1 - V(q)\Pi(\omega, q)}, \tag{10.12}$$

and the pole of the denominator, as usual, provides the dispersion relation for the corresponding mode. In the electron gas, the Coulomb potential $V(q) \sim 1/q^2$ was singular at small q, which produced plasma oscillations. For now, let us assume that it is nonsingular and even has a q-independent coupling constant, $\lambda = 4\pi a/m$, with a scattering length a.

The function Π is still the same Lindhard function that we calculated for two fermionic propagators. Introducing the notation $x = \omega/(v_F q)$ we can write the equation to be solved as

$$1 = \frac{4\pi a}{m} \frac{m p_F}{\pi^2} \left[\frac{x}{2} \log\left(\frac{1+x}{x-1}\right) - 1 + \frac{i\pi}{2}\theta(1 - |x|) \right]. \tag{10.13}$$

The weak-coupling case, small $a p_F \ll 1$, would require the quantity in square brackets to be large. This is possible when $x \to 1$ and $-\log(x-1)$ gets large. If we write $\omega = uq$, where u is the speed of the mode, then $x \to 1$ means $u \to v_F$, or more accurately,

$$\frac{u}{v_F} = 1 + 2e^{-\pi/(4a p_F)^{-2}}. \tag{10.14}$$

Note that $x > 1$ and thus the imaginary term is absent in this case. Note further that this velocity is higher than for ordinary sound and even higher than v_F!

However, if the coupling is strong and $a p_F \gg 1$, the quantity in square brackets in (10.13) should instead be small. Note further that at large x, we find

$$\left[\frac{x}{2} \log\left(\frac{1+x}{x-1}\right) - 1 \right] \approx \frac{x}{2}\frac{2}{x} - 1 \approx 0, \tag{10.15}$$

and so we have to expand the function further, to order $1/x^2$. This produces an even larger velocity

$$\frac{u}{v_F} \sim \sqrt{a p_F} \gg 1, \tag{10.16}$$

and again, $x > 1$, so there is no absorption! Figure 10.2 explains the kinematics of this result.

Remarkably, there is another important phenomenon we can obtain from this equation. To explain it, let us first simplify the kinematics and the Lindhard function, going to the one-dimensional Fermi gas problem. In one-dimension there is no Fermi sphere: it is reduced to two points, $\pm p_F$. Let us take a hole near the left end, $p_h(k) = -p_F + \delta k$, and a particle outside the right end, $p_p(k+q)$. Select a special q close to the "diameter" of

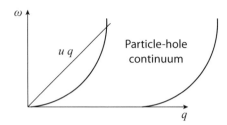

Figure 10.2. The zero sound and the particle-hole continuum on the energy-momentum plane. As the sound dispersion curve enters the particle-hole continuum its decays become possible.

the Fermi sphere $q = 2p_F + \delta q$, so that their energies are $\epsilon_F - v_F \delta k$ and $\epsilon_F + v_F(\delta k + \delta q)$, respectively.

Now the static Lindhard function is

$$\int \frac{dk}{2\pi} \frac{n_k - n_{k+q}}{\epsilon_k - \epsilon_{k+q}} \approx \int \frac{dk}{2\pi} \frac{0-1}{-v_F(2\delta k + \delta q)} \sim \frac{1}{v_F} \log(\delta q / p_F). \tag{10.17}$$

So, as $\delta q \to 0$ or $q \to 2p_F$, the function Π has infinite response, and thus any small coupling will lead to a new mode. This is called the *Peierls instability of a one-dimensional Fermi gas*, a signal that the gas basically wants to become a crystal with a period close to $1/(2p_F)$.

A specific form of it, the *spin-density waves*, is relevant for specific electronic materials, especially in the two-dimensional case, as originally proposed by Overhauser. Dense nuclear/quark matter may also undergo a transition to solid crystalline structure (see, e.g., [66]).

To finish our discussion of this topic, let us generalize the static Lindhard function to small but nonzero T. Thermal occupation numbers for particles and holes are now not 0 and 1, but are given by Fermi distributions, and very near ϵ_F these numbers instead are 1/2 and 1/2, so the numerator in expression (10.17) is zero. More accurately, we get

$$\int \frac{dk}{2\pi} \frac{\tanh(\epsilon_k / T)}{\epsilon_k - \epsilon_{k+q}} \sim \log(T/\epsilon_F).$$

This means that freezing of the gas happens at exponentially small $T/\epsilon_F \sim \exp[-O(1/\lambda)]$.

10.3 Superfluidity in ^{3}He

The thermodynamics of the fermionic liquid ^{3}He is of course very different from that of bosonic ^{4}He. In particular, there are no BEC-like transitions at temperatures of a few K, and the region of interesting phase transitions is shifted down by three orders of magnitude, $T/\epsilon_F \sim 10^{-3}$. Reaching that temperature was not easy, and the Nobel Prize was awarded to David Lee, Douglas Osheroff, and Robert Coleman Richardson, who discovered two phase transitions along the melting curve. Those were soon realized to be two distinct superfluid phases of helium-3 called ^{3}He-A and ^{3}He-B (see Figure 10.3).

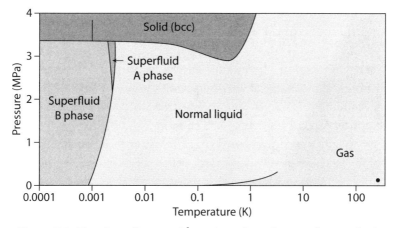

Figure 10.3. The phase diagram of ^{3}He. From the web page of LTL/Helsinki University of Technology (http://ltl.tkk.fi/research/theory/he3.html), an excellent source for many phenomena in liquid ^{3}He.

Cooper pairs of ^{3}He atoms, allowed by Fermi statistics, include the usual BCS pair, with the total spin S and orbital momentum L

$S = 0$, $\psi_\alpha \psi_\beta \epsilon_{\alpha\beta}$ antisymmetric, $L = 0$ symmetric, total = antisymmetric.

An alternative option is

$S = 1$ symmetric, $L = 1$, $\epsilon_{\alpha\beta\gamma} \psi_\alpha \psi_\beta$ antisymmetric, total = antisymmetric.

Looking at atomic scattering amplitudes, one finds that the spin-0 channel has a very strong repulsive interaction, while spin-1 is attractive, as needed for pairing. So pairing should occur in the second spin-1 channel.

Since a pair has $S = 1$ and $L = 1$, how are the directions of those in the condensate related? Obviously, one option is that they are opposite to each other, adding up to zero total angular momentum, $\vec{J} = \vec{S} + \vec{L} = 0$. This option is known as the "spin-orbital locked" phase. Another option is to have them parallel, producing an anisotropic or polarized condensate, with $J = 2$. While the exact structure of phase A (see Figure 10.3) is a bit complicated, it is indeed anisotropic and is (partly) this $J = 2$ option. For the discussion of the condensate structure, see the paper by Leggett [62]. His suggested structure was proved to be correct later, and he was awarded the 2003 Nobel Prize for it.

Without going into details, note that the anisotropic phase A has unusual rotation properties. Like all other superfluid phases, it can only rotate in the form of an ensemble of quantized vortices, since the order parameter—whatever complicated form it may—must be periodic on any loop going around the vortex core. But rotations of spin and angular momenta are different, which complicates the story, to the extent that vortices may lose axial symmetry or even split their cores into two subcores. (For details, see the web page mentioned in Figure 10.3.)

10.4 Pairing in Nuclei and Nuclear Matter

From the first days of nuclear physics, the simplest quantity to measure experimentally was the total ground-state energy of the isotopes with various Z and N. Plotted versus the total number of protons Z and neutrons N, one finds the volume and surface terms, the shell effects near the "magic numbers," and so forth.

Let us focus on only one term in such empirical mass formulas, the so-called pairing term. Plotting the total energy versus Z and/or N determines two different curves for even and odd values. The former nuclei are a little more deeply bound. Now that we are familiar with Cooper pairing and BCS quasiparticles, it is natural to relate these observations to the existence of the energy gap in quasiparticle energy spectra. Recall that the energy needed to break a Cooper pair is 2Δ.

To separate the pairing effect from background smooth variation with N, Z, one can define the pairing gap by using the binding energies $B(N, Z)$ of four subsequent isotopes, say, with the neutron number between $N-2$ and $N+1$, using the following combination,

$$\Delta_n = \frac{1}{4}[B(N-2, Z) - 3B(N-1, Z) + 3B(N, Z) - B(N+1, Z)], \tag{10.18}$$

in which all "smooth" dependences drop out, while the pairing remains. For heavy nuclei, its magnitude is about 1 MeV, both for pp and nn pairs.

This value should be compared not to the binding-per-particle energies themselves, which are of the order $B \sim 8$ MeV, but to the nuclear Fermi energy, which is of the order $\epsilon_F \sim 30$ MeV. Recall that ϵ_F is defined as the kinetic energy, counted from the bottom of the attractive mean-field potential. So the gap-to-Fermi-energy ratio is

$$\frac{\Delta}{\epsilon_F} \sim \frac{1}{30}, \tag{10.19}$$

Which is still small, yet much larger than in other examples considered so far.[2]

As usual, we start discussing possible channels for pairing that are allowed by Fermi statistics. Quantum numbers of nucleons are the isospin $I = 1/2$; spin $S = 1/2$; and on top of that, the orbital momentum L. Cooper pairs are subject to the following restrictions required by Fermi statistics.

When the total isospin of a pair is zero, $I = 0$, the isospin wavefunction $\psi_\alpha \psi_\beta \epsilon_{\alpha\beta}$ is antisymmetric, so it describes the pn channel only. If the total spin is zero, $S = 0$, as well, it contributes another -1 to the particle permutation factor. In contrast, Fermi statistics require the full permutation factor to be -1 for fermions. With two minuses already required, we need the antisymmetric coordinate wavefunction, $(-)^L = -1$. So, in this channel the orbital angular momentum L must be odd.

Similarly, when $I = 0$, and $S = 1$, then L must be even.

---

[2] A hypothetical metal with such a ratio would retain superconductivity not only at room temperature but even in an oven.

If $I = 1$ the isospin wave function is symmetric. So, compared to the $I = 0$ channel discussed above, the restrictions on L cases are interchanged. Now for $S = 0$ and $S = 1$ channels, L must be even and odd, respectively.

In general, many pairing channels are allowed by Fermi statistics. Which one succeeds in creating the condensate depends on dynamics. The most attractive channel is the one that creates the minimal free energy of the system. In practice the channel is decided by following the RG equations in all channels: the one that reaches condensation "earlier" than others, in the RG "time" $t = \log(p_F/|p - p_F|)$, is likely to be the operational channel.

To decide the outcome of such race, one needs to have some knowledge about the sign and magnitude of the nuclear forces in each channel. It is logical to start looking at these interactions in various channels, one by one, first considering the known scattering amplitudes in vacuum. The $I = 0$ pn channel is attractive; it has a bound state—the deuteron d—although with very small binding, $B_d \approx 2.2\,\text{MeV}$. The $I = 1$ nn channel is overall attractive as well, but the attraction is a bit weaker, unable to produce a bound state, but only a virtual[3] resonance in the nn scattering amplitude. And indeed, in the middle part of the isotope Z, N chart, where $N \approx Z$ and thus proton and neutron Fermi momenta are very similar, it is the pn pairing that dominates.

Nuclear matter inside neutron stars is very different, being far from $p \leftrightarrow n$ symmetric. Let us recall why this is so. Suppose normal matter, consisting of about equal numbers of electrons, protons, and neutrons, is strongly compressed. If the densities of these species are comparable, then the p_F values of all three species mentioned are comparable as well. If so, the electron energy is $E_e = p_F^2/2m_e$, while that for the nucleons is (for $N = p, n$) $E_N = p_F^2/2m_N$. Since the electron mass is much smaller than that of the nucleons, $m_e \approx m_N/2{,}000$, its Fermi energy is much larger. This means that electrons are too expensive to have! Inverse beta decay

$$e + p \rightarrow \nu_e + n$$

can obliterate nearly all electrons and protons. The neutrinos leave the star and then, to a good approximation, only neutrons remain.

Very different densities of p and n imply that the radii of their respective Fermi spheres are hugely different. Since pairing is a phenomenon in a thin shell on their surfaces, the pp and nn pairings must develop separately. In the nn race, the winner can be the $S = 1$, $L = 1$, $J = 2$ channel, due to a strong spin-orbit force voting for the largest value of J possible. Yet the actual winner seems to be the simplest: the $S = 0$, $L = 0$ channel, with an isotropic condensate.

Since neutrons have no charge, this dominant component of neutron stars is a superfluid. Because stars are rotating, recall that angular momentum can only be maintained by some ensemble—crystalline—of quantized vortices, as in a rotating cup of liquid ^{4}He discussed in Chapter 5.

---

[3] A side remark: if just a small modification of the nuclear forces would make a neutron pair bound, the primordial nucleosynthesis in the Big Bang would change our Universe beyond recognition. So we know for a fact that such small modifications were not allowed even 13 billion years ago!

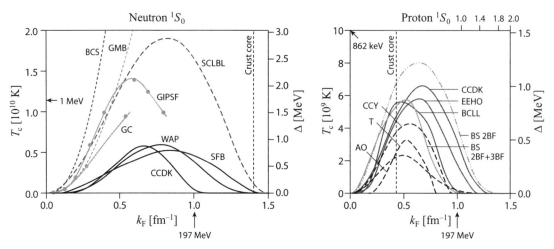

Figure 10.4. Pairing critical temperatures (proportional to the gaps) as a function of density (via Fermi momentum p_F) for neutrons and protons, respectively. Adapted from Hormuzdiar and Hsu [68].

The *pp* race for superconductivity is believed to be won by the simplest $S = 0$, $L = 0$ channel as well. The gap value Δ in the quasiparticle spectrum, and the critical temperature proportional to it, depend on the density: they reach $1-2$MeV at the star's center (for more details, see, e.g. [68] and references cited there). Since inside pulsars the more relevant quantity is the critical temperature rather than the gaps, these are shown in Figure 10.4. The units preferred in the astrophysics literature may be converted to more standard ones by noting that $1\,\mathrm{MeV} \approx 1.1 \times 10^{10}\mathrm{K}$, as also indicated in the left panel. Different curves compare different approximations used in the literature: their spread shows the theoretical uncertainty.

The density (or k_F) dependence, $T_c(k_F)$, in the Figure has a maximum at some density and decreases at high density. This is because the nuclear potential, attractive at large distances, becomes repulsive at smaller distances due to the repulsive cores of the nucleons. When the gaps and T_c of the $S = 0$, $L = 0$ superfluid vanish, superfluidity/superconductivity switches to another channel with larger L: the wavefunction of the pair at the origin vanishes as r^L, reducing the effect of the repulsive core.

These temperatures are of interest for the following reason. All neutron stars, observed by us as radio pulsars, are created in supernova explosions. So initially they are very hot, with $T \gg T_c$ in a nonsuperfluid normal phase. Then they cool down and eventually acquire both superconductivity and superfluidity. During this process, their thermodynamical properties (such as specific heat) change dramatically: a gapped Fermi surface carries many fewer thermal excitations compared to the ungapped normal phase. The transition may also affect the way rotational and magnetic fields are distributed inside the star, since cold stars should build ensembles of vortices.

Furthermore, since the proton fluid is charged, its condensate is superconducting, and its lattice of quantized vortices carry a certain amount of magnetic flux: they are very different from vortices in superfluids. We will discuss their differences in the next chapter.

11 Vortices and Topological Matter

With this chapter we start examining another central idea of the book: the "bodies" of many-body theory need not be some fundamental particles, that is, quanta of the fields that appear in the Lagrangian of the system. In particular, they can be a type of soliton, more specifically, topological solitons, which are stabilized by the topological quantum numbers and are not at all directly related to the quantum numbers of the fundamental fields.

Vortices are the simplest example of this kind of matter. While they are extended objects in three dimensions, they can be considered as localized excitations in two dimensions. Phases of vortex matter in two dimensions is the subject of the next section. We will discuss vortices in superconductors in Section 11.2. The interaction between them can be attractive or repulsive: the corresponding materials are called type I and II superconductors (see Section 11.3).

The next application is the phase transitions associated with the rotation of nuclei, as a function of the rotation angular velocity (Section 11.4). Similar physics, but on a much larger scale, is observed in neutron stars, in which the transitions are known as pulsar glitches (Section 11.5).

11.1 The BKT Phase Transition of Vortex Matter in Two Dimensions

The Berezinsky-Kosterlitz-Thouless (BKT) transition [71,72] can be found in several two-dimensional systems in condensed-matter physics that are approximated by the planar Heisenberg model, often called the XY model. These systems include Josephson junction arrays and thin disordered superconducting granular films. The transition is rather unusual, and the authors who discovered it were awarded the 2016 Nobel Prize.

The Heisenberg model describes the nearest-neighbor spin interactions. The spins themselves are placed on the sites of some lattices. In contrast to the Ising model, in this case spin is not a quantum variable, and its direction is described by some unit vector. The Hamiltonian of the model is

$$H_{XY} = -J \sum_{ij} \cos(\theta_i - \theta_j),$$ (11.1)

where J is a parameter, and the θ_i are the direction angles of these classical spins. The sum is done only over the pairs of spins on the two-dimensional cubic lattice that are nearest neighbors: so in the continuum formulation, this difference would be replaced by a derivative. The continuous version of the model has a Hamiltonian of the form

$$H_{XY} = \frac{J}{2} \int d^2x \left(\partial_\mu \theta(x)\right)^2.$$ (11.2)

(A small pedagogical digression. While the coordinates of the model are "planar" (two-dimensional), it is not necessary for the spin vectors to belong to the same two-dimensional plane. One may generalize the model, allowing spins to belong to some N-dimensional space $\sim \sum_{\mu,a}(\partial_\mu n^a)^2$, $\mu = 1, 2, a = 1, ..., N$. In this case it is known as the sigma model. Furthermore, one of the coordinates can be interpreted as the Euclidean time: in this case the model describes (1+1)-dimensional field theory. Note that it is not a Gaussian (noninteracting) theory. To see this, recall that for the unit vector, $\vec{n}^2 = 1$, its components are not independent. So one of them can be expressed in terms of the others, and if we do that, the action becomes nonlinear and thus nontrivial. This model in four dimensions is important for understanding chiral symmetry breaking in QCD; and in higher dimensions, it is an ingredient of string theory.)

A good pedagogical introduction to BKT transition in this model is given by Kogut [73]: here we outline only the main results. In one- and two-dimensional systems there can be no condensation, because fluctuations destroy the would-be condensate. So spins cannot be aligned on average: $\langle \vec{n} \rangle = 0$. As a result, there is no order parameter, and the transition is not the usual second order, as we discussed in Chapter 10 using the Landau mean-field model and Wilsonian indices. So, one may ask: Why can there be any phase transitions at all?

As usual, we begin with the perturbation theory. At low T the system is only slightly excited, so we can think of θ as varying little and expand cos to quadratic order. Then the correlator of the spin directions

$$\langle e^{i\theta(x)-i\theta(0)} \rangle = \frac{1}{Z} \int \prod d\theta_i \, e^{i\theta_n - i\theta_0 - (J/[2T])(\delta\theta)^2}$$ (11.3)

is Gaussian and can be evaluated by the usual rules involving Green functions.

Note that by inverting the Laplacian in two dimensions,

$$G(x) = \int \frac{d^2k}{(2\pi)^2} \frac{e^{i\vec{k}\vec{x}}}{k^2} = -\frac{1}{2\pi} \log|x|,$$

we find that—unlike in three or more dimensions—it actually grows with separation of the sources. Therefore, even weak interactions at the atomic scale in two dimensions get strong in the IR. (In two dimensions it is rather trivial; no renormalization group is needed.)

Completing the perturbative calculation at low T, we find the following result for the long-distance correlator of the spin directions:

$$\langle e^{i\theta(x)-i\theta(0)} \rangle \sim \left(\frac{a}{|x|} \right)^{\frac{T}{2\pi J}}. \tag{11.4}$$

(Here and below, a is the UV cutoff or the atomic scale.) A power dependence on the distance is natural in two dimensions, a feature that in three dimensions we can only achieve under the well-tuned conditions of the second-order phase transitions.

At high T the coefficient of the interaction term is small, $J/T \ll 1$, and we can expand the exponent. The so-called strong coupling-like expansion is based on the following observation: the average of $e^{i\theta_n}$ vanishes with integration over the angle. A nonzero average requires the exponent to appear with its conjugate. This means the result is zero unless we keep the Nth order in $J/(2T)$, where N is the minimum number of lattice links between the two ends (the site 0 and the site n), so it is the distance in units of a. This calculation of the correlator produces the following result:

$$\langle e^{i\theta_n - i\theta_0} \rangle \sim \left(\frac{T}{J} \right)^N \sim e^{-N\log(\frac{T}{J})}. \tag{11.5}$$

So, at high T, we expect to have a correlator decaying exponentially with the distance $r = Na$.

The BKT transition happens when one regime switches to the other. Nontrivially, the transition takes place between two disordered phases, which differ in the behavior of their correlation functions. Since the two regimes are qualitatively different, the transition point must still be a singularity. I have already explained that there cannot be an ordered phase with the usual second-order transition: and indeed, it is in fact a very smooth *infinite-order* transition.

The physics of the BKT transition can also be understood by using an ensemble of vortices. In two dimensions and at scales much larger than the lattice atomic scale a, these vortices can be considered to be a pointlike *topological defects*. Around their core, approximated by a point, the spins or the angle θ rotates, so that the circulation integral around it, known as vorticity,

$$v = \frac{1}{2\pi} \oint_C dx^\mu \partial_\mu \theta, \tag{11.6}$$

gets quantized for periodic θ on the contour C, as we discussed above.

The energy of the vortex is logarithmically divergent

$$E = \kappa \, \log \left(\frac{R}{a} \right), \tag{11.7}$$

where κ is some coefficient, while R and a are the IR and UV cutoffs—the size of the vessel and the atomic scale, respectively.

The positions of the vortices are arbitrary, so the number of corresponding states is $N_\text{states} \sim (R/a)^2$. The logarithm of the number of states is the entropy, and therefore the free energy of vortices is

$$F = E - TS = (\kappa - 2T)\log\left(\frac{R}{a}\right).$$ (11.8)

So, according to this argument, at high T the entropy term wins, and the free energy becomes negative. Its magnitude grows if R is as large as possible. This means the system becomes a random "plasma" of vortices.

Yet at T low compared to κ, the opposite happens: the energy wins over the entropy. The large value of the logarithm in this case forces F to be large and positive, so the individual vortices are strongly suppressed. The solution to this difficulty at low T is that positive and negative vortices are coupled into pairs, the "topologically neutral molecules." In this case the large IR cutoff R is replaced by the (much smaller) size of the molecules. The pairs are neutral dipoles of the charges, so their interaction is short range. The correlation functions in such a "molecular-gas" phase decay in the usual way, exponentially.

The (unusually simple) prediction for the critical temperature stems from the free energy (equation 11.8):

$$T_c = \frac{\kappa}{2}.$$ (11.9)

The importance of the BKT transition stems from its two main lessons. One is that the explanation of its very existence is highly nontrivial in terms of the original microscopic degrees of freedom, but it can be explained rather simply in terms of the effective topological excitations—the vortices. In the final chapters of this book, devoted to the description of the QCD vacuum (the ground state of the Universe we live in), we will meet the analog of the BKT transition, in which the molecular phase at low density of the topological objects—instantons or instanton-dyons—will be replaced by a disordered plasma of them at higher density.[1]

The second lesson stems from the rather unusual property of the high-T phase from the point of view of the RG. The field correlators in this phase are conformal, being the temperature-dependent powers of the distance x. Therefore, the discovery of the BKT phase strongly affected studies of the two-dimensional conformal symmetries and theories in general.

Jumping from the 1970s to today, let us consider the recent interesting study of the ensemble of two-dimensional vortices [75]. The system of vortices form a kind of many-body system of the plasma type, interacting by means of a two-dimensional Coulomb law. Onsager [74] suggested the use of the mechanics of vortices to understand turbulent systems of this type. He also pointed out that systems of a very large number vortices can be treated as a fluid, which itself can be described in terms of hydrodynamics.

Wiegmann and Abanov [75] studied whether the quantization of vorticity can lead to some unusual macroscopic behavior of such "vortex fluids." These authors considered a chiral setting of the problem, including vortices with only one sign of the vorticity.

---

[1] Ironically, in this case the low density of the topological solitons occurs at high T, and low density at low T, but these phenomena are naturally explained by the general properties of QCD at high T.

Let me mention some of their results. There appear anomalous viscous terms in the stress tensor,

$$\tau_{xy} = \tau_{yx} = \eta(\nabla_x v_x - \nabla_y v_y)$$

(11.10)

$$\tau_{xx} = -\tau_{yy} = -\eta(\nabla_x v_y + \nabla_y v_x),$$

plus an addition to the pressure

$$\delta p = 2\eta|\omega| + 2\frac{\eta^2}{\sqrt{|\omega|}}\Delta\sqrt{|\omega|},$$

(11.11)

where ω is vorticity of the flow (or the density of vortices) and Δ is its Laplacian.

The anomalous viscosity value has a very peculiar value

$$\eta = \frac{1}{4}\gamma,$$

(11.12)

where γ is the single vortex strength.

Some unusual properties of these terms can be seen from the fact that unlike Euler's inviscid incompressible fluid, this fluid does exert a force on a shear flow. Yet the force acts normally to the velocity and produces neither work nor any dissipation (!), unlike the usual Navier-Stokes viscosity. Curious features also appear if the two-dimensional manifold is curved. The modification of the density of vortices turns out to be proportional to modifications of the scalar curvature R.

These effects are called "anomalous" because they do not originate from the action in the path integral (as for the Feynman diagrams we have discussed in all applications so far), but instead from the volume element (i.e., the Jacobian) of the space of collective coordinates. We will meet some other anomalies of a similar nature in Chapter 12, related to the topology of the gauge theories.

11.2 Vortices in Superconductors

So far we have considered vortices in superfluids. They are rather different from vortices in superconductors that we now discuss. The difference stems from the fact that in the present case, the Cooper pairs are electrically charged, while in superfluids the basic boson is not. Rotating current creates magnetic fields, which lead to dramatic changes in the interactions of such objects.

In superfluids the wave function of the collective condensate needs to be periodic, and therefore

$$\hbar\int_{\text{loop}} \vec{\nabla}\phi\, dl = \hbar 2\pi n,$$

(11.13)

and since $\hbar\int_{\text{loop}} \vec{\nabla}\phi = 2m\vec{v}$, we find a very slow decrease in velocity with distance, $v(r) \sim 1/r$. As a result, the total energy of a vertex (per length) $\sim \int v^2 r\, dr \sim \log(r_{\text{max}}/r_{\text{min}})$ needs to be regulated in IR by the size of the vessel.

In superconductors the presence of an Abelian electromagnetic field leads to an additional term with the vector potential

$$\hbar\vec{\nabla}\phi = 2m\vec{v} + 2e\vec{A} \qquad (11.14)$$

because of the covariant (long) derivative in the action, needed for gauge invariance. Note that we put here $2e$, the charge of a Cooper pair, in front of $\vec{A}$. And therefore we have another option: to make $\vec{A}$ be decreasing slowly with distance. Specifically, in cylindrical coordinates, let us take

$$A_\phi = \frac{\hbar 2\pi n}{2e}\frac{1}{r}, \qquad (11.15)$$

with the other components being zero. This option leads to much smaller energy. Indeed, unlike velocity, which enters the energy density directly (squared), the vector potential contributes to it via the magnetic field (squared), its curl.

In cylindrical coordinates,

$$\text{curl}\vec{A} = \left[\frac{1}{r}\frac{\partial A_z}{\partial \phi} - \frac{\partial A_\phi}{\partial z}, \frac{\partial A_r}{\partial z} - \frac{\partial A_z}{\partial z}, \frac{1}{r}\frac{\partial(r A_\phi)}{\partial r} - \frac{1}{r}\frac{\partial A_r}{\partial \phi}\right], \qquad (11.16)$$

and thus we find that for our choice of A_ϕ, the curl is zero in bulk. So we "do not have to pay" for the long-range pure gauge tail of the vortex: it disappears for free: However, a more accurate calculation shows that it is not zero but a delta-function contribution, $B_z \sim \delta(r)$, since there is a nonzero quantized flux of the magnetic field

$$\Phi_0 = \frac{2\pi\hbar}{2e} \approx 2.07 \times 10^{-15}\,\text{T/m}^2 \qquad (11.17)$$

going through the vertex core. Note that this contribution is defined by a combination of two fundamental constants, the Plank constant and the electron charge. Note also that the flux quantum is very small: for example, if we have a superconducting magnet $1\,\text{m}^2$ in cross section holding a field of $B = 2\,\text{T}$, the number of vortices in it will be $\sim 10^{15}$!

11.3 Type I and II Superconductors

The Landau-Ginzburg [56] phenomenological theory of superconductivity uses the same quartic potential as the Landau model of second-order phase transitions (3.20), but it modifies the kinetic term, replacing the usual derivatives by the covariant ones

$$\partial_\mu \to \partial_\mu - 2ei\,A_\mu,$$

including the electromagnetic field. It is required because the order parameter (which we now will call ψ rather than ϕ) corresponds to the wavefunction of the Cooper pairs, with the electric charge $2e$.

Using this theory, we can quantify the vortex solution, as was done by Abrikosov [69]. From the action we get the following two equations for the Cooper pair condensate wavefunction ψ and vector potential $\vec{A}$:

$$\frac{1}{2m}\left(-i\hbar\vec{\nabla} - 2e\vec{A}\right)^2 \psi + \alpha\psi + \beta\psi|\psi|^2 = 0 \tag{11.18}$$

$$\frac{1}{\mu_0}\text{curl}(\text{curl}\,\vec{A}) = -\frac{ie\hbar}{m}(\psi^*\nabla\psi - \psi\nabla\psi^*) - \frac{4e^2}{m}|\psi|^2\vec{A} = 0. \tag{11.19}$$

The first is the (nonlinear) Schrödinger-like equation, containing phenomenological coefficients (here called α, β), of the quadratic and quartic terms in the action. Note the main element of (11.18), a covariant or "long derivative," which provides gauge invariance, over simultaneous change of the phase of ψ and adding its gradient to $\vec{A}$. The second equation is the Maxwell equation, curl $\vec{B}$ equal to the electromagnetic current on the r.h.s.. The expression for the current follows from the Landau-Ginzburg action, by its standard differentiation over $\vec{A}$.

To understand how the vortex solution works, it is instructive to consider first a simpler problem, with a flat boundary between normal and superconducting phases. The normal phase has a nonzero field $\vec{B} \neq 0$ that penetrates into the superconducting part by a certain length λ: in the bulk of the superconductor, the magnetic field is expelled. The superconducting phase has nonzero condensate $\psi \neq 0$, penetrating into the normal phase by some other length ξ. The derivatives are only one-dimensional, normal to the plane of phase separation.

It is a good simple exercise to obtain the two penetration lengths from the equations above. They are

$$\lambda = \sqrt{\frac{m\beta}{4e^2\mu_0|\alpha|}}, \quad \xi = \sqrt{\frac{\hbar^2}{2m|\alpha|}}. \tag{11.20}$$

Their ratio $\lambda/\xi = \kappa$ is the crucial parameter of the problem, because when one calculates the surface energy of the phase boundary, it turns out that for $\kappa > 1/\sqrt{2}$ the ratio is negative; otherwise it is positive.

When κ is positive, typical for pure metals, is called the *type-I superconductor*: in this case the system tries to have as little phase boundary as possible. If a type-1 superconductor is placed in a magnetic field, and for geometric reasons expelling the field is too costly, complicated patterns appear. Two examples are shown in Figure 11.1: as indicated, the scale is 1 mm, so these regions are macroscopically large. The exact pattern depends on geometry and the sample history: the so-called Prozorov pattern on the left has been observed only recently (in 2007 [57]), while the pattern that Landau suggested on the right was assumed to be the case since 1930s.

In contrast, when κ is negative, the material is called a *type-II superconductor* and possesses negative boundary energy. It is then beneficial to have as much boundary as possible. Thus any magnetic flux passing through the sample is split into the smallest allowed units with one quantum of flux Φ_0, the Abrikosov vortices . As we already know, at large r the gauge field is topologically nontrivial with circulation proportional to integer n, but it is "pure gauge," producing zero (bulk) $\vec{B}$.

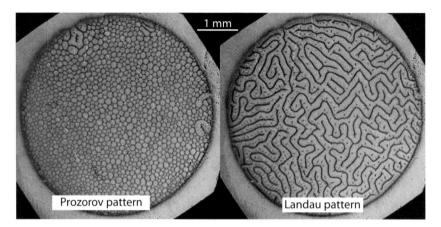

Figure 11.1. Patterns of a type-I superconductor in a magnetic field depend on the sample history, such as whether the magnetized sample was cooled (a) or was a superconducting one applied to a field (b). From Prozorov [57]. Copyright 2007 by the American Physical Society.

Now, curl(curl) in cylindrical coordinates is

$$-\frac{\partial}{\partial r}\frac{1}{r}\frac{\partial}{\partial r}(r A_\phi) = \frac{4e^2}{m}|\psi|^2 A_\phi,$$ (11.21)

and on the r.h.s. I have ignored the gradients of ψ, assuming that ξ is small and it is mostly constant. This equation also has (on top of the $1/r$ tail) an exponential core solution $\sim \exp(-r/\lambda)$ with $\lambda^{-2} = 4e^2/m$. The magnetic field is

$$B_z = \frac{\Phi}{2\pi\lambda^2} K_0(r/\lambda)$$ (11.22)

with Bessel function K_0. At large r it decays exponentially, but at small r it has a logarithmic singularity. But here the equation changes: gradients of ψ appear and regulate B at small r to be constant.

Unlike vortices in superfluids such as He-II, the interaction between vortices in superconductors is exponentially small at large distances. (This is because the long-range field is now delegated to a pure-gauge vector potential and does not lead to a magnetic field in the bulk.)

Many of these vortices can be packed together, which is what applications demand. Figures 11.2 shows a calculation for a densely packed triangular[2] lattice. The bottom image shows the experimental study of the mixed phase using external particles that become pinned to vortices and then are sucked out of them by an electric field. Type-II superconductors are widely used in devices, as they allow both supercurrent and magnetic fields to coexist.

The near-ideal lattice of vortices is of course not always the case: their pinning at the ends and production history often create much more complex vortex matter, ranging from

---

[2] In fact Abrikosov's original paper considered a rectangular lattice, which was a bit easier to calculate, but it was later found that the lowest energy state is the triangular one.

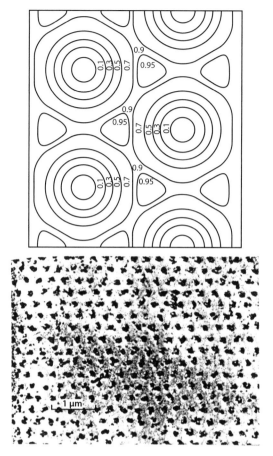

Figure 11.2. Triangular vortex lattice, theory and experiment. The former is from W. H. Kleiner, L. M. Roth, and S. H. Autler, "Bulk Solution of Ginzburg-Landau Equations for Type II Superconductors: Upper Critical Field Region," *Physical Review* 133 (1964). Copyright 1964 by the American Physical Society. The latter is for the superconducting alloy $Pb_{96}In_4$ at $T = 1.1$ K, and is from U. Essmann and H. Träuble, "The Direct Observation of Individual Flux Lines in Type II Superconductors," *Physics Letters* A 24 (1967).

liquid to solid. What matters, however, is that rather high permanent magnetic fields can be created and supported, and the production of type-II superconductors is a big industry.

11.4 Phase Transitions in Strongly Rotating Nuclei

If a (deformed) nucleus gets into rotational states, the rotational band of states is observed. The characteristic feature is that the excitation energies depend on the angular momentum L via the usual expression

$$E_L = \frac{L(L+1)}{2\bar{J}}, \tag{11.23}$$

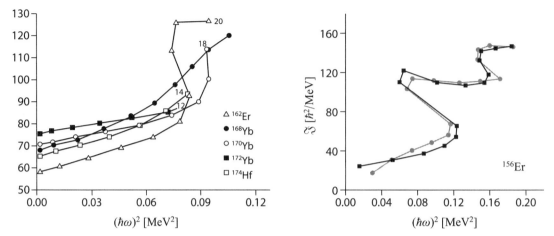

Figure 11.3. Moment of inertia $2\bar{J}/\hbar^2$ (MeV^{-1}) versus $(\hbar\omega)^2$ (MeV^2). The left plot (from the 1970s; taken from [76]) indicates existence of the rotational phase transition. In the right plot (from [81], the squares are subsequent states in the rotational band predicted by the theory, while the circles are the excited states seen in experiment) shows the current status of such experiments. The specific nuclei discussed are indicated on the plots.

where $\bar{J}$ is the nuclear moment of inertia. The matter distributions in the ground state of nuclei are known (e.g., from the formfactors of electron scattering). Straightforward calculation of $\bar{J}$ predicts the solid-body moment of inertia, corresponding to the picture that nuclei rotate as a whole (i.e., as a solid body).

However, this simple theory was found to be in strong disagreement with experimental data, which had shown significantly smaller values of $\bar{J}$ for the lowest part of the rotational band, $L = 2, 4, 6$. However, when subsequent experiments were able to reach much higher L values, a dramatic change of $\bar{J}$ was observed. Curiously, the value of $\bar{J}$ seemed to come close to the originally predicted solid-body value of the moment of inertia. The physics of these interesting phenomena, displaying variable moments of inertia of rotating nuclei, is the subject of this section (for a review, see, e.g., [76]).

Small moments of inertia naturally suggest that only a part of the matter is rotating. It is not difficult to imagine various scenarios: for instance, most of the matter could be at rest, with a kind of wave traveling on the surface of the nucleus. However, since these rotational states exist for a long time (in the relevant nuclear time units, of course), we then need to explain how there seems to be no friction between the stationary and moving parts of the nucleus.

After having discussed experiments with a rotating glass containing superfluid helium in Chapter 5, the reader is well prepared for the idea that nuclear matter can be partially a superfluid, and thus the rotational motion can be a supercurrent, occurring without friction. If the supercomponent ρ_s remains at rest, one expects the moment of inertia $\bar{J}$ to be significantly reduced, as observed.

Such an idea, suggested by Bohr, Mottelson, and Pines [77] soon after the BCS paper, indeed related pairing in nuclei with the BCS pairing in superconductors.

Migdal [78] soon developed this idea further, and Belyaev [79] had developed the *pairing plus quadrupole* model, which quantified the effects very successfully, combining rotational and vibrational types of motion into one theory.

Mottelson and Valatin [80] took the theory one step further, arguing that a high degree of rotation would invoke Coriolis forces against pairing, eventually killing the pairing and superfluidity. Indeed, a rotation and a magnetic field should have a qualitatively similar effect. Suppose we would like to describe nuclei in a rotating frame (just as when we calculate in the frame of rotating Earth). The Hamiltonian in the rotating frame is corrected by a term

$$H_{\text{rotated}} = H_{\text{rest}} - \vec{\omega} \vec{J}, \qquad (11.24)$$

where $\vec{\omega}$ is the rotation frequency, and $\vec{J}$ is the total angular momentum. The last term acts as a magnetic field, in the sense that it tries to force all nucleons to have the same direction as $\vec{J}$, minimizing the energy. This is in conflict with the pairing: for $S = 0$ Cooper pairs, the two spins must have opposite directions. So a strong enough magnetic field, or a large enough rotational frequency ω would kill the Cooper pairs! As we will discuss below, while the main idea proved correct, the exact behavior as a function of L turned out to be more complicated than initially thought.

Assuming that ωJ must exceed the Δ value of the pairing, which is on of a scale of MeV or so, one finds that the total angular momentum of the nuclei must be $L > O(10\hbar)$. It is not a trivial experimental task to excite nuclei to a state of such rapid rotation without destroying them, but such experiments were eventually done.

In them it was indeed observed that the moment of inertia $\bar{J}$ changes to its solid-body value for very rapid rotation (see Figure 11.3). The left plot shows moment of inertia versus the rotational frequency squared. One can see that there exists some critical value of the rotational frequency, and that several levels correspond to it. At the same rotational frequency one finds increasing values of the moment of inertia: this region describes de-pairing of Cooper pairs, one by one. When the nuclear matter becomes "completely normal," its moment of inertia reaches its solid-state value, and in the upper right corner of the plots it does not grow any more!

This is not the end of the story, as shown in the right plot in Figure 11.3. Jumping over many years of developments, let me just comment that excitation to much higher rotational states (up to about $60\hbar$) can now be detected with modern gamma quanta detectors.

The most prominent feature of the plot is the appearance of the *second phase transition* at higher rotational frequency. Indeed, these heavy nuclei have different numbers of protons and neutrons, so there are two different Fermi spheres. The element Er (shown in the right panel) has $Z = 68$ protons, and the isotope used has $A - Z = 156 - 68 = 88$ neutrons, 20 more than its number of protons.[3] Thus the Fermi surface area—included in the BCS-pairing Lagrangian—is different for the two types of nucleons. The second

---

[3] The protons and neutrons on the Fermi surface should have the same Fermi energy, but protons have an electromagnetic Coulombic potential that is missing for the neutrons.

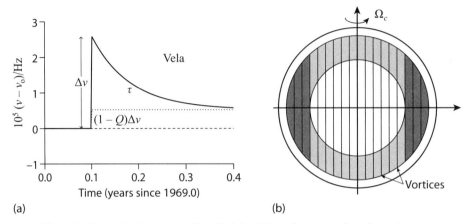

Figure 11.4. Qualitative picture of a glitch by Vela pulsar (a) and a schematic picture of rotational vortices (b). (The actual number of vortices is of course astronomically large and cannot be drawn.)

transition is thus nothing other than proton de-pairing. The corresponding calculations reproduce the data quite well.

Note also that these data show clear backbending at the critical regions. Such curves and the so-called hysteresis phenomena associated with them are familiar features of first-order transitions, observed in finite systems. Well, nuclei are finite systems as well.

In summary the ground state of slowly rotating nuclei are made of superfluid and rotate differentially. Rapidly rotating nuclei proceed through one or two de-pairing phase transitions until they become normal fluids, rotating as solid bodies.

11.5 Glitches of Pulsars

In this section we discuss basically the same phenomena as in the previous one, but at a much larger scale, replacing nuclei by neutron stars (pulsars). While nuclei are too small to possess vortices inside them, neutron stars are macroscopically large, of radius $R \sim 10$ km. Their rotations must therefore be associated with large ensembles of microscopic vortices, both in neutron superfluids and proton superconductors, as we discussed above. Since neutron stars, seen as pulsars, are "dead" (i.e., no longer possess any source of energy) they constantly cool down as the originally stored heat is radiated away. It takes millions of years to cool them beyond radio detection, and therefore we fortunately can observe hundreds of pulsars at various stages of this process.

Since their rotational frequencies are normally very stable and are measured extremely accurately to many digits, it was noticed that at some times there are small jumps $\Delta\omega$ upward, with subsequent relaxation. A *glitch* is a sudden change (increase) in the rotational frequency of a pulsar. For the Crab pulsar, for example, glitches happen about once per year, and $\Delta\omega/\omega \sim 10^{-8}$. A Schematic picture of this process is shown in Figure 11.4a.

Glitches themselves and their magnitudes are not particularly puzzling: they are believed to be similar to earthquakes and constitute events in which the thin solid crust breaks and settles into a new configuration as the star cools down. The interesting feature we will focus on is related to the relaxation process and the very long *relaxation times* observed: $\tau \sim$ month. The normal hydrodynamical response time $\sim R/c_s \sim$ $10 \, \text{km}/10^5 \, \text{km/s} \sim 10^{-4}$ s is completely incompatible with this value. So why does it take so long?

The answer was investigated for a long time (for references, see, e.g., an early review [82] on superfluidity in neutron stars and a more recent review [83] of the glitches). In short, it takes such a long time for relaxation because rotational vortices are strongly pinned to some underground inhomogeneities of the crust, and it takes a very long time for multiple repinning events. This motion of vortices is necessary, since changes in the global rotational frequency ω of the star directly imply changes in the number of vortices. The process of eliminating excess vortices is difficult, which is why the relaxation time is so long. Vortices can only be killed at the surface (the crust of the star). Although some forces push the vortices there, they resist these forces by pinning to whatever they can hold on to, namely, the defects of the crust or its crystalline structure. Vortex repinning leads to the "slow creep" process [70]. The overall mechanisms of pinning and repinning are complicated, but their studies have explained the long relaxation times observed.

Among speculations about what can trigger glitches, there is the recent science-fiction-like idea [84] that it can be induced by a chunk of dark matter, perhaps even a macroscopic dark comet/planet/star of some kind, flying through the pulsar. Of course, we still do not know much about dark matter, except something about its overall mass density in galaxies and some loose limits on its interactions with visible matter.[4]

Let me add another point, which is under consideration now. Since the famous recent observation of the gravitational waves by LIGO in the fall of 2015, produced by coalescing massive black holes, an observation of waves emitted by coalescing neutron stars happened in 2017. During the spiral rotation preceding the merger, the tidal forces are expected to deform the stars substantially, so their sizes and thus equations of state can be measured. Reheating of the stars by tidal forces and subsequent merger causes massive x-ray emission, coincidental with the emission of the gravitational waves: both were in fact detected. We would be able to learn the temperature history of the event, sensitive to the stars specific heat and perhaps even to the gaps at the corresponding Fermi spheres.

[4] If such dark chunks exist, one may also ask how frequently they fly through Earth (and us). Since the usual earthquakes are spherical waves emanating from a certain point, while those from a propagating dark matter chunk should rather be a cylindrical wave emanating from a line, one can in principle tell them apart. But since Earth is geologically noisy, perhaps it is very hard to notice such cylindrical waves, even if they exist.

12 | Strongly Coupled Fermionic Gases

Trapped ultracold fermionic atoms, unlike bosonic ones, cannot of course be Bose condensed, but they should display a condensation of Cooper pairs and thus superfluidity. These phenomena were predicted, and when sufficiently low temperatures were reached in laboratory, they were indeed observed. For an experimental review, see [91].

However, a somewhat unexpected bonus of these studies was the possibility of tuning the scattering length a to any desired value using Feshbach resonances. We will discuss the tuning method in the next section.

The limit $a \to \infty$, also known as the *unitary limit*, turned out to be especially interesting, because in this case the scattering cross section is as large as it can possibly be. The term "strongly coupled" in the chapter title refers to this case. In the unitary limit, any information about the particular atoms is gone, so there is no scale associated with the atoms. As a result, a universal scale-invariant (conformal) regime appears, which is of significant theoretical interest.

We discuss the global phase diagram of fermionic gases in Section 12.2. It is also known as the BCS-BEC crossover, in which weakly coupled BCS Cooper pairs existing at the Fermi surface gradually change into a BEC of strongly bound atomic pairs, or molecules. This transition turns out to be surprisingly smooth, related to the fact that no symmetry is changed during the process.

The hydrodynamics and kinetics at strong coupling is the subject of Section 12.3. The main question here is whether, in the limit $a \to \infty$, one can define the Fermi sphere or quasiparticles, and, if not, what is the (analog of) the particle mean free path. The kinetic quantities—viscosity, diffusion constant, and the like—must be studied, and indeed they show some universal features.

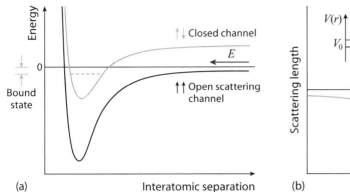
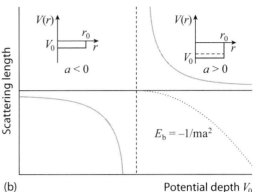

Figure 12.1. Tunable scattering length via Feshbach resonance. (a) Feshbach resonance. (b) Scattering length for a square well potential and the appearance of a bound state of energy E_b past a critical well depth. Adapted from Randeria and Taylor [86].

12.1 Weak and Strong Coupling Regimes in Trapped Fermionic Gases

In quantum mechanics the low-energy scattering amplitude has the following general form:

$$f(k) = -\frac{1}{1/a + ik},$$
(12.1)

where a is the scattering length, and k is the (relative) momentum. In Chapters 4 and 7 we studied the so-called *weakly coupled* gases, with small $f \approx -a$, using a as an expansion parameter.

Now we are interested in the opposite limit: the strongly coupled regime, in which $(1/a) \to 0$. Therefore the scattering amplitude is $f = i/k$ at small k. This leads to the largest possible cross section in the s-wave channel $\sigma \sim 1/k^2$ allowed by unitarity, so this limit is also called the unitarity limit.

Historically, the issue appeared in the framework of nuclear physics. The scattering lengths of two neutrons is $a_{nn} = -18.9 \pm 0.4$ fm, and for neutron-proton it is even larger: $a_{pn} = -23.740 \pm 0.020$ fm. These values, compared to the proton/neutron size of about .7 fm, are indeed huge. This happens because the nuclear forces produce bound/virtual states at a near-zero energy. In our discussion of nuclear matter, this fact played no role, but it becomes relevant for the discussion of (very dilute) "neutron skins" on the surfaces of heavy nuclei.

Today strongly coupled gases are studied using trapped ultracold fermionic gases. Any value of a can be experimentally obtained using the Feshbach resonance. Different atomic states have different magnetic moments, so their hyperfine splitting can be changed by applying an external magnetic field. One can move the levels to induce a resonance, between the bound closed channel state and the zero energy open (scattering) channel in Figure 12.1a. The corresponding change of the scattering length a as a function of the potential depth V_0 (basically, the external magnetic field) is shown in Figure 12.1b. As

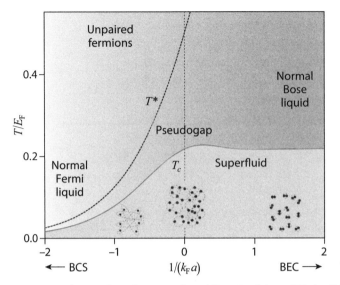

Figure 12.2. Qualitative phase diagram. Adapted from Randeira and Taylor [86].

explained in quantum mechanics textbooks, the appearance of a new level corresponds to a jumping from $-\infty$ to ∞.

(For a reader who would like to have a concrete example, consider the Li^6 atom sometimes used in such experiments. The nucleus has six nucleons (three protons and three neutrons), and the ground state spin $S_{\text{nucl}} = 1$. Three electrons are in [He]$2s^1$ configuration, the electron spin is $S_e = 1/2$, and the total angular momentum of the atom thus can have two values: $j = 3/2, 1/2$. These are the two channels under consideration.)

From the experimental perspective, one difficulty was cooling fermionic trapped gases (Li^6, K^{40}) to a temperature much smaller than the Fermi energy ($T \ll \epsilon_F$), to see whether a degenerate Fermi gas at strong coupling has a Fermi sphere. When this was achieved, around 1999, the quest continued to reach even lower temperatures, of about $T \sim 10\,\text{nK}$ or so, at which a superfluid phase could be formed. This goal was achieved around 2006.

12.2 Global Phase Diagram

Figure 12.2 from [86] shows the so-called global phase diagram of the BCS-BEC crossover as a function of temperature T/E_F and coupling $1/(k_F a)$. This diagram shows schematically the evolution from the BCS limit, with weak attractive interaction between atoms, large Cooper pairs, and exponentially small critical temperature (the far left edge of the plot), to the BEC limit on the right side, with Bose-condensed tightly bound molecules.

The vertical axis is T/E_F: the upper right region is the high-T normal phase, and the region below is a superfluid region. At small negative a (the left side of the plot) one has weakly attractive fermions, which generate the exponentially small critical temperature

$$T_c \sim \exp\left(-\frac{\pi}{2|a|\,p_F}\right) \tag{12.2}$$

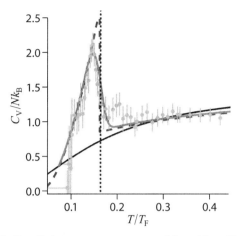

Figure 12.3. Specific heat versus temperature. Adapted from Ku et al. [87].

predicted by the BCS theory of superconductivity. At small positive a (the right side of the plot) the atoms are slightly repulsive. This happens because a bound state of spin-zero diatomic molecules exists, a descendent of the Cooper pair. The superphase is a BEC of such molecules, and $T_c/E_F \approx .22$ in such regimes (which agrees with Einstein's original expression for this ratio). We have already studied both of these regimes in Chapters 4 and 9.

The most interesting "unitarity gas" is in the middle of the plot, where $1/(k_F a) = 0$. This regime has only one dimensional parameter, the density or Fermi momentum k_F. The unitary gas is not just strongly interacting: it has no scale whatsoever and thus is conformally invariant. Its Cooper pairs correspond to strongly interacting pairs with size comparable to $1/k_F$, a gap $\Delta \sim k_F$, and a Fermi sphere with radius k_F. So, is there a Fermi sphere? Is it a Fermi liquid? If it is a superfluid, what is its critical temperature? What is Δ/E_F, if the pairs exist? Do any diagrammatic calculations make sense in this regime, in the region where weak coupling expansion is clearly inadequate?

Numerical simulations of the fermionic paths (I do not describe them here, sorry) established the existence of a superphase at zero T, with a large gap comparable to but smaller than the Fermi energy:

$$\frac{\Delta}{E_F} \approx 0.44.$$

Then experiments have found the phase transition: the specific heat of the unitary Fermi gas as a function of T/E_F exhibits a peak at

$$\frac{T_c}{E_F} \approx 0.18$$

(see Figure 12.3, from [87]). The dots denote experimental points; the curves are various theoretical results for comparison. Below this temperature several features characteristic of a superfluid were observed, in particular a lattice of quantized vortices (see Figure 12.4 from the MIT group).

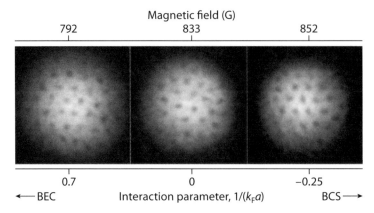

Figure 12.4. Quantized vortices in rotating Fermi gases. The upper scale shows the external magnetic field shifting Feshbach resonance: the three plots are for all three regimes possible with greatly changed scattering lengths. From Zwierlein et al. [88].

12.3 Thermodynamics and Kinetics at Strong Coupling

Apart from first-principle numerical simulations, which for fermions are not straightforward to do, there are no good means to study the strongly coupled system. As we will see, even some issues of principal importance remain unresolved.

Let us recall the main feature of the unitary gas. The scattering length a, the only parameter that includes the atomic scale, is gone when $a \to \infty$. Thus the system has no length scales of its own. There is the only external scale—the interparticle distance—and it should therefore be described by its trivial powers, defined by dimension multiplied by some universal numbers.

We do not have good ways to calculate these numbers, but they of course can be extracted from experiment. In particular, the so-called Bertsch parameter [85] is the ratio of the total energy at zero temperature E_0 to that of the ideal fermi gas:

$$\xi = \frac{E_0}{(3 N E_F / 5)}. \tag{12.3}$$

Experiment gives $\xi = 0.376(5)$, compared to various theoretical approaches in [86]. Similar relations for other quantities, such as the gap Δ / E_F and the critical temperature T_c / E_F, should be universal as well. As far as I know, extrapolation of the BCS theory to its boundary is in qualitative agreement with these measured values.

(There is an interesting detail related to the measurements of ξ. Bertsch defined the energy in a standard setting, assuming a homogeneous fluid in a container of some volume V. The measurements are, however, done for gases trapped in harmonic potential wells. For an ideal gas, repeating the original Einstein calculation but instead using a trap, one finds that instead of the original three-dimensional momentum integral, one has a similar six-dimensional phase-space integral. The question is how exactly can ξ be measured.)

Figure 12.5. Time snapshots of the system's shape (radial cross-section), demonstrating elliptic flow in a unitary Fermi gas. Adapted from Gelman et al. [89].

Conformal invariance of the unitary gas also has many other consequences that can be falsified. For example, it implies that the bulk viscosity should be zero, since this coefficient is related to modification of the spatial scale of the system. (I am not aware of any experimental tests of it so far.)

One feature of interest is whether the unitary gas displays the hydrodynamical properties characteristic of other strongly coupled systems, such as quark-gluon plasma we will discuss later on in Chapters 15 and 16. When the magnetic field is away from the Feshbach resonance, the scattering length a is of the atomic size, while the distances between atoms are 100–1,000 times larger. The system is very dilute, and if the trap is switched off, atoms simply ballistically fly away with the momenta they already have: rescattering is negligible. For such a dilute gas, the angular distribution of particles is isotropic.

However, when the trap is switched off while the Feshbach resonance exists and $a \to \infty$, the system displays a completely different flow. The mean free path is small, and one finds a hydrodynamical elliptic flow, shown in Figure 12.5. The pressure gradient is larger along the shorter direction, so it explodes more rapidly in this direction.

Small mean free path means low shear viscosity of the fluid: a near-ideal liquid, as it is called [89]. It was argued in [89] that, say, at zero temperature there should be quantum viscosity, and at finite temperature (and entropy), it would be of the form

$$\eta = \alpha_\eta \times \hbar n + \beta_\eta \hbar s(T), \tag{12.4}$$

where n and s are particle and entropy density, respectively, and α_η, β_η are two dimensionless constants of the order 1. Later studies have shown that

$$\alpha_\eta \approx 5, \beta_\eta \approx 0.2. \tag{12.5}$$

The viscosity value can be inferred from the hydrodynamic expansion, such as that shown in Figures 12.5 and 12.6. This sequence of snapshots shows that the system is initially cigar-shaped and is elongated in the horizontal direction. But because the pressure gradient is larger in the other two directions, in which the system is the shortest, the hydrodynamic explosion proceeds mostly in these other directions. In a limit of a very long cigar it, becomes a cylindrical explosion that is independent of the distance along the cigar. This elliptic flow is completely analogous to that in heavy ion collisions, which we will discuss in connection to the viscosity of QGP below.

Viscosity also can be measured from the damping of various oscillation modes of the trapped gas. Since it has a cigar shape, there are various modes of oscillations

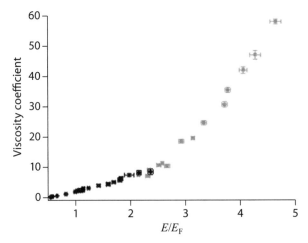

Figure 12.6. Shear viscosity versus the per-particle energy of the gas in units of Fermi energy (the experimental substitute for the temperature). Note that for most points $E/E_F > 1$, corresponding to a nondegenerate Fermi gas. Adapted from Gelman et al. [89].

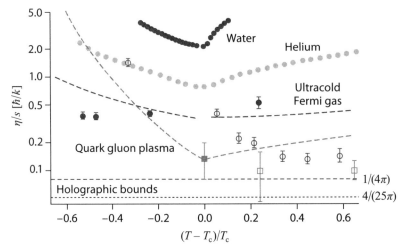

Figure 12.7. Shear viscosity divided by the entropy density, which in units $\hbar = k_B = 1$ have the same dimensions. This ratio is shown as a function of the temperature near respective phase transitions for the ultracold quantum gases (filled circles) and some other substances, including quark-gluon plasma (open points). The dashed lines in the lower part of the plot are bounds corresponding to holographic models at infinite coupling. Adapted from Adams et al. [90].

around different directions, so in [89], our group solved the hydrodynamical equations of motion and checked whether the two observed damping rates of different modes do in fact correspond to the same viscosity. A summary plot, relating viscosity divided by the entropy density of ultracold quantum gases to that of other substances (water, liquid helium I, and a quark-gluon plasma) is shown in Figure 12.7.

Note that this small viscosity (and thus short mean free paths or large scattering cross sections) is very surprising, and there is no theory to explain it, just measurements. These measurements are in direct contradiction to any diagrammatic calculation of scattering rates. In particular, when the fermionic atoms were in the degenerate Fermi gas regime, their collision rate must be significantly lowered by the Pauli blocking, which should lead to a suppression factor of about $(T/T_F)^2 \sim 10^{-3}$ in the experimental conditions. If true, the oscillations then would be basically collisionless and no hydrodynamic phenomena would be present. In a picture of BCS-type pairing, with relatively small modification of Fermi sphere, T in the above ratio is to be replaced by a gap, so the scattering suppression would be on the order of $(\Delta/\epsilon_F)^2$. We must then conclude that $\Delta/\epsilon_F \sim O(1)$, and basically there remains no Fermi sphere as such.

13 | Numerical Evaluation of Path Integrals

The foundation of this book is the Euclidean path integral formulation of quantum and statistical mechanics. So far we have used a very limited toolbox, considering only integrals that can be viewed as weakly perturbed Gaussian integrals, calculable via Feynman diagrams.

In this chapter we look at the Euclidean path integral formulation from a different perspective:[1] as a basis for certain "brute force" numerical methods, allowing the evaluatation of such integrals with more or less arbitrary integrands.

We start the chapter with a brief overview of some simple tactics used for ordinary integrals and then discuss subsequently the path-integral Monte Carlo methods for distinguishable particles, bosons in Section 13.3 and fermions in Section 13.4. This chapter can be considered as a prerequisite for the next one, where we venture into numerical studies of quantum field theories.

13.1 Numerical Evaluation of Multidimensional Integrals

Numerical evaluation of integrals may be considered as a rather technical issue that is not worth discussing in a book devoted to quantum many-body theory. Theorists and mathematicians generally tend to look down on such issues as belonging to "applied mathematics." But, as long as experience shows, success in science is not determined by how great the problem you study is, but by how deeply you understand the problem you consider. Computational methods is a vast field, which continues to produce new algorithms and spectacular results based on them. Their role in overall scientific progress has grown spectacularly, and not just because of increasing computer power.

Everybody who comes across an integral starts with a few obvious textbook methods. One such method uses the classical definition of an integral as the limit of a sum. The

---

[1] One more general method based on the Euclidean path integral formulation is its semiclassical expansion. We will discuss some applications of those at the end of the book.

textbook definition of the integral is based on a grid with step a and the limit $a \to 0$. Using $N \sim 10^3$ or even 10^6 points for not too complicated functions is a trivial task for any computer, so one-dimensional integrals can usually be done in this way without difficulty.[2] Yet, in D dimensions the number of points needed grows exponentially $\sim N^D$. For example, for $D = 3$, using 10^6 points ($N \sim 10^2$ in each dimension) usually works. But it is clearly impossible to use the grid methods for, say, $D \sim 10$ or more.

The second obvious method is known as *naive Monte-Carlo method*. One can select some reasonably large (e.g., $N \sim 10^6$) number of randomly placed points inside the D-dimensional box and average the function over these points. Naive expectation is that the statistical accuracy is $\sim 1/\sqrt{N}$ for any D. In fact, it would be true if the function has relatively small variation, with comparable magnitudes of the function at any point.

Unfortunately, most functions with many variables are not like that: they vary greatly and peak only in some small subvolumes in their space of definition. Therefore most of the points sampled randomly miss those important regions, and the efforts made for evaluation of the function there are wasted. Playing with the subvolumes helps sometimes, provided we know where the function is peaked. In practice, however, a lot of people has wasted an enormous amount of computer time using this method, with a sizable fraction obtaining wrong (or very inaccurate) answers, often without even suspecting that this is so.

Let us estimate how many variables we need for statistical mechanics on the lattice. The simplest case, the two-dimensional Ising model on, say a 100×100 lattice, already has $2^{10,000}$ configurations. For the Heisenberg model with continuous spin, one needs to evaluate integrals in these $D = 10^4$ dimensions. Therefore, an application of the naive Monte-Carlo method to any such problem is completely hopeless.

At this point people look into books on numerical methods or standard math libraries, but—for reasons unclear to me—they can hardly find anything useful there.[3]

However, there are very general algorithms, based on a few distinct ideas, that can successfully evaluate integrals with dimensions of, say, millions of variables. Some have been known for decades and are widely used by the professionals—and yet they still have not made their way into popular culture and textbooks.

Fortunately, it is possible to explain how these algorithms work in physics terms for our problems at hand. And, of course, once understood, one may then use them in contexts that have nothing to do with statistical mechanics.

The main difference these methods have with the naive Monte Carlo method is that the points at which the function is evaluated are not selected randomly. Instead a process generating a path in "configuration space" (as we will call our set of variables) is chosen. It can be stochastic (Monte Carlo methods), a solution of some differential equation (molecular dynamics), or the solution to stochastic equations (the Langevin equation).

---

[2] There are always exceptionally difficult cases (e.g., those with strongly oscillating integrands). We do not discuss such cases here.

[3] Even the popular (and otherwise excellent) book *Numerical Recipes: The Art of Scientific Computing*, by W. H. Press et al. only mentions the Metropolis algorithm (discussed below) in a vague form rather than include a detailed description of it.

The main point is that these algorithms automatically find where the function is the largest and go there, without wasting time elsewhere. They also maintain the correct balance between the energy and entropy, which is needed to keep realistic statistical fluctuations in a system.

Let me enumerate some methods that are used, with brief comments:

- **The Heat Bath** is very direct and powerful but can be implemented only in special cases.
- **The Metropolis algorithm** [92][4] is very general and has no error except the statistical one. This is my favorite.
- Approaches based on solving the **Langevin stochastic equation** update all variables (not just one) and allow for analytic treatment but the small-step limit should be taken with care.
- **Molecular dynamics** algorithms substitute solution of certain classical mechanics equations of motion in place of the statistical/quantum mechanics equations. However, the method assumes that the system in question is chaotic enough, which may not work in some cases when the system is integrable or has "islands of stability," separated from the main distribution.
- **Hybrid** algorithms, alternate between the stochastic Metropolis algorithm and molecular dynamics. These may be the most efficient methods of all, if tuned well. They are especially useful for complicated nonlocal actions when evaluation of the action is very costly. Their use is standard in many lattice gauge simulations.

There is no need to go into the details of each method, as one can find descriptions online. Let me just give some general advice. When one needs to do an integral with several variables, $D > 3$ or so, do not use the simple lattice or the naive Monte Carlo methods. Yet very often the integrand can be split into $F(x) W(x)$, where the "weight" W is positive ($W > 0$) and relatively simple, while F is complicated (e.g., oscillating) or computer costly. It is optimal if the integral of W alone can be done exactly. If so, run a Metropolis generator with the weight W, produce an ensemble of points, and then calculate the integral by averaging F:

$$\frac{\int d^d x \, F(x) \, W(x)}{\int d^d x \, W(x)} = \langle F \rangle_W$$

13.2 First Steps: The Path-Integral Monte Carlo Method for Few-Body Systems with Distinguishable Particles

The initial attempts to apply computational methods to address quantum mechanics problems used the path-integral Monte Carlo (PIMC) method. The first paper (known to me) in which a quantum mechanical problem was treated by numerically evaluating

---

[4] In fact this algorithm was proposed by five authors, including the very famous physicist Edward Teller. One can only speculate on what problem this team was studying.

a path integral was the pedagogical paper by Creutz and Freedman [93]. It considered one particle moving in two one-dimensional quartic potentials, with an anharmonic well $V(x) = \lambda x^4$, as well as in a double-well potential $V(x) = \lambda(x^2 - f^2)$. I recommend that every student reads this paper. It was followed by truly revolutionary work by Creutz [94], which was the first application of computers to the study of confinement in gauge theories.

It made so great an impression on me, that I instantly decided to test for myself whether it really works. It indeed passed the tests, and those efforts resulted in another methodology paper [95] in which a couple of "realistic" few-body systems, such as (1) the He atom, with two electrons in three dimensions (total dimension = 6); and (2) He nuclei, with two protons and two neutrons in three dimensions (total dimension = 12) were solved. (Yes, one can subtract the three coordinates of the center of mass in the second problem, to make it nine-dimensional, but we did not even bother to do so.) Both problems had a certain history (existing approximate methods, etc.), but to use *a brute force first-principle* approach looked inviting. It all worked, and we got very reasonable ground-state energies.

(Assuming that computer speed doubles every 2 years, and noting that this work was done more than 30 years ago, one can estimate our computer then was 2^{15} times slower than they are today. This comment is made to encourage readers to try some similar calculation now!)

There is no point in showing any results of that ancient work, but it is instructive to explain why precisely these two problems were selected. In problem (1) we dealt with two electrons: they have spin 1/2, and one can think of the two of them as having two different spin directions. This makes them nonidentical and not subject to quantum statistics problems. For the nucleons there are spins and isospins: thus we have four nonidentical nucleons. To study larger systems, fermionic antisymmetrization of the paths would become unavoidable: and of course we did not know then how to deal with that.

13.3 Liquid ^{4}He PIMC Simulations: Bose Clusters, Condensation, and the Critical Temperature

The problems that arise in quantum many-body applications stem from the fact that the particles we want to study are identical. Therefore, the paths with permutations, in which they exchange places after a Matsubara time τ completes its period β, need to be included in the calculations. For bosons all paths with possible permutations should simply be added to the path integral; for fermions one further needs to include those notorious minus signs.

So, for the bosonic case, one has an algorithm with permutations, which is a bit more complex: yet otherwise it is not a serious problem. In the case of Fermi statistics, each permutation should be accompanied by negative sign: this is a very serious problem, since the Metropolis and similar algorithms cannot be used for nonpositive weights.

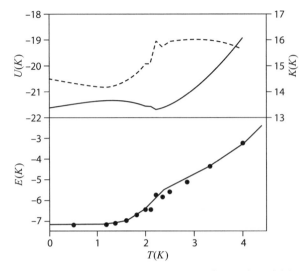

Figure 13.1. Bottom panel: The energy per atom of He. The solid lines show experimental results at saturated vapor pressure; closed circles show the PIMC calculations (D. M. Ceperley and E. L. Pollock, "Path-Integral Computation of the Low-Temperature Properties of Liquid ^{4}HE," *Physical Review Letters* 56, 351 [1986]). Upper panel: The potential energy (solid line and left scale) and kinetic energy (dashed line and right scale).

For Bose systems it soon became possible to treat real systems as complicated as liquid ^{4}He. Let me describe some aspects of these calculations, using as illustrations the plots from the excellent review by Ceperley [96]. The upper part of Figure 13.1 shows the kinetic and potential energy per atom as a function of the temperature. The effect of condensation at the critical temperature is hardly seen. The sum of these energies, shown in the lower plot, hints at the presence of the lambda point. Note however, that the experiment is shown here by a line, while the "theory" (that is, the PIMC measurements) are shown by points. The conclusion is that to detect phase transitions numerically is not a trivial task.

Precise determination of the critical point from free energy or specific heat (its second derivative) is very hard to do. Significant efforts are required from computational physicists to locate it. For a given binary interaction known as the Aziz potential, the results are shown in Figure 13.2 from Boninsegni et al. [104]. The computed value $T_c = 2.19\,\mathrm{K}$ is shifted up from the experimental one by about a percent, presumably because only the two-body interaction potential was included in the simulation.

Instead of calculating global thermodynamical observables, such as energy, another strategy is to try to detect the BEC condensate itself. In the language of paths, it can be done using the notion of *Bose-clusters* introduced by Feynman in two classic papers of 1953 [97]. The Euclidean path integral is done over the periodic paths; which means that at $\tau = 0$ and $\tau = \beta = 1/T$, the particles are at the same locations. Yet for bosons it may not be the same particles! Figure 13.3 shows paths of the particles in a periodic box (dashed box at the center): the paths continued outside the box can also be put inside it by using periodicity. Circles are the initial locations at $\tau = 0$ (the same as $\tau = \beta$, since the Euclidean

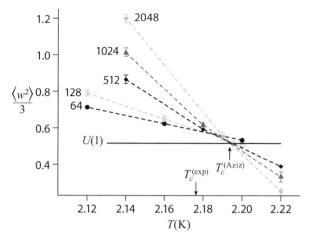

Figure 13.2. Finite size scaling determination of the critical temperature T_c. Numbers near the lines are the number of atoms used in each simulation. Adapted from Boninsegni et al. [104].

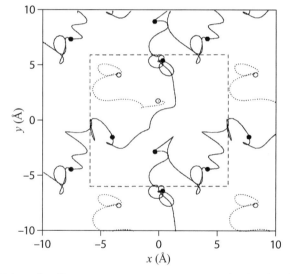

Figure 13.3. Paths of He atoms at temperature 0.75 K. The dotted path is that of the cut polymer, the one that is not periodic in imaginary time. Its end-to-end distribution is used to calculate the momentum distribution. The other four atoms are involved in an exchange that winds around the boundary in the x and y directions. The dashed square represents the periodic boundary conditions. The paths have been Fourier smoothed before plotting. Reprinted from [96].

paths must be periodic). One can see on this plot a cluster of four particles exchanging their places.[5]

The density of k-particle clusters n_k increases as T decreases (the Euclidean time duration $\beta = \hbar/T$ gets longer, and the opportunity to get mixed with neighboring atoms

---

[5] Their behavior resembles the popular kid's game of musical chairs, occupied with permutations.

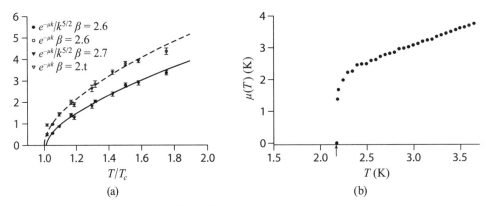

Figure 13.4. The effective chemical potential μ versus temperature T for lattice simulation of (a) gauge theory monopoles and (b) liquid ^{4}He. The arrow shows the experimentally located critical temperature T_c. Adapted from D'Alessandro et al. [99] and Ramamurti and Shuryak [100].

increases). Feynman pointed out that the critical temperature T_c is defined by the singularity of thermodynamical quantities, stemming from certain divergences in the partition function. Those divergences come about because the Bose clusters had reached a macroscopically large size, forming an "infinite cluster."[6] So, BEC can be detected by counting clusters!

It turns out that counting clusters is a rather practical way to measure T_c. Defining an *effective chemical potential* μ by the expression (for an ideal Bose gas)

$$n_k = \frac{\exp(-k\mu/T)}{\lambda_B^3 k^{5/2}}, \tag{13.1}$$

and using n_k from the simulation, one can plot $\mu(T)$ and find where it reaches zero. This idea had first been used for monopoles in the QCD vacuum by D'Alessandro, D'Elia, and myself [99] (see Figure 13.4). For liquid ^{4}He, Adith Ramamurti and I [100] recently simulated the paths, determined the cluster probabilities, and plotted the resulting $\mu(T)$ defined in (13.1). The results are shown in Figure 13.4b. As you can see, the zero of μ is right where the critical temperature (for this potential) needs to be, at 2.19 K. The lesson is that this method is a very sensitive way of locating T_c.

The amplitudes for clusters, Feynman argued, consist of dynamical and combinatorial parts. For the straight part he proposed the amplitude and the extra action is given by the following simple expression:

$$A_{\text{exchange}} = e^{-S_{\text{exchange}}} = \exp\left(-\frac{mTd^2}{2\hbar^2}\right), \tag{13.2}$$

where S_{exchange} is the extra action needed for the exchange of two particles separated by distance d during Euclidean time $\beta = \hbar/T$. The action is the time integral of a kinetic energy $\int_0^\beta \frac{m}{2}\frac{d^2}{\beta^2}\,d\tau = \frac{md^2T}{2}$. This simple formula is of course an estimate, corresponding to

---

[6] The word "infinite" here means of course a number of particles which scales as the total number used in a simulation.

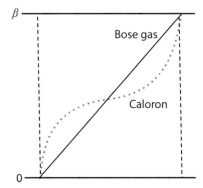

Figure 13.5. The paths on (Euclidean) time versus space plot. The dashed lines correspond to two atoms at rest. The diagonal straight line is Feynman's path, by which a particle can exchange locations. The dotted line marked "caloron" is the path including the interparticle potentials. The straight solid or dotted lines show the path of one particle jumping to the location of the other: where the other one jumps is not indicated here. Adapted from Cristoforetti and Shuryak [98].

an ideal gas made of noninteracting particles, moving along the straight diagonal line shown in Figure 13.5, with the exponent being the kinetic energy times time β.

If a cluster includes k particles, their paths form a closed polygon with k "jumps." The combinatorial factor is thus the number of such polygons. Feynman noticed that this number depends on k in a predictable way. Using it together with the action estimate (13.2), Feynman concluded that the statistical sum over k starts to diverge at some *universal value of the action* (for a particular space dimension d). Feynman's estimate for the critical value was

$$S_c \approx 1.655. \tag{13.3}$$

Numerical studies by Elser in the three dimensional ($D = 3$) case provided a numerical value for this universal critical action that is a bit smaller, $S_c \approx 1.44$. If one uses the mean interparticle separation $d = 3.5\,\text{Å}$ and the ^{4}He mass, then Feynman's formula and Elser's action would predict $T_c = 2.72\,\text{K}$, clearly in the ball park but not too close to the actual critical value $T_c = 2.17\,\text{K}$. Feynman's value is even further away. Feynman explained the difference by the "helium effective mass," and the issue was left there for many years.

Cristoforetti (then a visiting student from Trento, Italy) and I [98] revived Feynman's ideas 50 years later. By this time, PIMC simulations for liquid ^{4}He had been done, and the helium binary potentials were well known and tested. We decided to test whether Feynman's idea of a universal action—independent of particular forces between atoms— is correct.

It was obvious that Feynman's estimate was very naive: his straight path was clearly unrealistic, since He atoms are not free but move in the collective potential of all neighboring atoms. However, one can correct Feynman's estimate by a more realistic calculation of the action needed for a "jump." In the case of a solid lattice of He atoms, the result is shown in Figure 13.6a. A single elementary jump occurs from one minimum to the next, and the path can easily be calculated once the potential is known (see the

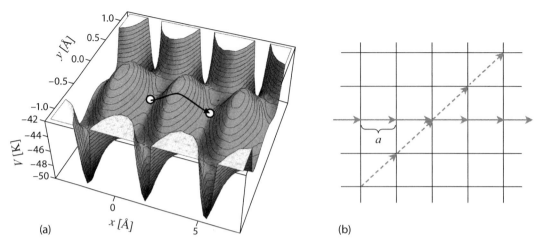

(a)   (b)

Figure 13.6. Sketches of the Feynman's jumps, from locations of one atom to the next. (a) The figure illustrates a single jump, in the potential created by all other stationary atoms. (b) Linear sequences of multiple correlated jumps, generating a "supercurrent." From [98].

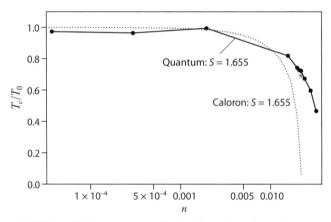

Figure 13.7. The critical temperature obtained from the caloron/instanton solution (dashed line) and from the one-dimensional PIMC simulation (points). The star shows the experimental value of the lambda point. Adapted from [98].

dotted line in Figure 13.5). Now if the cluster polygon is large, and its shape locally is just a straight line, or a "supercurrent line" of atoms, each jumping to the next location (see Figure 13.6b).

Evaluating the action per jump for variable density, we get the prediction for T_c shown in Figure 13.7, stemming from Feynman's critical action (13.3). Surprisingly, the star—the observed density and temperature—turned out to be right on the calculated line!

(As a by-product of that study, we investigated a question that resurfaces time and again: Why is it that liquid He-II may have a superflow, but solid He may not? Direct simulations by Ceperley and others indeed did not find superclusters in the solid phase. From our calculation we also see that and concluded that this is simply because the solid phase has a bit higher density. The interparticle potential is rather sharp, and the action

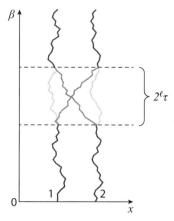

Figure 13.8. Two particle paths, with and without permutation from [100].

for a jump simply exceeds the value needed for a supercluster, even at zero T. It would be fun to have a supercurrent in a solid body show up in the simulations in some modified toy models. It did not work out in solid He just because of the numbers involved.)

Feynman was right in 1953, when he predicted the existence of the *universal action value per particle permutation*, leading to a supercurrent. All one needed to do—but it waited for 60 years—was to include the potential energy and to calculate the correct value for the action needed for two atoms' permutation. Even Feynman's value for it (given in equation (13.3)) turned out to be correct.

13.4 Path Integrals for Fermions

The fermionic amplitudes are supposed to change sign when the number of fermionic loops is changed by one. When two fermions interchange their locations, there appears one common path (loop) instead of two separate paths, so a minus sign appears in the amplitude. In terms of the paths shown in Figure 13.8, we can write the amplitude as

$$(Amplitude) = (Direct) - (Exchange). \tag{13.4}$$

As already mentioned, the weight being negative is a very serious problem, because standard Monte-Carlo methods (e.g., Metropolis algorithm) cannot be used.

In the 1980s Oleg Zhirov (then my graduate student) and I decided to start with fermions moving in one dimension only. This case is special, because one can always enumerate fermions along the line and thus pretend that the "exchange" never happens. For a small time step a, the one-particle amplitude is

$$U(x^f, x^i, a) \sim \exp\left[-\frac{m(x_f - x_i)^2}{2a} - a V\left(\frac{x_f + x_i}{2}\right)\right], \tag{13.5}$$

where the superscripts i and F stand for "initial" and "final," respectively. For two particles it can be written as

$$U(x_1^f, x_1^i, a)\, U(x_2^f, x_2^i, a) - U(x_1^f, x_2^i, a)\, U(x_2^f, x_1^i, a) \approx U(x_1^f, x_1^i, a)\, U(x_2^f, x_2^i, a)\, \exp[-a\, V_{\text{Pauli}}]$$

with the Pauli potential defined as

$$V_{\text{Pauli}} = -(1/a) \log\left[1 - \exp\left(\frac{-m(x_1^f - x_2^f)(x_1^i - x_2^i)}{a}\right)\right]. \tag{13.6}$$

Note that when two particles get close, the exponent becomes small, and so the argument of the log approaches zero and the potential becomes very large. So, a node of the amplitude can be viewed as a repulsive potential. It is going in the right direction: indeed, fermions must have a larger energy than distinguishable particles have in the same setting.

If particles never jump over the node-generated barrier, their order along the line remains preserved, and if the Pauli potential is included, the simulation can be done by PIMC. We checked it for several ($n = 3$, 4, and 5) particles put in a harmonic potential: for distinguishable particles, the ground-state energy is $\hbar\omega(1/2 + 1/2 + \cdots)$, but for fermions it should be $\hbar\omega \times (1/2 + 3/2 + 5/2 + \cdots)$, since each must be put into the next available level. So our algorithm with this Pauli potential correctly reproduced that! Unfortunately, due to certain (unrelated) events, I left for the United States at the time of that work, and it was concluded with another student Tursunov without me [101]. A description of the method and its use is also given in [102], in which its many successful tests are discussed.

This is not the end of the story. The decisive step forward, allowing use of the idea in any dimension, was made by Ceperley [103]. It has been applied to fermionic problems, including liquid ^{3}He. The main idea can be explained by considerings the various paths of one fermion, keeping all other fermion paths frozen. The one-dimensional node of the amplitude discussed above gets promoted to a *nodal surface*, which surrounds each fermion, keeping it inside a *nodal cell*. Paths that are not allowed to leave the nodal cell are called "restricted": a sum over the restricted paths obviously has no sign change in the amplitude.

How to find the nodal surface? It can be inferred from the simplest case of noninteracting fermions. After all, the kinetic energy needs to be the same as for an ideal Fermi gas, $(3/5)(p_F^2/(2M))$. So the cell is inferred to be approximately a sphere of some radius, which can be in turn be determined from the energy. Or alternatively, it can be inferred from the exact noninteracting density matrix for the ground state (GS in the equation)

$$\rho_{\text{GS}}(\vec{r}_1 \ldots \vec{r}_N) \sim \det[e^{i\vec{k}_i \times \vec{r}_j}], \tag{13.7}$$

where the momenta $\vec{k}_j$ are run over all occupied states in the box for which the simulation is done.

The simulations reported were done for 38 and 66 fermionic He atoms and take an order of magnitude longer than for bosons. It is done for $0 < T < 4\,\text{K}$ and fixed density.

It has been found that using ground state nodes, one finds the correct T dependence, but the energy is systematically shifted upward by about 1.5 K. (Experiment shows that at $T = 0$, $E \approx -2.5$ K while for $T = 3$ K, the energy $E \approx 0$ K.)

In summary, the fermions are surrounded by some nodal surfaces, which hardly depend on the potential but do depend on the local density (which is a function of the location of all other fermions). One can reformulate the fermionic path integrals in terms of restricted paths, which are then sign-definite.

14 | Quantum Field Theories on the Lattice and Supercomputers

Historically, when QFTs were initially developed, there was a rather significant gap between the methods used for their description and many-body theory. While this gap partially remains in the relevant courses and textbooks to this day, there is no real reason for it any more.

This book has been based on the Euclidean path integrals from the very beginning, and there are no barriers between the two disciplines—zero-temperature QFTs and the many-body theory—anymore. While the quantum theory of particles is described by integrals over particle paths $x_i(\tau)$, $i = 1, \ldots, N$, the fields are described by their field histories $\phi(\tau, \vec{x})$. The finite-temperature formulations of both require the Euclidean time τ to be periodic—defined on a circle—with period $\beta = \hbar/T$. Numerical methods described in Chapter 13 use discretized time, defining a one-dimensional lattice. Their extension to QFTs simply changes the setting to a $D = (d + 1)$-dimensional lattice, where d is the number of spatial dimensions. This may increase the number of variables to be integrated over, but even for the paths of many particles, this number was quite large already. So many-body path simulations and those of QFTs use essentially the same tools, which we discussed in Chapter 13.

The reader who has progressed this far into this book thus already knows that the same numerical tools are applicable, and is at this point perhaps puzzled as to why one even needs a chapter introducing QFTs on the lattice. Indeed, it would not be needed if I were to describe only scalar fields (e.g., the so-called Higgs sector of the Standard Model). This chapter allows us to switch to non-Abelian gauge theories, which are the key element of the Standard Model of particle physics. These theories are of course part of standard QFT courses, and readers familiar with them can skip most of this introduction. We still have to include it, following, as always, Einstein's guiding principle: to be as simple as possible but not simpler.

We start the chapter with some brief comments on the history of this subject. Then we introduce lattice formulation of the gauge fields, first Abelian and then non-Abelian.

The aim is to design a formulation that possesses *exact* gauge invariance, at any value of the lattice spacing a, and not just in the continuum limit $a \to 0$. This task is nontrivial, and the corresponding design—due to K. Wilson and others—needs to be carefully explained.

Practical simulations, starting with the pioneering papers by M. Creutz in 1979, had the hard task of proving that—notwithstanding a relatively small dynamical range in which the simulations can be done—the results obtained do depend on the lattice inputs in a way fully consistent with the predictions of the renormalization group (RG). Simply put, they had to show that the results obtained do have a physical meaning, independent of the lattice discretization and parameters used. A demonstration that it is the case is the central achievement of this effort, allowing the creation of a whole industry dedicated to calculating the features of QCD not directly observable in experiments, and even exploring a much wider set of gauge theories. We will continue to use lattice data in the chapters to follow: nowadays the data from numerical experiments significantly complements the "real" data from actual experiments.

Since we seek the non-Abelian gauge fields for the first time in this book, we start in Section 14.1 with its history. Sections 14.2 and 14.3 formulate Euclidean gauge-invariant lattice formulations in Abelian and non-Abelian versions. Section 14.3.1 describes the phenomenon of confinement as it was discovered on the lattice.

14.1 History

Non-Abelian gauge theories were discovered by C. N. Yang and R. Mills in their famous 1954 paper [105]. The main idea is to generalize gauge invariance of QED to more general non-Abelian gauge groups. Two decades later it was found that both the weak and strong interaction parts of the Standard Model are based on it, as are the many extensions beyond the Standard Model that have been proposed.

The Yang-Mills construction later was traced to some deep mathematical constructions in differential geometry and it also has significant connections to Einstein's general relativity. It took a lot of effort to build proper QFTs containing the Yang-Mills gauge fields. Unfortunately, in this book we have to skip all these beautiful connections and many other technical achievements, for the simple reason that they are described in standard QFT textbooks. We only enumerate the main points, before engaging with our main goal: a description of its formulation on the lattice, the one actually used in practical applications.

Much theoretical work was done in the 1960s and 1970s. Derivation of the Feynman rules for this theory took significant effort. After it was completed, this theory was shown to be "renormalizable" in the same sense as QED: all divergences can be systematically eliminated by the counter-terms in the Lagrangian, and the physical effects of the regularized theory can be reduced to charge and (fermion) mass renormalizations.

Treating non-Abelian gauge theories by the RG has lead to the discovery of "asymptotic freedom" by Politzer, Gross, and Wilczek in 1973. We have already discussed it briefly, as examples of different beta functions in Chapter 13. Unlike that for QED and many other known theories in $D = 4$ dimensions, the QCD-like theories (with not too many fermions) have the *opposite sign* for the beta function, so that the charge strength versus distance scale moves in the opposite direction. In QED the charge grows as one approaches it in space (in UV), but in the non-Abelian theories of electroweak and strong sectors of the Standard Model, it grows as one moves *away* from the charge (toward IR).[1] Politzer, Gross, and Wilczek also related this feature with the experimentally observed weakly interacting point-like quarks, thus creating the fundamental theory of strong interactions—quantum chromodynamics (QCD).

Due to asymptotic freedom, QCD is a much better theory than its "parent," QED. While QED has only a perturbative formulation, lattice QCD (which we are going to discuss below) provides a much more complete definition of this theory, describing its behavior in the weak as well as the strong coupling regime. It works as follows. One starts by defining a weakly coupled Lagrangian at some small lattice scale a and then studies (numerically) what happens at much larger distances (smaller momenta, or IR). It is found that, while the effective charge keeps growing in this direction, certain nonperturbative phenomena occur, or even phase transitions[2] take place to qualitatively new phases of the theory. We will discuss these phenomena in the remainder of this book.

In QED the "physical charge of the electron" mentioned in various tables and textbooks is defined at large distances, as is indeed relevant for atomic and condensed-matter applications. It is famously rather small, $\alpha = e^2/(\hbar c) = 1/137$, which allows the use of expansion in its powers. We know that the electric charge is partially screened by vacuum $e^+ e^-$ pairs, but as one approaches the charge, its unscreened value appears larger and larger. Eventually, at (extremely) small distances, the QED coupling becomes large, $\alpha_{QED} \sim O(1)$, so that the perturbative treatment no longer makes sense. We have no "nonperturbative QED" definition, so it remains unknown what to do next.

Perhaps one may argue that RG flow can still be followed by making finer and finer lattices. Attempts to do so have produced unacceptable answers, describing the formation of certain quark-pair condensates, which we know do not exist in real-world QED. But because the scale at which $\alpha_{QED} \sim O(1)$ is many orders away from the momentum scale of any physics experiments, very few people worry about that. Most likely QED is part of a larger theory, and many new phenomena outside QED may come into play on the road to such a theory.

QCD was then developed in two directions. The first deals with small distances of "hard processes" (meaning the coupling is small and the perturbative treatment via Feynman diagrams can be used). It is called "perturbative QCD," and we will discuss it in connection with the high-T quark-gluon plasma in Section 14.3.1.

The other direction, loosely called "nonperturbative QCD," considers "soft phenomena" (those with relatively small momenta but strong coupling). It now includes several

---

[1] We will discuss this effect in Chapter 15, devoted to the discussion of hot quark-gluon plasmas.
[2] The reader has already seen examples of similar behavior earlier in this book. For example, the RG flow of interactions producing Cooper pairing and superconductivity is of the same kind.

subcommunities such as (1) theorists doing QFTs, (2) numerical experimentalists doing lattice gauge theory simulations, (3) experimentalists studying hadronic spectroscopy and various reactions, (4) particle and nuclear theorists making models of nucleons and other hadrons, and (5) people studying high-energy collisions in the hope of learning about properties of very dense hadronic matter. We will discuss hadronic matter in Section 15.9.

14.2 Gauge-Invariant Abelian Theory on the Lattice

Before we start describing lattice gauge theories, let me say a few words about why such a special formulation is needed. Let us take as an example QED. Its (Euclidean) action depends on gauge field A_μ and fermionic fields of electrons ψ

$$
S = \int d^4x \left[\frac{1}{4}(F_{\mu\nu})^2 + \bar\psi(\gamma_\mu i D_\mu + m)\psi \right]
$$

(14.1)

$$
F_{\mu\nu} \equiv \partial_\mu A_\nu(x) - \partial_\nu A_\mu(x), \quad i D_\mu \equiv i\partial_\mu + e A_\mu.
$$

The first term is quadratic in the vector potential A_μ, and without the second (electron) term,[3] the theory would be noninteracting, describing free photons.

Note that even at this point we may ask some obvious questions: are four components of the A_μ field really necessary? Free photons have only two transverse polarizations, so can we eliminate two of these four fields? The answer to this is "no": by keeping free photons only, we would lose, for example, the Coulomb field of a static charge. We can, however, eliminate one component of A_μ, because it is possible to add to A_μ a gradient of any function

$$
A_\mu \to A_\mu + \partial_\mu f
$$

(14.2)

without a change in the field $F_{\mu\nu}$ (since the double derivatives cancel).

Let us now look at the electron term. In it the complex field ψ is multiplied by its conjugate $\bar\psi = \psi^+\gamma_4$. So, we may multiply it by some constant phase $\psi \to e^{if}\psi$, which is irrelevant, as it would be cancelled by that of the conjugate field. We thus have learned that there is the so-called *global* $U(1)$ symmetry corresponding to phase rotation of ψ. We have already discussed such situations in Chapter 4 for complex scalar fields.

QED has in fact much higher symmetry. An appropriate combination of the two transformations discussed can be generalized into the main property of QED, its *local gauge invariance*. In fact the phase of the electron wavefunctions and gauge fields do not have any direct physical meaning by themselves. Gauge invariance of this theory allows a

---

[3] The reader at this point may worry about several technical details I have not explained, such as the exact Euclidean definition of the spinor fields and gamma matrices in this term. These details are not important for understanding the main arguments being made, and their discussion would be just a distraction.

simultaneous transformation of both

$$\psi \to e^{if(x)}\psi; \quad A_\mu \to A_\mu + \frac{1}{e}\partial_\mu f(x), \tag{14.3}$$

with an arbitrary "gauge" function of spacetime point $f(x)$, complemented by a shift of A_μ by its gradient. After such transformation, the two terms in the so-called gauge-invariant or long derivative in the action cancel each other, so D_μ is gauge invariant. The gauge invariance of QED is local, not global. Out of two fields—the gradient part of A_μ and the phase part of ψ—only one combination is physical. The other is redundant; it can be ascribed any value (e.g., it can be eliminated via an appropriate choice of the gauge).

Let us now address the discretization issue. Naively, one can take a four-dimensional lattice with a step a, define $A_\mu(x)$ by its values at the lattice sites, and describe derivatives as nearest-neighbor differences, as is done for a scalar fields. But by doing so, one would loose the exact local gauge invariance, which would be recovered only in the continuum limit $a \to 0$. It is in principle possible to work with such discretization, but practice had shown that it creates significant problems.

Fortunately, an alternative gauge-invariant lattice formulation exists, proposed by K. Wilson in 1974 [106]. Its main idea is that while the charged field ψ is defined at the sites of the lattice, the gauge field is defined at the *links* connecting[4] two neighboring sites.

When we discussed the BKT transition in Chapter 11, I introduced a phase field $\theta(x)$, with the action $\sum_{n,\mu} \cos(\theta(n) - \theta(n+\mu))$ depending on the phase difference between two neighboring sites (the lattice step is $a = 1$ for now). Let us now think of this field as the phase of ψ. Note that it should have the property of changing sign if one goes backward on the link:

$$\theta_{-\mu}(n+\mu) = -\theta_\mu(n). \tag{14.4}$$

The next step is to introduce one more field $\theta_\mu(n)$ defined on the links, going from site n in the direction $\mu = 1, \ldots, 4$. It will appear in the action

$$S = J \sum_{\text{plaquettes}} [1 - \cos(\theta_{\mu\nu})] \tag{14.5}$$

via a "discrete curl" defined at the elementary squares (or plaquettes, as they are usually called):

$$\theta_{\mu\nu} = \theta_\mu(n) + \theta_\nu(n+\mu) + \theta_{-\mu}(n+\mu+\nu) + \theta_{-\nu}(n+\nu). \tag{14.6}$$

The theory just defined has local gauge invariance under the following transformation. For any function of the coordinates $\chi(n)$, the field transformation

$$\theta_\mu(n) \to \theta_\mu(n) - \chi(n) + \chi(n+\mu) \tag{14.7}$$

leaves the action unchanged. (The reader should check that additions at each corner of the plaquette cancel out.)

---

[4] In fact, the proper name of such fields in mathematics is "connections."

These lattice-based fields $\theta_\mu(n)$ and $\theta_{\mu\nu}(n)$ can now be identified with the vector potential A_μ and the field strength $F_{\mu\nu}$ of electrodynamics; the arbitrary function $\chi(n)$ can be identified with the phase of the electron's wavefunction. We have thus accomplished our goal, defining QED on the lattice in a way that keeps local gauge invariance exact.

To complete this section, we need only add one more set of notations, to smooth the transition to the next section. We can change variables, from our real field $\theta_\mu(n)$ to a "unitary one," its exponent:

$$U_\mu(n) = \exp(i\theta_\mu(n)). \tag{14.8}$$

The exponent of $\theta_{\mu\nu}$, using its definition (14.6), can be written as

$$\exp(i\theta_{\mu\nu}(n)) = \prod_{\text{plaquettes}} UUUU, \tag{14.9}$$

where I have omitted the indices and locations of the U variables, and just note that they represent subsequent links along the plaquette. Notice that the cosine in the action can be written as $2\text{Re}(\exp(i\theta_{\mu\nu}(n)))$.

One more important comment. After exponentiation, the fields θ can be recognized as angles, which means they have certain periodicity properties. So, the version of QED defined in terms of u_μ is known as *compact QED*. The two versions differ, for example, in terms of their possible topological configurations.

14.3 Non-Abelian Gauge Theory on the Lattice

Now, turning to QCD, we need to consider quark fields instead of electrons. Since the former have internal quantum numbers called "colors," the quark wavefunction should have a color index $\psi_i(x)$, $i = 1, 2, 3$. The direct analog of the QED gauge transformation (14.3) is

$$\psi_i(x) \rightarrow U_{ij}(x)\psi_j(x), \tag{14.10}$$

where $U(x)$ is a 3×3 unitary matrix,[5] being an arbitrary function of spacetime coordinates. In other words, we want the particular quark orientation in the color space—instead of the phase—to become redundant and have no physical meaning. Another way to put this is to suggest that we are free to use independent color coordinates at any site. Going to another (equally arbitrary) set of coordinate would require a modification of the gauge fields. However, since unitary matrices do not commute, this transformation of A_μ or U_μ will now be a bit more complicated.

In retrospect, it is in fact more convenient to discuss a proper nonperturbative definition of the Yang-Mills theory on a discretized spacetime (Euclidean) lattice [106], than it is in the original continuum formulation. While the lattice sites are reserved

---

[5] The reader is reminded that a sum over a repeated index (in this case j) is always implicitly assumed. We will also often skip writing all indices altogether, writing $\hat{U}\psi$ instead.

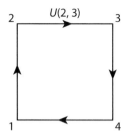

Figure 14.1. A small plaquette with a matter field at its corners. The link variables U have arrows indicating their directions.

for the matter fields (quarks), the connecting gauge fields are defined on "links." For simplicity, let us use a cubic four-dimensional lattice; we start with only one small square (or plaquette), shown in Figure 14.1.

There are four (vector-valued) matter fields $\psi_i(x_A)$ defined at its corners, numbered by $A = 1, \ldots, 4$. The gauge field in this mini-universe is described by four unitary "relative orientation matrices"[6] $U_{ij}(L, M)$, associated with the links between the two sites (e.g., the one for the (2,3) link shown in the figure). Since the purpose of U is the same as that of the vector field, to undo (or compensate for) arbitrary rotations of matter fields at all corners, the obvious transformation law for the $U(L, M)$ is

$$U(L, M) \rightarrow \Omega(L) U(L, M) \Omega^{-1}(M). \tag{14.11}$$

This construction makes it extremely easy to formulate many gauge-invariant quantities. For example, a combination of two quark field and two links

$$\psi^+(2) U(2, 3) U(3, 4) \psi(4)$$

would then be independent of any possible rotations at point 3. It is clear that one can make open strings of any length from a product of such link variables, and its gauge transformation would be the rotations at its ends, compensated by rotations of the quark fields there. Finally, it is the "closed strings"—the trace of the product of link matrices over any closed path—that are gauge invariant. One obvious example is $\text{Tr}[\hat{U}(12)\hat{U}(23)\hat{U}(34)\hat{U}(41)]$.

How are the link variables related to vector potentials? The formal relation is

$$U(1, 2) = \text{Pexp}\left(ig \int_1^2 A_\mu^a(x) T^a dx_\mu\right), \tag{14.12}$$

where the definition includes the *path-ordered exponent (Pexp) of a matrix-valued variable* defined as follows. If the length Δx_μ of the path 1-2 is small, we can use an expansion of the exponent as $(1 + ig A_\mu^a(x) T^a \Delta x_\mu + \cdots)$. If it is not small, it can be divided into a series of small steps, with multiplication of the rotation matrices at each step. The path-ordered exponential Pexp is the limit of such a procedure as the step size goes to zero.

---

[6] For clarity, here L, M are not the indices of the matrix (which I imply without showing them below) but just a label of the link connecting two points.

Let us consider some simple examples of gauge fixing, to clarify the similarity and differences of the Abelian and non-Abelian cases. Let us take one plaquette with arbitrary U on its links and ask, how many of them can be eliminated (set to $\hat{U} = \hat{1}$)? Note that we can do it subsequently for three links, but at least the last one it will remain not equal to 1, unless the configuration is pure gauge. The same is true for the non-Abelian case, although there is one difference.

It is easier to follow the argument if $A_\mu(x)$ is independent of x: in the Abelian case the product of all U over the contour is trivial, because phase differences just cancel out. Thus, this configuration is physically identical to the no-field $A_\mu = 0$ one. This is not true for the non-Abelian case, for which one finds instead

$$UUUU \equiv \hat{U}(1, 2)\hat{U}(2, 3)\hat{U}^+(1, 2)\hat{U}^+(2, 3).$$

It is still true that $U^+(1, 2)$ is the inverse of $U(1, 2)$, and the same is true for $U^+(2, 3)$ and $U(2, 3)$, but now the combination of four Us is still not trivial, because in general two matrices do not commute. The lesson: in the non-Abelian case even constant potentials can lead to nonzero gauge-invariant (that is, physical) consequences.

It would be useful to keep this lesson in mind while thinking about the proper definition of the field strength. The Yang-Mills definition (again related to the Riemann curvature tensor in terms of Chrystoffel symbols in differential geometry) takes the form

$$G_{\mu\nu} = \partial_\mu A_\nu - \partial_\nu A_\mu + ig[A_\mu A_\nu], \tag{14.13}$$

where the coupling g is introduced to separate the nonlinear commutator term from the others. Note that, due to the last commutator term, even a constant A_μ can lead to a nonzero field strength (unlike in QED).

Understanding the relation between continuum and lattice notations can be helped by using the Campbell-Baker-Hausdorff formula

$$e^\lambda e^\mu = \exp\left(\lambda + \mu + \frac{1}{2}[\lambda, \mu] + \frac{1}{12}[(\lambda + \mu), [\lambda, \mu]] + \cdots\right), \tag{14.14}$$

from which a non-Abelian version of the Stokes theorem for small contours follows:

$$e^\lambda e^\mu e^{-\lambda} e^{-\mu} = \exp([\lambda, \mu] + \cdots). \tag{14.15}$$

Exercise *Calculate the (no-trace) $UUUU$ product over plaquette to order a^2, and find how it is related to the field strength $G_{\mu,\nu}$ defined in (14.13). Consider a to be small compared to the scale at which A_μ changes and expand it to the first power of the gradient and commutator.*

Exercise *Continue the calculation in the previous exercise to the traced product $\text{Tr}(UUUU)$ up to order a^4. Since it is gauge invariant, and the only gauge-invariant operator of dimension 4 is $G^a_{\mu,\nu}G^a_{\mu,\nu}$, they must be proportional. Verify this statement and find the coefficient of proportionality.*

Exercise *If you use Maple or Mathematica, attempt to calculate $\text{Tr}(UUUU)$ to even higher order a^6. Try to identify all gauge-invariant operators and their coefficients.*

Exercise *Add to a single plaquette product by Wilson the set of all double plaquettes and* Tr($UUUUUU$), *known as "strip," "chair," and "butterfly" plaquettes, to the action. Tune the coefficients of these terms to cancel unwanted higher-order operators on the lattice, which (we think) are not present in continuous QCD. The results are known as "improved actions" among lattice practitioners.*

With this latticized definition of variables, let us now turn to quantization of this theory, which means performing the integral over Us at all links and ψs at all sites. Wilson has provided the following definition:

$$Z = \int (\Pi_{\text{links}} DU) \exp\left[-\frac{1}{g^2} \sum_{\text{plaquettes}} \left(1 - \frac{1}{N_c} \text{Tr} UUUU\right)\right],$$

where N_c is the number of colors, the sum is made over all plaquettes, and all products $UUUU$ are traced. The coupling g defines the absolute magnitude of the action.

This complicated integral, provided we are able to calculate it, represents properties of the "pure glue" QFT, called $SU(N_c)$ *gluodynamics*. Note that no attempt was made to fix the gauge: in fact, the partition function defined above includes the integral over all gauges.

It is a somewhat unusual arrangement, and so let us discuss how it works in an example. Suppose we define the so-called *Wilson contour*, take a closed rectangle of the size R in spatial and T in (Euclidean) time directions, multiply all U along it, and trace the result at the end:

$$W \equiv \frac{1}{N_c} \text{Tr}[UUU \cdots U]. \tag{14.16}$$

Since it is a closed loop, it is gauge invariant.

Its vacuum expectation value is defined by the ratio

$$\langle W \rangle \equiv \frac{\int W e^{-S} DU}{\int e^{-S} DU}, \tag{14.17}$$

where the action S is as defined by the exponent in the partition function above.

While W itself is the same for all gauge choices, the integral includes all gauges possible. The volume of the gauge group appears in the power equal to the number of sites on the lattice. Since the same integral appears in the numerator and denominator, it cancels out. For a gauge-invariant observable, like W, it does not matter whether we fix the gauge or integrate over all of them.

Suppose we do not want to integrate over all gauges and insist on doing the calculations with physical degrees of freedom only, eliminating all the redundant ones. The reader is perhaps familiar with the practice of using some gauge conditions imposed on A_μ. Popular examples are

- $A_0 = 0$ (the temporal or the Weyl gauge),
- $\partial_m A_m = 0$ (the Coulomb gauge, $m = 1, 2, 3$),
- $\partial_\mu A_\mu = 0$ (the Landau gauge),

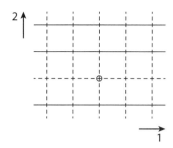

Figure 14.2. Example of a gauge-fixing tree in the two-dimensional lattice. The solid lines have the nontrivial link variables U, while the dashed links indicate a "tree" starting from a root point (the small circle), where one can set $U = 1$ by using the available gauge freedom at the appropriate sites.

- $n_\mu A_\mu = 0$ (the axial gauge, where n is some unit 4-vector), and
- $x_\mu A_\mu = 0$ (the Fock-Schwinger gauge).

Indeed, one can use gauge freedom (e.g., for putting $A_0 = 0$), as for all temporal plaquettes, it is possible to set two links going in the time direction to zero. However, the preceding discussion should have prepared the reader to realize that these gauge conditions, although named after famous physicists, actually do not do their job properly, because not all gauge freedom is completely eliminated by any of these conditions. Indeed, even for the single plaquette discussed above this is obvious, since all of these conditions can only set two out of four links to zero, but we now know that gauge freedom allows us to eliminate three of them. The next exercise asks the reader to invent the *complete* gauge fixing.

Exercise *Define the so-called gauge-fixing tree on the lattice, starting from a particular point and fixing as many links as possible. So, for lattice configurations there is no problem fixing the gauge completely, and in fact it can be done in many ways. A simple example of it is shown in Figure 14.2: in this case $A_1(x) = 0$ as in some axial gauge with $n = (1, 0)$, but some residual gauge transformation makes it possible to put $A_2(x = 0) = 0$ as well.*

Unfortunately, there is no practical way to complete the gauge fixing in a continuum description, so in fact only partial gauge fixing, using conditions like those given above, is actually done in practice.

14.3.1 The Confining Potential

In the limit of very large time T of the Wilson contour its expectation value with the partition function Z weight defined above (14.16) is related to the interaction potential between two external static charges

$$\lim_{T \to \infty} \langle W \rangle \sim e^{-V(R)T}. \tag{14.18}$$

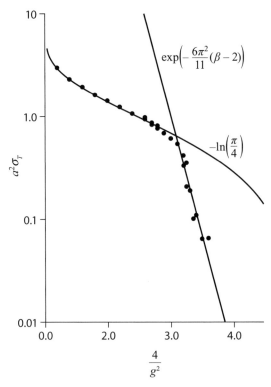

Figure 14.3. The lattice spacing squared times the string tension as a function of bare coupling $\beta \equiv 4/g^2(a)$. The points are the results of Monte-Carlo calculations, while the lines are the expected weak and strong coupling behavior. Adapted from Creutz [94].

At present nobody understands how to calculate it analytically, but a numerical integration performed by Creutz in 1979 [94] indicates that the result seems to be a linear confining potential

$$V(R) = \sigma_T R. \tag{14.19}$$

Furthermore, it was also demonstrated that the coefficient σ_T—known as *the string tension*—correctly depends on the lattice spacing a in the continuum limit $a \to 0$. The dimension of the tension $\sigma_T \sim$ energy/length $\sim 1/a^2$, and the asymptotic freedom (RG expression for the $SU(2)$ pure gauge theory) is

$$g^2(a) = \frac{12\pi^2}{11\log(1/\Lambda a)} \tag{14.20}$$

Inverting this expression, we find the scale a as a function of the input coupling $g(a)$ at this scale

$$a \sim \exp\left(-\frac{12\pi^2}{11g^2(a)}\right). \tag{14.21}$$

So, performing a series of calculations using different bare couplings $g^2(a)$, one can measure the string tension and test whether the results agree with this prediction. The results, shown in Figure 14.3, clearly demonstrate a rapid transition from strong coupling behavior at small values of

$$\beta \equiv 4/g^2(a)$$

to weak coupling behavior, at larger β. Note another important fact that this plot conveys: the transition between the two regimes happens at

$$\alpha_s \equiv \frac{g^2}{4\pi} \approx 0.16. \tag{14.22}$$

Knowing experimental value of the string tension $\sigma_T \approx (420\,\mathrm{MeV})^2$, one can use it to define the lattice spacings a for any of these calculations in MeV or fm units. Of course, this particular calculation was done in 1979 for the simplest $SU(2)$ gauge theory without quarks, but since then it has been repeated hundreds of times for many other settings, including QCD with physical quark masses. Knowing the absolute scale of a for each calculation, one can now predict in absolute units all other observables, such as hadronic masses.

15 | Theory of Quark-Gluon Plasma

The attentive reader has probably already noticed that in the case of non-Abelian gauge theory, I changed the standard order: we did not start with the Lagrangian and the perturbation theory for it, as we did for all other systems discussed previously. The reason is that gauge symmetry is much more tricky than other symmetries, and it is essential to demonstrate first how it can be implemented in its full glory. The lattice formulation shows very clearly that one can proceed even without gauge fixing.

Unfortunately, in perturbation theory, gauge fixing is unavoidable, and it brings with it a large number of technical questions. Standard QFT courses pay a lot of attention to these issues, teaching the details of quantization in covariant gauges and discussing Faddeev-Popov ghosts and Gribov copies. It is all needed for certain goals, but not so far for ours. As we will see shortly, the main perturbative results for finite-temperature QCD were in fact first obtained in a Coulomb gauge, in which these issues do not arise.

We start in Section 15.1 by defining three scales of quark-gluon plasma: the thermal, electric, and magnetic screening scales. At high enough T these scales are separated by powers of the coupling, small due to asymptotic freedom: $g^2(T) \ll 1$. The first diagrams are evaluated in sections 15.2 and 15.3.

In Section 15.4 we depart from hot QCD to discuss cold quark matter. In this case the large parameter, ensuring smallness of the coupling, is the chemical potential instead of the temperature.

Some resummation of the diagrams, similar to what was done for an electron gas and nuclear matter, is discussed in Section 15.5. In Section 15.6, we discuss the infrared divergences and show that, even at very high T, there remains a nonperturbative magnetic sector of the theory, for which the perturbative approach cannot be used. It can be investigated using dedicated lattice studies of a three-dimensional magnetic effective theory. In Section 15.7 we discuss convergence of the finite-T perturbative series. In Section 15.8 we come to the so-called hard thermal loop (HTL) resummation. Finally, in Section 15.9 we discuss results of the finite-T lattice simulations.

15.1 Overview and Scales of Quark-Gluon Plasmas

Perturbative calculations for high-T quark-gluon plasmas were started by me in 1976 [108] and Kapusta [109]. These papers include the first diagrams and their resummation. First reviews were those by me [110] and by Gross et al. [112], establishing this new field.

Ignoring all small quark masses (of u, d, s quarks), and ignoring effects associated with heavy c, b, t quarks completely, one has a simplified situation in which T is the only scale in the problem. Thus the (free) energy density and pressure are, to zeroth order, on dimensional grounds,

$$\epsilon \sim p \sim T^4. \tag{15.1}$$

In the zeroth order in the coupling, the quark-gluon plasma is just an ideal gas of quarks and gluons, so the coefficients in the expressions above follow, using the usual Bose-Einstein distributions for gluons and Fermi-Dirac distributions for quarks and antiquarks.

The dimensionless coupling g is often expressed as $g^2/4\pi = \alpha_s$, and so the first corrections to thermodynamical observables are $O(g^2)$ corrections to the free gas expressions. Since it is straightforward to calculate the lowest-order diagrams, I will skip this in the interest of time and move to the next issue, diagram resummations.

The first physical issue was screening or antiscreening, related to RG running of the coupling. Cold quark matter has the same diagrams as QED, so screening is obvious. The gluonic part—appearing at finite T only—was new. In the mid-1970s, when I first approached this problem, the following facts were known:

- in QED plasma, the charge is screened by matter particles;
- in QED vacuum, the charge is screened by vacuum polarization; and
- in QCD vacuum, the charge at small distances is anti-screened (this is of course the celebrated asymptotic freedom).

The remaining question was then: What happens in a QCD heat bath at high enough T at large distances? The answer came from a calculation [108] of the polarization tensor $\Pi_{\mu\nu}(T, q)$:

- In QCD plasma, the charge is screened at large distances.

That is why the quark-gluon plasma was called a plasma. Schematic behavior of the effective electric charge is shown in Figure 15.1: so a combination of screening and anti-screening keeps it small at all distances, provided T is high enough.

What this means is that, apart from the obvious scale provided by the temperature

- energy-momentum scale $p \sim T$,

there is also the so-called

- electric or Debye scale $p_E \sim g(T)T$,

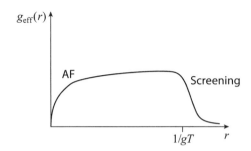

Figure 15.1. Schematic behavior of the effective electric interactions in the quark-gluon plasma as charge versus distance.

at which the static electric charge is screened. Significant progress on understanding what happens at this scale has been obtained by Braaten and Pisarki [115], who made complete resummations of hot thermal loops. We will not discuss its later form, as an effective Lagrangian, but proceed instead to a somewhat more intuitive kinetic formulation due to Blaizot and Iancu [116].

Furthermore, I [108] found that a static magnetic field is not screened: thus infrared divergences and other nonperturbative phenomena survive in the magnetic sector. First resummation of the so-called ring diagrams produced a finite plasmon term [108,109], but higher order diagrams are still infrared divergent. Polyakov [117] argued that there should be a new scale at which magnetic fields are also screened

- magnetic mass scale $p_E \sim g^2(T)T$.

Linde [118] was quick to suggest a monopole-related screening mechanism. Many years later Kajantie and collaborators have explicitly calculated the couplings of the corresponding effective action for the magnetic sector of QCD. It includes the zero Matsubara frequency sector of the vector potential $\vec{A}$ and the temporal component A_0, which plays the role of an adjoint Higgs field. So, this model is a three-dimensional version of the Georgy-Glashow model, with the dimensional coupling $g_3 = g^2 T$. Two other couplings are the coefficient of mass term $\mathrm{Tr}(A_0)^2$ and of the quartic coupling $(\mathrm{Tr}(A_0)^2)^2$. The A_0 mass term, from the four-dimensional calculation outlined in Section 15.2, is positive. Therefore this magnetic theory is not in the broken phase, $\langle A_0 \rangle = 0$, and thus it has no magnetic monopoles in it. However, this theory is a confining one, with a mass gap: it has been studied by using three-dimensional lattice simulations (see [111] for details).

Brute-force perturbative QCD evaluation of multi-loop diagrams such as in [119] have been done: they have demonstrated that convergence of perturbative QCD series is not reached in the region of interest. Multiple different resummations also have been done [122,123].

Of course, thermodynamical quantities are not the only ones we would like to know. Kinetic quantities like viscosity are crucial to understanding equilibration and matter evolution in heavy ion collisions: we will turn to their discussion in Section 16.3. Jumping ahead of the main logic of the book, Table 15.1 shows the main scales (normalized to T) for weak and strong coupling regimes.

Table 15.1. Comparison of the basic scales in high-T gauge theories in the weak coupling and the strong coupling limit of the $N = 4$ supersymmetric(SUSY) gauge theory.

Quantity	Weak Coupling (Small g, N)	Strong Coupling Limit (Large $g^2 N$)
Interparticle $n^{-1/3}$	~ 1	$\sim 1/N^{2/3}$
Debye screening radius R_D	$\sim 1/g$	~ 1
Magnetic screening radius R_M	$\sim 1/g^2$	~ 1
Sound attenuation Γ_s	$\sim 1/g^4 \log(1/g)$	$1/3\pi$

Note: All constants are ignored except in the last entry.

15.2 The Perturbative Formalism of QCD

The continuous version of the Yang-Mills action in perturbative normalization

$$S = \int d^4 x \frac{1}{4}(G^a_{\mu\nu})^2$$

$$G_{\mu,\nu} = \partial_\mu A_\nu - \partial_\nu A_\mu + ig[A_\mu A_\nu]$$

(15.2)

looks similar to the gauge part of the action of electrodynamics. Yet there is a crucial difference here, because the field strength contains the commutator we discussed previously in Chapter 14. So it is not just Gaussian path integrals due to terms of order $\sim g A^3, g^2 A^4$. Recall the derivation of the Feynman diagrams for the quartic potential in the quantum mechanical example in Chapter 3 from which we started this course. We should now do basically the same steps, expanding $\exp(-S)$ in powers of the nonlinear terms, or powers of the coupling g. As usual, the quadratic part of the action defines the Gaussian operator and propagators. Since it has two derivatives (or two momenta in momentum representation) we find $G \sim (numerator)/k^2$, as usual. The numerator depends on the gauge choice: these and other details of Feynman rules for gauge theory I do not want to cover at present. The cubic and quartic vertices can be read off the Lagrangian directly.

To master this formalism, it is helpful to consider some simple examples. The single-body observable $O_{mn}(\omega, k)$ (such as energy) of the transverse gluons in the $A_0 = 0$ gauge can be written as

$$\epsilon = T \sum_n \int \frac{d^3 k}{(2\pi)^3} O_{mn}(\omega, k) D_{mm}(\omega_n, k),$$

(15.3)

where $D = (\delta_{mn} - k_m k_n / k^2)/(\omega_n^2 + k^2)$ is the usual zeroth-order gluonic propagator. Evaluating the sum first, one gets the famous "bosonic summation" formula we used before

$$\sum_n \frac{1}{x^2 + n^2} = \frac{2\pi}{x} \left(\frac{1}{2} + \frac{1}{\exp(2\pi x) - 1} \right), \tag{15.4}$$

which leads to the following result:

$$\epsilon(T) = 2(N_c^2 - 1) \int \frac{d^3 k \, k}{(2\pi)^3} \left(\frac{1}{2} + \frac{1}{\exp(k/T) - 1} \right) = 2(N_c^2 - 1) \frac{\pi^2 T^4}{30}. \tag{15.5}$$

The first term—the divergent energy of zero-point oscillations of all oscillators in the vacuum—is omitted as usual, while the second (thermal bath) term is kept.

It is also worth repeating the evaluation for the zeroth-order single-body observables for fermions, again by using the propagator $S = (\gamma_\mu p_\mu + m)/(p^2 + m^2)$. Recall the following generic fermionic identity we derived using the Sommerfeld Watson trick:

$$\sum_n \frac{1}{x^2 + (2n+1)^2} = \frac{\pi}{2} \left(\frac{1}{2} - \frac{1}{\exp(\pi x) + 1} \right) = \frac{\pi}{2x} \tanh\left(\frac{\pi x}{2} \right). \tag{15.6}$$

In our context the r.h.s. has some rather transparent physical meaning that we have already discussed in connecion to the electron gas. Note also how an extra an factor 2 (from the two poles of the l.h.s.) appears, accounting for equal number of quarks and antiquarks in the heat bath.

Note also the following curious detail: unlike the bosonic case, with its standard combination familiar from harmonic oscillator $\langle n_B(\text{quanta}) + 1/2 \rangle$, the fermionic one is $\langle 1/2 - n_F(\text{quanta}) \rangle$. Furthermore, at very high temperatures $T \to \infty$, the exponent $\exp(\omega_k / T) \to 1$ and therefore $\langle 1/2 - n(\text{quanta}) \rangle \to 0$. This term will have an important effect on high-T behavior of the fermionic propagator.

Explicit calculation of all closed 2-loop diagrams including all qq, qg, gg scattering processes to order g^2 are straightforward, and we will not discuss them here.

15.3 The Polarization Operator

The heat bath defines a preferred Lorentz frame, in which matter is at rest. So Lorentz symmetry is broken anyway, and thus the zeroth component of the vector potential A_0 is not the same as the others. The calculations are easier to perform in the "physical" noncovariant gauges, rather than in the covariant ones that are used in QFT textbooks. The calculations were first done in [108] using the Coulomb gauge. In this case not only are there no Faddeev-Popov ghosts, but also one can show that there is no vertex renormalization of the static charge. As a result, all one has to do in this gauge is to calculate the polarization tensor $\Pi_{\mu\nu}(\omega, k)$, much like what was done for the electron gas in Chapter 7.

As usual, the polarization tensor has two Lorentz structures corresponding to the electric (Coulomb) and the magnetic (transverse) fields. The former is described by

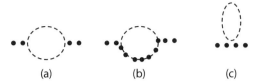

Figure 15.2. Three diagrams in the Coulomb gauge for polarization operator. The dashed lines indicate the transverse (magnetic) gluon propagators, dotted lines are for $\langle A_0 A_0 \rangle$ propagator.

$\Pi_{00}(\omega, k)$ and the latter by the spatial part of the polarization tensor

$$\Pi_{mn}(\omega, k) = \left(\delta_{mn} - \frac{k_m k_n}{k^2} \right) \Pi_\perp(\omega, k) + \frac{\omega^2 k_m k_n}{k^2} \Pi_{00}(\omega, k). \tag{15.7}$$

These two functions enter the corresponding "dressed" (resummed) Green functions as follows:

$$D_{00} = \frac{1}{\mathbf{k}^2 + \Pi_{00}(\omega, k)} \qquad D_{mn} = -\frac{\delta_{mn} - k_m k_n / k^2}{\omega^2 - \mathbf{k}^2 - \Pi_\perp(\omega, k)}. \tag{15.8}$$

The logarithmic contribution from the three diagrams shown in Figure 15.2 is sufficient to obtain the invariant charge renormalization, as in this gauge there is no renormalization of the Coulomb vertex function. Including the usual contribution from quarks (analogous to that in QED) plus these three gluon diagrams, we get the following contributions to the gluon polarization operator:

$$\Pi_{00}(\vec{q}) = \frac{g^2 \vec{q}^2}{\pi^2} \log(\vec{q}^2/\Lambda^2) \left[+\frac{1}{2} - 6 + 0 + \frac{N_f}{3} \right]. \tag{15.9}$$

(The fourth term is a contribution of the quark loop, added for completeness.) Note that the only negative contribution comes from the diagram in Figure 15.2b. Indeed it is only possible for this diagram because it has no imaginary part—the Coulomb line shown by dotted line is instantaneous and has no pole in energy, so the sign of the log is not restricted by the unitarity cut. Explicit calculation shows that it has a log term, $\log(\vec{q}^2)$, which contains not the usual four-dimensional momentum squared but its spatial components only.

Thus, resumming the Coulomb field of a charge via a sequence of polarization operators, we find

$$(1/\vec{q}^2) + (1/\vec{q}^2)\Pi_{00}(1/\vec{q}^2) + \cdots = \frac{1/\vec{q}^2}{(1 + \frac{g^2}{\pi^2} \log(\vec{q}^2/\Lambda^2)[(N_f/3) + (1/2) - 6]}, \tag{15.10}$$

which agrees with the asymptotic freedom formula by Gross, Politzer, and Wilczek. This derivation is in fact much simpler than theirs, done in the covariant gauge, and is due to Khriplovich [124], as early as 1968.[1]

---

[1] Khriplovich was a good friend and for few years even a roommate of mine: so I had a full 5 years before Gross, Politzer, and Wilczek to connect this sign of the correction to the parton model and the weakness of strong interaction at small distances, known since the famous SLAC experiment of 1969. I can only blame myself for knowing at the time next to nothing about the RG from QFT courses. Later David Gross told me that it was K. Wilson's lectures about indices that were crucial to their thinking at the time.

I mention all this for completeness. But in this book we have never discussed the UV completion of the theory, leaving these issues to QFT courses. Our focus should always be on the behavior of the dressed propagators in the opposite limit of soft momenta $Q \to 0$, that is, the IR.

Let me start with the electric polarization tensor in the soft limit $\omega, k \ll \mu$ for cold quark matter, in which case it comes from the fermion loop. We can call it the relativistic version of the Lindhard function

$$\Pi_{00} = -\frac{g^2}{2\pi^2} \left(\sum_f \mu_f^2 \right) \left[1 - \frac{\omega}{2k} \log \left| \frac{\omega+k}{\omega-k} \right| + \frac{i\pi\omega}{2k} \theta(\omega)\theta(k-\omega) \right], \qquad (15.11)$$

where the sum runs over all quark flavors if the chemical potentials for them are all different.

Let us try to understand what this expression means, starting with its real part. Note that the expression in square brackets is a function of ω/k, so there are different long-wavelength limits depending on this ratio. One is *the static limit* $\omega = 0, k \to 0, \omega/k = 0$, in which

$$D_{00}(\omega, k) \to -\frac{1}{k^2 + \kappa_D^2}, \qquad \kappa_D^2 = \frac{g^2}{2\pi^2} \sum_f \mu_f^2 \qquad (15.12)$$

The Fourier transform of this Coulomb-field propagator is the screened potential between two charges $V(r) \sim \exp(-\kappa_D r)/r$, and so the inverse $R_d = 1/\kappa_D$ is the *Debye screening radius*.

The opposite limit $k = 0, \omega \to 0, k/\omega = 0$ is called *the plasmon limit*. Using expansion at small k

$$1 - \frac{\omega}{2k} \log \left(\frac{\omega+k}{\omega-k} \right) \approx -\frac{k^2}{3\omega^2} + \cdots,$$

we find

$$D_{00}(\omega, k) \to \frac{1}{k^2(1 - \omega_D^2/\omega^2)}, \qquad \omega_D^2 = \frac{g^2}{6\pi^2} \sum_f \mu_f^2 = \frac{\kappa_D^2}{3}. \qquad (15.13)$$

The position of the pole, ω_D, is therefore the frequency of the longitudinal long-wavelength plasma oscillations. I do not present here the analogous expression for the magnetic polarization tensor, and only note that in this case the most interesting physical limit is the *gluon effective mass*, defined by the third on-shell limit $k = \omega \to 0, k/\omega = 1$:

$$m_g^2 = \Pi_\perp|_{\text{onshell}} = \frac{g^2}{4\pi^2} \sum_f \mu_f^2 = \frac{\kappa_D^2}{2}. \qquad (15.14)$$

The last important point is that the static magnetic field in a quark Fermi gas is not screened because in the static limit $\omega = 0, k \to 0, \omega/k = 0$, the magnetic polarization $\Pi_\perp \to 0$ vanishes.

The imaginary part of the tensor originates from the log in (15.11), when its argument is negative. The physics of this term is the *Landau damping* of waves in a plasma, which may occur if their ω/k match the longitudinal velocity of some particles inside the Fermi sphere. If this happens, the particles may "surf the wave," continuously draining its energy.

Now is time to present some results of my calculation for the hot QGP. Rather surprisingly, they lead to the polarization tensor being the same function of ω, $\vec{k}$ as for the cold quark matter: the difference is only the coefficient! The Debye screening parameter obtains the following contributions from the three diagrams Figure 15.2a–c plus that of the fermionic loop:

$$\kappa_D^2 = g^2 T^2 \left(\frac{1}{2} + 0 + \frac{1}{2} + \frac{N_f}{6} \right). \tag{15.15}$$

Note that the unusual diagram Figure 15.2b, with a transverse-Coulomb loop (which is responsible for the unexpected sign and the asymptotic freedom) does not contribute at all!

The relation between this parameter and the plasma oscillation frequency and effective on-shell gluon mass are the same as above, at $T = 0$. Very importantly, perturbatively there is, again no static magnetic screening , as in QED.

15.4 Cold Quark Matter and Perturbation Theory

To zeroth order in coupling, cold quark matter is just the relativistic Fermi gas of quarks. I omit expressions for it as trivial.

The order g^2 [150, 151] has the same two diagrams as for the electron gas and nuclear matter, the dumbbell and the sunset. Recall from Chapter 7 that in the electron gas, the former did not contribute because it was canceled by "jellium," a compensating positive charge. In the quark matter this contribution is also zero, but for a different reason: including all three quark colors makes the matter colorless. Technically, each loop of the dumbbell diagram contains the color matrix $\hat{t}_a$, $a = 1 \cdots, 8$ in the vertex, and $\text{tr}[\hat{t}^a] = 0$.

The remaining sunset diagram is the quark loop with a gluon exchanged. We do not want to calculate the divergent vacuum part, and we apply the modified $i\epsilon$ prescription to each quark propagator, putting both quarks on shell $\omega_k^2 = k^2 + m^2$, with $\omega_k < \mu$ and k inside the Fermi sphere for each flavor. The physical meaning of it is that it is part of forward scattering amplitude due to the exchange interaction, in which two quarks have to exchange all quantum numbers (namely, their momenta, colors, and spins) to find an empty place inside the Fermi sphere. The corresponding correction to thermodynamical potential is [108]:

$$\delta\Omega = \frac{g^2}{4\pi^4} \sum_f \left\{ \frac{3}{2} \left[\mu_f p_f - m_f^2 \log\left(\frac{\mu_f + p_f}{m_f} \right) \right]^2 - p_f^4 \right\}, \tag{15.16}$$

where as before the sum runs over the flavors f. The reason I give this expression is that in nonrelativistic and ultrarelativistic plasmas it leads to corrections that look very similar,

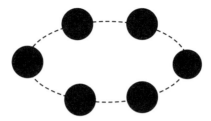

Figure 15.3. Example of a ring diagram with $n = 6$ gluon polarization operators in one single gauge field loop.

but closer inspection shows that they have the opposite sign:

$$\delta\Omega|_{\mu_f \gg m_f} = \frac{g^2}{8\pi^4} \sum_f p_f^4 \qquad \delta\Omega|_{\mu_f \ll m_f} = -\frac{g^2}{4\pi^4} \sum_f p_f^4. \tag{15.17}$$

The sign of the first (nonrelativistic) expression corresponds to attractive electric interaction. As for the latter (ultrarelativistic) case, magnetic interactions are as large as the electric ones, and the sign is no longer obvious. But one may ask how it is possible to violate the well-known quantum-mechanical theorem, according to which the ground state energy of any system can only decrease in the second order of the perturbation theory. The answer is that this theorem is not applicable in the relativistic case: the necessity of renormalization—subtraction of the divergent vacuum contribution—voids this argument.

15.5 Ring Diagram Resummation

In higher orders at high T there appear soft (IR) divergences of the perturbation theory, which are absent in vacuum. The simplest examples of these divergences are the ring diagrams exemplified in Figure 15.3. Their electric contribution to the Matsubara frequency zero is

$$\delta\Omega|_n = \frac{1}{n} \int \frac{d^3k}{(2\pi)^3} \left(\frac{\Pi_{00}}{k^2}\right)^n, \tag{15.18}$$

and for $n > 2$ these are clearly divergent at small k, at which $\Pi_{00} = \kappa_D^2$.

However, following the example of Gell-Mann and Bruckner [50] in their discussion of a cold QED plasma of electrons, we can sum all such diagrams over n from 2 to infinity first [108, 109], obtaining

$$\delta\Omega^{\text{ring}} = \frac{1}{2} \int \frac{d^3k}{(2\pi)^3} \left[\log\left(1 + \Pi_{00}/k^2\right) - \Pi_{00}/k^2\right], \tag{15.19}$$

which is IR-finite. The resulting correlation energy or "plasmon term" at high T and μ is equal to

$$\delta\Omega^{\text{plasmon}} = -\frac{N_c^2 - 1}{12\pi} g^3 T^4 \left[\frac{N_c}{3} + \frac{N_f}{6} + \sum_f \frac{\mu_f^2}{6\pi^2 T^2}\right]^{3/2}. \tag{15.20}$$

This nice example of resummation shows that sometimes the divergence of diagrams is just the way Nature tells us that the answer is not necessarily an analytic function of g^2, but the g^3. Note that the result is still small as soon as g is small.

The physical nature of this correlation energy is the interaction of a charge with its Debye cloud, and its magnitude per particle is $\sim g^3 T$, as we get by dividing this contribution by the particle density. The same magnitude has a correlation energy for a static quark in a plasma. Following Debye, we can derive it much more simply by writing the potential as

$$4\pi V(r) = -\frac{4}{3}\frac{g e^{-\kappa_D r}}{r} \approx \frac{4}{3}\frac{g}{r} - \frac{4}{3}g\kappa_D + \cdots \tag{15.21}$$

and noticing that the second constant potential times g of a point static charge is the correlation energy

$$E_{\text{corr}} = -\frac{4}{3}\alpha_s \kappa_D = O(g^3). \tag{15.22}$$

15.6 IR Divergences and Magnetic Sector

Unfortunately, the resummation of IR-divergent diagrams to a finite answer just demonstrated is not the general case. The bosonic nature of the gluon field together with its Abelian (or "charged") properties leads to IR problems that do not appear in the QED plasma for example. Its presence can in fact already be seen on dimensional grounds. Each loop at the nonzero temperatures contains the factor T and a sum over discrete Matsubara frequencies. Consider only the sector with zero frequencies (which corresponds to the classical limit of the vanishing Plank constant, or three-dimensional gauge theory). The factor T enters any loop and in fact makes the coupling dimensional, its power growing with the number of loops. Something should compensate for it in the denominator, and that can only be the particle momenta. And growing powers of momenta in the denominator must eventually produce IR divergences. In fact, the 4-loops or the eighth-order diagrams already have such power divergences. Therefore

$$\delta\Omega^{(8)} \sim g^6 T^4 \left(\frac{g^2 T}{q_{\text{min}}}\right), \tag{15.23}$$

where a cutoff has been explicitly put in.

Polyakov [117] has suggested that this disease can only be cured by the development of some magnetic screening length for the gluomagnetic field of the order of $\kappa_M \sim g^2 T$. If so, using it as the IR cutoff in (15.23) we would conclude that all diagrams above the eighth order may contain a contribution of the same magnitude, namely $\delta\Omega \sim g^6 T^4$. So, terms of such order cannot be perturbatively calculated.

We can compare the situation in high-T QCD with that of a star (e.g., the Sun). If we want to evaluate its total thermal energy, normal plasma physics would be enough. It is true that the Sun has a very complicated magnetic field structure, with spots of opposite polarities and magnetic fluxes protruding from them into space. We do not know how to calculate the energy of these spots, but of course however interesting they may be, their total energy is negligible.

One can separate, by the so-called dimensional reduction, soft fields of a *magnetic scale* $g^2 T$ from all harder scales. It has been shown in [113] how, in principle, an effective theory could be constructed to deal with this particular problem by marrying analytical techniques (to determine the coefficients of the effective theory) and numerical ones (to solve the nonperturbative three-dimensional effective theory). The resulting effective theory is a three-dimensional theory of static fields, with Lagrangian:

$$L_{\text{eff}} = \frac{1}{4}(F_{ij}^a)^2 + \frac{1}{2}(D_i A_0^a)^2 + \frac{1}{2}m_D^2(A_0^a)^2 + \lambda(A_0^a)^4 + \delta L, \tag{15.24}$$

with $D_i = \partial_i - ig\sqrt{T}A_i$. This strategy has been applied recently to the calculation of the free energy of the quark-gluon plasma at high temperature [114]. This technique of dimensional reduction puts special weight on the static sector (it singles out the contributions of the zero Matsubara frequency), and a major effort is being devoted to the calculation of the coefficients of the effective Lagrangian (which contain the dominant contribution to the thermodynamical functions).

15.7 Where Do the Perturbative Series Converge?

In Chapter 16 we will discuss QGP produced experimentally, and the region of temperatures we can hope to reach is never larger than three to four times the critical temperature T_c. The corresponding momenta are about $3T$, but $\alpha_s \sim 1/3$ or $g \sim 2$, and one can ask how good are perturbative series in this regime.

Due to the efforts of many people over many years, all the calculable terms are now known. Following [119, 120], I present the results for the free energy $F(T)$ (the same as Ω at zero density or μ):

$$
\begin{aligned}
F = d_A T^4 \frac{\pi^2}{9} \Bigg\{ &-\frac{1}{5}\left(1 + \frac{7d_F}{4d_A}\right) + \left(\frac{g}{4\pi}\right)^2 \left(C_A + \frac{5}{2}S_F\right) \\
&- \frac{16}{\sqrt{3}}\left(\frac{g}{4\pi}\right)^3 (C_A + S_F)^{\frac{3}{2}} - 48\left(\frac{g}{4\pi}\right)^4 C_A(C_A + S_F)\log\left(\frac{g}{2\pi}\sqrt{\frac{C_A + S_F}{3}}\right) \\
&+ \left(\frac{g}{4\pi}\right)^4 C_A^2 \left[\frac{22}{3}\log\left(\frac{\bar{\mu}}{4\pi T}\right) + \frac{38}{3}\frac{\zeta'(-3)}{\zeta(-3)} - \frac{148}{3}\frac{\zeta'(-1)}{\zeta(-1)} - 4\gamma_E + \frac{64}{5}\right] \\
&+ \left(\frac{g}{4\pi}\right)^4 C_A S_F \left[\frac{47}{3}\log\left(\frac{\bar{\mu}}{4\pi T}\right) + \frac{1}{3}\frac{\zeta'(-3)}{\zeta(-3)} - \frac{74}{3}\frac{\zeta'(-1)}{\zeta(-1)} - 8\gamma_E + \frac{1759}{60} + \frac{37}{5}\log 2\right] \\
&+ \left(\frac{g}{4\pi}\right)^4 S_F^2 \left[-\frac{20}{3}\log\left(\frac{\bar{\mu}}{4\pi T}\right) + \frac{8}{3}\frac{\zeta'(-3)}{\zeta(-3)} - \frac{16}{3}\frac{\zeta'(-1)}{\zeta(-1)} - 4\gamma_E - \frac{1}{3} + \frac{88}{5}\log 2\right] \\
&+ \left(\frac{g}{4\pi}\right)^4 S_{2F}\left[-\frac{105}{4} + 24\log 2\right] + O(g^5) \Bigg\}.
\end{aligned}
\tag{15.25}
$$

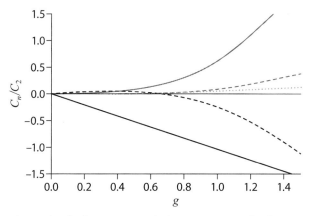

Figure 15.4. Ratio of subsequent perturbative corrections to the first one $O(g^2)$: g^3, $g^4\log(g)$, g^4, g^5, $g^6\log(g)$ are shown versus g by the soild thick, dashed thick, dotted, soild thin, and dashed curves, respectively.

Evaluated numerically for QCD with n_f quark flavors, this expression simplifies to the numerical form

$$
F = -\frac{8\pi^2 T^4}{45}\left\{1 + \frac{21}{32}n_f - 0.09499\,g^2\left(1 + \frac{5}{12}n_f\right) + 0.12094\,g^3\left(1 + \frac{1}{6}n_f\right)^{3/2}\right.
$$
$$
+ g^4\left[0.08662\left(1 + \frac{1}{6}n_f\right)\,\log\left(g\sqrt{1 + \frac{1}{6}n_f}\right) - 0.01323\left(1 + \frac{5}{12}n_f\right)\left(1 - \frac{2}{33}n_f\right)\,\log\left(\frac{\bar{\mu}}{T}\right)\right.
$$
$$
\left.\left. + 0.01733 - 0.00763\,n_f - 0.00088\,n_f^2\right] + O(g^5)\right\}.
\tag{15.26}
$$

For QED with n_f massless charged fermions with charges $q_i e$, the free energy is

$$
F = -\frac{\pi^2 T^4}{45}\left\{1 + \frac{7}{4}n_f - 0.07916\,e^2\sum q_i^2 + 0.02328\,e^3\left(\sum q_i^2\right)^{3/2}\right.
$$
$$
\left. + e^4\left[\left(-0.00352 + 0.00134\,\log\frac{\bar{\mu}}{T}\right)\left(\sum q_i^2\right)^2 + 0.00193\sum q_i^4\right] + O(e^5)\right\}.
\tag{15.27}
$$

With the coefficient of the perturbative results in hand, we can investigate where the perturbative series for the free energy is convergent. The first perturbative corrections $O(g^2)$, calculated by me in 1970s, were an order of magnitude smaller than the zeroth order, so the beginning of the perturbative series looked promising. The next corrections of the series $p/T^4 = \sum_n C_n g^n$ normalized to the first one C_2, are shown in Figure 15.4. The reader can see that the convergence happens at couplings $g < 0.8$, $\alpha_s = g^2/(4\pi) < 1/20$ or so, smaller than the values we need to compare to experiment, such as $g \sim 2$, $\alpha_s \sim 1/3$. So, as for many other real-world problems, the perturbation theory for weakly coupled QGP serves as a qualitative rather than a quantitative tool.

15.8 Hard Thermal Loop Resummations and the Quasiparticle Gas

The ring diagram resummation just discussed is an example in which hard momenta $p \sim T$ inside the small loops (polarization tensor) are different from softer scale $p \sim gT$ in the large loop. The program initiated and carried out by Braaten and Pisarski [115] was to identify all diagrams containing hard thermal loops (HTLs) and resum them, obtaining some effective Lagrangian (Frenkel and Taylor [115]) to be consistently used at this softer scale.

Instead of pursuing this rather technical discussion, note that it can be alternatively explained by classical kinetics. The fluctuations at the soft scale $k \sim gT \ll T$ can be described by classical fields, because the associated occupation numbers $N_k \sim T/E_k \sim 1/g$ are large. As emphasized by Blaizot and Iancu [116], these soft excitations can be described in terms of average fields that obey classical equations of motion. In QED these equations are the Maxwell equations:

$$\partial_\mu F^{\mu\nu} = j^\nu_{\text{ind}} + j^\nu_{\text{ext}} \tag{15.28}$$

with a source term composed of an external perturbation j^ν_{ext}, and an extra contribution j^ν_{ind} (referred to as the *induced current*), can be obtained using linear response theory

$$j^\mu_{\text{ind}}(x) = \int dy\, \Pi_{\mu\nu}(x-y)\, A^\nu(y), \tag{15.29}$$

which generalizes the usual dielectric and diamagnetic constants. To leading order in weak coupling, this polarization tensor is given by the one-loop approximation we discussed above, $\Pi \sim g^2 T^2 f(\omega/p)$.

In a non-Abelian theory, linear response is not sufficient: constraints due to gauge symmetry force us to take into account specific non-linear effects. The relevant generalization of the Yang-Mills equation reads [116] :

$$D_\nu F^{\nu\mu} = \Pi^{ab}_{\mu\nu} A^\nu_b + \frac{1}{2} \Gamma^{abc}_{\mu\nu\rho} A^\nu_b A^\rho_c + \cdots, \tag{15.30}$$

where the induced current on the r.h.s. is nonlinear: when expanded in powers of A^μ_a, it generates an infinite series of HTLs not only in self-energy but also vertices with any number of soft gluonic legs.

Physically, the need for resummation arises from the existence of collective excitations in the system, whose properties are not well captured by perturbation theory. One might hope that a "gas of quasiparticles" would do a better job of reproducing the results. The main spirit of such treatments is that quasiparticles have effective masses that are functions of g and T, which one could first evaluate perturbatively, and only then include their motion (e.g., [121]) and interactions [122].

As an example of such an approach I present a calculation by Blaizot et al. [123], which will be compared to lattice data. For technical reasons, they prefer to focus on entropy $s(T)$ rather than on, say, $p(T)$, which at least is not defined up to an arbitrary constant (the bag term).

The entropy density of their quasiparticle gas is defined as the entropy of the ideal gas of quasiparticles:

$$s = \int \frac{d^3k}{(2\pi)^3} \left[(1 + N_k) \log(1 + N_k) - N_k \log N_k \right], \tag{15.31}$$

where $N_k = \frac{1}{e^{E_k/T}-1}$, and $E_k = \sqrt{M^2(T) + k^2}$. Note that at this stage, the (T-dependent) effective quasiparticle mass $M(T)$ is not yet defined.

The pressure p and the energy density ϵ are then given by the corresponding ideal gas expressions, corrected by a function $B(T)$, that has been adjusted to satisfy thermodynamical identities. That is, one defines the pressure and energy density as

$$p = -T \int \frac{d^3k}{(2\pi)^3} \log\left(1 - e^{-E_k/T}\right) - B(T) \tag{15.32}$$

$$\epsilon = \int \frac{d^3k}{(2\pi)^3} N_k E_k + B(T). \tag{15.33}$$

Such a parametrization obviously fulfills the identity $\epsilon + p = Ts$. The function $B(T)$ is then determined by requiring that $s = \frac{dp}{dT}$.

The effective mass used is the HTL one. The quasiparticle gas can be viewed as an approximation to a more general approach known as *hard thermal loop perturbation theory* [122], in which one adds the HTL mass term to the tree-level Lagrangian and then subtracts it from the perturbation part.

Blaizot and Iancu noticed that for a special case of the entropy, the calculations can be pushed one order further, to make better "Φ-derivable" self-consistent approximations. The stationarity property of the free energy entails important simplifications in the calculation of the entropy. Indeed, we have

$$S = -\frac{dF}{dT} = -\frac{\partial F}{\partial T}\bigg|_D, \tag{15.34}$$

where the last derivative is at fixed propagator. Due to that expression the temperature dependence of the propagator can be ignored here. Explicitly we get:

$$S = -\int \frac{d^4k}{(2\pi)^4} \frac{\partial N(\omega)}{\partial T} \operatorname{Im} \log D^{-1}(\omega, k) + \int \frac{d^4k}{(2\pi)^4} \frac{\partial N(\omega)}{\partial T} \operatorname{Im}\Pi(\omega, k) \operatorname{Re} D(\omega, k) + S' \tag{15.35}$$

where $N(\omega) = 1/(e^{\beta\omega} - 1)$, and

$$S' \equiv -\frac{\partial (T\Phi)}{\partial T}\bigg|_D + \int \frac{d^4k}{(2\pi)^4} \frac{\partial N(\omega)}{\partial T} \operatorname{Re}\Pi \operatorname{Im} D.$$

At two-loop order in the skeleton expansion, the entropy then takes the simple form

$$S = -\int \frac{d^4p}{(2\pi)^4} \frac{\partial N}{\partial T} \left\{ \operatorname{Im} \log D^{-1} - \operatorname{Im}\Pi \operatorname{Re} D \right\}. \tag{15.36}$$

The simplifications discussed here have led to important cancellations, leaving an expression for the entropy that is effectively a one-loop expression, thus emphasizing the direct relation between the entropy and the quasiparticle spectrum. Residual interactions start contributing at order three-loop. This expression is also manifestly UV-finite.

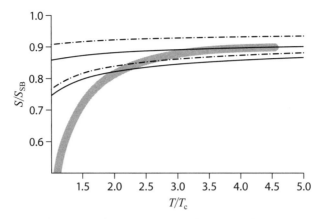

Figure 15.5. The entropy of pure $SU(3)$ gauge theory normalized to the Stephan-Boetzmann ideal gas entropy S_{SB}. Solid lines: S_{HTL} from Blaizot et al. [123]. Dashed-dotted lines: S_{NLA}. For each approximation, the two lines correspond to the choices $\bar{\mu} = \pi T$ and $\bar{\mu} = 4\pi T$ of the running coupling constant $\alpha_s(\bar{\mu})$. The dark gray band represents the lattice results.

In the regime where the loop momenta are soft, as an approximation for Π we can use the corresponding HTL. As an illustration of the quality of the results obtained in this case, Figure 15.5 shows the entropy of pure $SU(3)$ gauge theory (normalized to the ideal gas limit). The agreement with the parameterized lattice results at large T ($T \sim 2.5 T_c$) is quite good.

However, a radically different approach—the strong coupling limit of $N = 4$ SUSY theory [197]—gives the value $s/s_0 = 3/4$, which is also not very far off. So it is difficult to conclude at the moment which of the two limits is more appropriate at $T \sim (2 - 3) T_c$.

15.9 Lattice Gauge Theory Simulations at Finite Temperature

We have already discussed the basic lattice settings and Wilson action: their main feature is exact gauge symmetry. Let me briefly go though the main scales of the problem. The lattice input is the value of the coupling $g(a)$ at its UV-end scale a, the lattice step. (Typically $a \sim 0.1\,\mathrm{fm} \sim 1/(2\,\mathrm{GeV})$.) It is assumed that in the UV regime the theory is weakly coupled, and so the standard gauge action is adequate. The size of the lattice $L = Na$ provides the IR cutoff: typically $\sim 3\,\mathrm{fm}$, $N \sim 30$. It is assumed that the scales of the physics phenomena lie in between the two:

$$a \ll \text{physical scale} \ll L. \tag{15.37}$$

From the numbers I have mentioned, one can see that "much less" in reality means a factor of $\sqrt{N} \sim 5$.

In this relatively narrow interval of scales, the effective coupling constant g runs, as prescribed by the RG flow, from weak in UV to strong in IR. As we discussed for

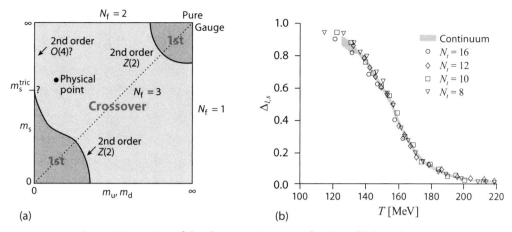

(a)   (b) T [MeV]

Figure 15.6. (a) Map of the phase transitions as a function of light and strange quark masses. (b) The temperature dependence of the quark condensate in high-T QCD. Adapted from Soltz et al. [125].

the original Creutz string-tension calculation (see Figure 14.3), it is indeed possible[2] to see a transition from the weak to the strong coupling regime and have a reasonably accurate representation of nonperturbative physics, such as confinement and hadron formation.

Numerical simulations of lattice gauge have become a large enterprise, occupying a good portion of the largest supercomputers. It took decades to develop adequate algorithms, correct quark fields respecting the chiral symmetry, and perform actual calculations. One major achievement is that by now these simulations reproduce most of the hadronic phenomenology, such as hadronic masses, reasonably well. However, discussion of these simulations will take us too far from our main interests. Therefore, I only briefly discuss the finite-T results (for a review, see [125]).

The main fact is the presence of the QCD transition, from the hadronic to the QGP phase, at $T_c \approx 155 \pm 5$ MeV. Figure 15.6a shows the order of this transition as a function of quark masses. The physical point is outside all regions of the singular phase transitions. Thus it is a smooth crossover. This is also seen in Figure 15.6b which shows the dependence of the quark condensate on T. It shows that the transition is a smooth curve.

Figure 15.7 shows a compendium of the thermodynamical quantities, energy density, pressure, and entropy. The first two are also shown as the "interaction measure" $\epsilon - p$. The inflection point of the $p(T)$ curve (or the maximum of the specific heat, the second derivative of p) is usually used as a definition of T_c.

These lattice results are used as the equation of state in heavy-ion physics applications. Unfortunately, direct lattice simulations with nonzero chemical potentials are not

---

[2] Compare this range to that for Cooper pairing, in examples we studied before: it takes typically many decades of scale change in the RG flow to make the interaction strong enough for pair binding. Fortunately, for the gauge theories the rate of the RG flow is enhanced by the rather large coefficients involved, and therefore less than one order of magnitude is enough.

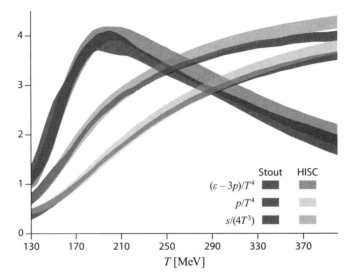

Figure 15.7. The continuum-extrapolated thermodynamical observables versus temperature, from lattice gauge theory simulations. Different shades of gray correspond to different lattice discretisations used: their spread gives an idea of the current accuracy of these calculations. HISC and stout are two technical terms describing different formulations of the quark action. Adapted from Soltz et al. [125].

possible, because the path integral would then have to be taken using a complex weight. The current practical solution is calculating several (currently six) derivatives over μ at $\mu = 0$ and then using a Taylor series in μ / T. The coefficients of the expansion—higher susceptibilities—are related with many-body event-by-event fluctuations, so they can be compared individually to the actual data.

16 | Quark-Gluon Plasma in Experiment

Quark-gluon plasma (QGP) is the densest form of matter that has been created in the laboratory. It has not been seen in our Universe since milliseconds after the Big Bang. While the QCD phase transition is perhaps not the earliest in cosmology—it is expected to be preceded by the electroweak phase transition at temperatures three orders of magnitude higher—it is the only one we can study experimentally. As we will discuss, this form of matter has rather unusual properties.

And still, from the theoretical perspective, it is the simplest form of strongly interacting matter, its normal high-T phase. The quasiparticles in it have quantum numbers of quarks and gluons, the same fields that appear in the QCD Lagrangian. The super phases, possessing nontrivial condensates, are the ones at $T < T_c$, and those we will discuss in Chapters 17–20.

In the next section we briefly discuss the history of the field, its main facilities and collaborations. Then we revisit the phase diagram in Section 16.2, this time from the experimental point of view. In Section 16.3 we focus on collective flow phenomena, which convinced the community that a tiny exploding QGP fireball, of the size of only a few femtometers, is truly a macroscopic object, and that the laws of thermodynamics and hydrodynamics can indeed be used for its description. Relativistic hydrodynamics is the subject of Section 16.4.

Section 16.5 considers perturbations propagating on the surface of the fireball (sound waves) and compares them with the waves that formed just after the Big Bang. The cascading sounds in the Big Bang and possible observations of them via gravitational waves are discussed in Section 16.6.

16.1 Experimental Quest for QGP

In the 1990s, an experimental program, using high-energy heavy ion collisions for the production of highly excited hadronic matter, had been formed, at the intersection of high-energy and traditional nuclear physics. It originated from such well-established centers of nuclear physics as Lawrence Berkeley Laboratory (Berkeley, California) and Gesellschaft für Schwerionenforschung (Darmstadt, Germany), but its main centers became two laboratories: Brookhaven National Laboratory (BNL) in Upton, New York, and the European Center for Nuclear Research (CERN) in Geneva, Switzerland.[1]

The first dedicated facility, the Relativistic Heavy Ion Collider (RHIC), had its first run in the summer of 2000. It is a collider with $100 + 100$ GeV/N in the heavy ion mode and up to 250+250 in the pp mode. The two main experiments are PHENIX and STAR.[2] The community of heavy ion physicists is now also working at the world's largest collider, the LHC at CERN. Two high-energy physics experiments—CMS and ATLAS—both have their heavy ion subgroups, and there is also a dedicated heavy ion experiment, ALICE. The LHC heavy ion mode is PbPb , started in 2010. The field thus has a firmly established experimental base at least for another one or two decades or so.

After having considered the properties of highly excited hadronic matter theoretically in the previous chapter, in this one we review which of these properties have already been observed in real-time experiments. But before we get to specifics, we have to answer the following general questions:

- Is hadronic matter really produced in heavy ion collisions? Or, to put it more quantitatively, how do we know that a fireball produced by heavy ions is large enough to be treated macroscopically? Another take on this is: If one switches to smaller nuclei, or pA, or pp collisions with variable multiplicity bins, at what point do we see a transition to a microscopic system? What is going on during "elementary" pp collisions at comparable energies? What is the smallest drop of QGP that is still QGP and not just a collection of a few gluons and quarks?

- Does the macroscopic treatment of the explosion confirm the theoretical predictions for the thermodynamics of QGP? Is it true that it is near-scale independent, or $\epsilon \sim p \sim T^4$, as one finds on the lattice for $T \gg T_c$? The answer is definitely "yes," hydrodynamics based on lattice equations of state works very well.

- Does the macroscopic treatment of the explosion confirm the perturbative predictions for QGP kinetics? Note that lattice simulations are restricted to Euclidean time and have problems deriving kinetic (time dependent) quantities. So there is no first-principle nonperturbative derivation of the viscosity. Diagrammatic derivation of it has failed dramatically.

For a general review of the current status of heavy ion physics see, for example [147].

---

[1] Both were home to many Nobel-Prize-level discoveries in high-energy physics, and heavy ion physics initially had difficulties establishing itself at these institutions. CERN for example, now has a dedicated 1-month-a-year slot, which is quite adequate. Taking into account the high cost of the colliders and detectors, using them for diverse science objectives is amply justified.

[2] Originally PHENIX was supposed to focus on the electromagnetic probes (photons and dileptons), and STAR on hadrons. But both can do hadronic physics, although differently, and lately STAR has added a large electromagnetic calorimeter. A new version of PHENIX is under construction now, to focus on jets.

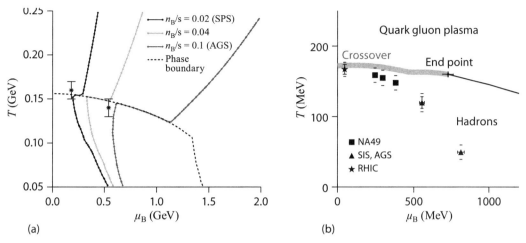

Figure 16.1. (a) Example of the adiabatic cooling path. (b) Compilation of experimental data on chemical freezeout parameters from different experiments, together with the phase transition line and the hypothetical tricritical point.

16.2 Mapping the Phase Diagram

We already have discussed the QCD phase diagram in a $T-\mu$ plot. Two views of it, plotted on the temperature T–chemical potential for baryon charge μ_B plane, are shown in Figure 16.1.

The first of them is the theoretical view of how the cooling paths should look in this plane. If the resultant system is macroscopically large and its evolution is sufficiently slow on the scale of microscopic reactions, its paths on the phase diagram follow *adiabatic* lines, on which entropy is kept constant. Since the baryon number is also conserved, the ratio of both densities, n_B/s, is constant along its adiabatic line. Examples of such lines are shown in Figure 16.1a. Note the characteristic zigzag shape, which appears because the lines with the same n_B/s value cross the phase transition line at two different points.[3] The equation of state used in hydrodynamical models should be calculated along such adiabatic paths.

In real collisions, these adiabatic lines should have well-defined starting and ending points. Let me start with the latter, related to the "freezeout" phenomenon. Both in heavy ion collisions (the "little bang") and in cosmology (the Big Bang), expansion of the system causes densities to decrease to the point where certain reaction rates no longer keep up with the expansion rate: at that point the reaction products are "frozen." In heavy ion collisions about a dozen different species of secondaries are produced, mesons $\pi, K, \rho, \omega, ...$, baryons $N, \Lambda, \Sigma, \Xi, \Omega, ...$, and antibaryons, plus light nuclei $d, t, {}^3\text{He}, {}^4\text{He}, ...$, hypernuclei, and antinuclei. Without going into any detail, let me just state that rather accurate parameterizations of all these particle yields can be

---

[3] One may then naively worry that they got mixed up inside the line: this is not so, the constant ratio n_B/s distinguishes curves from each other in the mixed phase, although all of them collapse to them same line on this particular plane.

well described by three parameters: T_{ch}, μ_{ch}, V, where the subscript means chemical freezeout, and V is the effective volume of the system.

As a result, we are quite confident about mapping the resulting data on the phase diagram; some points are shown in Figure 16.1b, together with lattice-based estimates about the crossover line. One can see that all points are safely on the hadronic side of the transition, although quite close to it.

Chemical freezeout determines the observed particle composition, but in fact elastic collisions continue, and the so-called "kinetic freezeout" happens later. The largest systems have $T_f \approx 90 - 100\,\text{MeV}$, approximately half the critical temperature T_c.

Now, let us return to the question of the beginning of the phase diagram paths. Marking them is not so well defined, but it is clear that one needs a certain accuracy in the definitions of T and μ to mark the point on the phase diagram. Basically all we know depends on the applicability of hydrodynamics as a description of the explosion: a consensus is that the paths begin (proper) time $\tau \sim 0.5\,\text{fm}$, at which the initial T is about twice the critical temperature.

16.3 Collective Flow and Viscosity of QGP

For heavy ion collisions at RHIC and LHC, we do have impressive evidence for macroscopic behavior. This behavior manifests as *flows*, which are various forms of a collective expansion. Flows are observed at all energies, and the first data from RHIC, since its start in 2000, have demonstrated this behavior in most spectacular way. The accuracy of its hydrodynamical description is getting even better at the highest energies of LHC, which started operations in 2010.

There are three basic forms of these flows:

- *the radial flow*, in the direction normal to the beam, it is maximal for central collisions (which are axially symmetric, or zeroth harmonics in ϕ),
- *the elliptic flow* or the second harmonics v_2, and
- *higher harmonics of the flow*, $v_m = \langle \cos(m\,\phi) \rangle$ with $m > 2$, where ϕ is measured around the beam axis with respect to the impact parameter direction $\vec{b}$.

The radial flow in Pb-Pb (SPS, LHC) or Au-Au (RHIC) collisions is well documented and is explained by ideal (inviscid) hydrodynamics. Viscosity plays very little role. Matter emerges from the collision with a collective transverse velocity of up to $.8\,c$. Its profile and magnitude is now firmly established from a combined analysis of particle spectra, femtoscopy correlations, and deuteron coalescence, among other observables.

The elliptic flow in noncentral collisions develops for geometric reasons. The ellipticity of the almond-shaped overlap region defines the initial ellipticity ϵ_2, and hydrodynamics yields v_2/ϵ_2 predictions. So nuclear geometry defines uniquely the centrality dependence of the elliptic flow. It is a very important QGP signature.

The higher harmonics of the flow are produced by fluctuations of the initial matter distribution. Since fluctuations of individual nucleons at different locations in the transverse plane cannot be correlated, the problem reduces to the sound waves from local—delta-function like—perturbations.

In short, the sound waves looks like bunch of circles on the surface of a pond when a dozen pebbles are thrown into it. As everyone knows, the circle radii (linearly) increase with time, and they eventually disappear. Imagine, however, that suddenly the weather changes, the temperature drops rapidly, and the pond freezes so quickly that the the circles on its surface remain imprinted in ice. This sounds like mere fantasy, but it does happen in reality, and not only in heavy ion collisions but also in Big Bang cosmology as well. With proper instruments, one does observe frozen sound circles in the sky!

Any v_m are measured as a function of transverse momenta, particle type, impact parameter, and collision energy. Furthermore, each can be obtained from 2, 4, 6, or 8 particle correlations. This wealth of data described by the theory left no doubt that there is collectivity for all distributions, meaning all secondaries have those harmonics. This fact severely constrains viable models of heavy ion collision, excluding many exotic possibilities. Hydrodynamic explosion remains the only option. The description of higher-order moments v_m, $m = 3, 4, 5, 6$ by hydrodynamics is its true triumph.

I will not discuss the details. Let me just mention that higher harmonics are basically sound modes, and from their damping we can deduce the value of the shear viscosity, which is in the range[4]

$$\frac{\eta}{s} \approx \left(\frac{1}{4\pi}\right)(1.75 \pm 0.25). \tag{16.1}$$

Perturbative kinetics states that viscosity (as for many other kinetic coefficients) is proportional to the mean free path $l_{\mathrm{mfp}} = 1/(n\sigma) \sim (1/T)(1/\alpha_s^2)$, and thus

$$\frac{\eta}{s} \sim \frac{1}{\alpha_s^2} \gg 1 \tag{16.2}$$

for weak coupling $\alpha_s \ll 1$, which is in clear disagreement with the data (16.1).

Note the similarity of the situation for QGP and trapped unitary Fermi gases: both are scale invariant, both exhibit strong coupling, both have thermodynamics well described by nonperturbative simulations, and yet their kinetic coefficients are not described by ordinary kinetics. Even the values for η/s are similar!

16.4 Relativistic Hydrodynamics

If in high-energy collisions of hadrons and/or nuclei a macroscopically large, excited system is produced, its expansion and decay can be described by relativistic hydrodynamics. Its first application for such purposes was made in the classical work by L. D. Landau in 1953, following the pioneering statistical model by Fermi of 1951 and an important paper by Pomeranchuck in 1952 (who has pointed out that even if Fermi is right and the

---

[4] The value in the first bracket, $1/(4\pi)$, is the infinite coupling limit in certain supersymmetric theories that have gravity dual. We will not discuss this duality in this book, so it is to be taken just as some suggested theoretical limit.

conditions he predicted do happen, it cannot be the final stage of the collisions, since strong interaction in the system will persist. Pomeranchuck proposed some universal freezeout temperature T_f.)

Initially, Laudau theory was applied to pp Collisions, But after disscussion of transverse flow [148], it became clear that one needs to collide larger systems to see whether it is applicable. Often hydrodynamics is considered to be a consequence of the kinetic equations. In fact, the stronger the interaction in the system is, the more accurately hydrodynamics models the system. The kinetic approach (using the kinetic equations), on the contrary, was never formulated except for weakly interacting systems.

The conceptual basis of hydrodynamics is not the energy-momentum-particle number conservation, as is often stated: those conservation laws are always valid. In contrast, hydrodynamics may also be valid for sufficiently large systems with sufficiently strong interactions and small mean free path. It assumes that the stress–energy and particle currents can be expressed in *gradient expansion* form, with the zeroth order—ideal inviscid hydrodynamics—using local expressions only, without any gradients. Then a set of local conservation laws for the stress tensor ($T^{\mu\nu}$) and for the conserved currents (J_i^μ)

$$\partial_\mu T^{\mu\nu} = 0 \tag{16.3}$$

$$\partial_\mu J_i^\mu = 0$$

is used for the thermally equilibrated $T^{\mu\nu}$ and J_i^μ, which are related to the bulk properties of the fluid by the relations

$$T^{\mu\nu} = (\epsilon + p)u^\mu u^\nu + pg^{\mu\nu} \tag{16.4}$$

$$J_i^\mu = n_i u^\mu.$$

Here ϵ is the energy density, p is the pressure, n_i is the number density of the corresponding current, and $u^\mu = \gamma(1, \mathbf{v})$ is the proper velocity of the fluid. In strong interactions, the conserved currents are isospin (J_I^μ), strangeness (J_S^μ), and baryon number (J_B^μ). For hydrodynamic evolution, isospin symmetry is assumed and the net strangeness is set to zero; therefore, only the baryon current J_B is considered below.

To close this set of equations, we also need the equation of state $p(\epsilon)$. We will also use two thermodynamic differentials,

$$d\epsilon = T ds \qquad dp = s dT, \tag{16.5}$$

and the definition of the sound velocity,

$$c_s^2 = \frac{\partial p}{\partial \epsilon} = \frac{s}{T}\frac{\partial T}{\partial s}, \tag{16.6}$$

and that $\epsilon + p = Ts$. Using these equations and the thermodynamic relations in the form

$$\frac{\partial_\mu \epsilon}{\epsilon + p} = \frac{\partial_\mu s}{s}, \tag{16.7}$$

one may show that these equations imply another nontrivial conservation law, namely, the conservation of entropy:

$$(s u^\mu)_{;\mu} = 0, \tag{16.8}$$

where the semicolon indicates a covariant derivative.

Realistic solutions of hydrodynamic explosions has been found numerically, and they describe very well the observed particle spectra and even the harmonics of the angular correlations. I will not describe here any specific examples.

Instead I focus on two exercises, which have some pedagogical value.

Let me remind the reader of a few general formulas valid in an arbitrary coordinate system. These equations of motion can be written as

$$T^{mn}{}_{;m} = 0, \qquad j^m{}_{;m} = 0. \tag{16.9}$$

For tensors of rank 1 and 2 these equations read explicitly

$$j^i{}_{;p} = j^i{}_{,p} + \Gamma^i_{pk} j^k, \tag{16.10}$$

$$T^{ik}{}_{;p} = T^{ik}{}_{,p} + \Gamma^i_{pm} T^{mk} + \Gamma^k_{pm} T^{im}, \tag{16.11}$$

where the comma denotes a simple partial derivative, and the Christoffel symbols Γ^s_{ij} are given by derivatives of the metric tensor $g^{ab}(x)$:

$$\Gamma^s_{ij} = (1/2)g^{ks}\left(g_{ik,j} + g_{jk,i} - g_{ij,k}\right). \tag{16.12}$$

In high-energy collisions, we should first worry about the longitudinal motion, so we discuss the one-dimensional case first. As an example, let us do the following transformation from Cartesian to light-cone coordinates:

$$x^\mu = (t, x, y, z) \longrightarrow \bar{x}^m = (\tau, x, y, \eta) \tag{16.13}$$

$$t = \tau\cosh\eta \qquad \tau = \sqrt{t^2 - z^2}$$

$$z = \tau\sinh\eta \quad \eta = (1/2)\log\frac{t+z}{t-z}. \tag{16.14}$$

In the new coordinate system, the velocity field (after inserting $v_z = z/t$) is given by

$$\bar{u}^m = \bar{\gamma}(1, \bar{v}_x, \bar{v}_y, 0), \tag{16.15}$$

with $\bar{v}_i \equiv v_i\cosh\eta$, $i = x, y$, and $\bar{\gamma} \equiv 1/\sqrt{1-\bar{v}_x^2-\bar{v}_y^2}$.

Now we transform the metric

$$ds^2 = g_{\mu\nu}dx^\mu dx^\nu = dt^2 - dx^2 - dy^2 - dz^2$$

$$= d\tau^2 - dx^2 - dy^2 - \tau^2 d\eta^2 \tag{16.16}$$

and therefore

$$g_{mn} = \begin{pmatrix} 1 & 0 & 0 & 0 \\ 0 & -1 & 0 & 0 \\ 0 & 0 & -1 & 0 \\ 0 & 0 & 0 & -\tau^2 \end{pmatrix},$$

(16.17)

The only nonvanishing Christoffel symbols are

$$\Gamma^{\eta}_{\eta\tau} = \Gamma^{\eta}_{\tau\eta} = \frac{1}{\tau}, \qquad \Gamma^{\tau}_{\eta\eta} = \tau.$$

(16.18)

Bjorken flow. The solution is independent of x, y, η and depends on τ only. It can be easily calculated using such coordinates. The coordinates are co-moving with the fluid, $u^\mu = (1, 0, 0, 0)$, so entropy conservation has only a time covariant derivative (with $\Gamma^{\eta}_{\tau\eta}$)

$$\frac{ds}{d\tau} + \frac{s}{\tau} = 0,$$

(16.19)

from which one gets the solution $s\tau = const.$

Another useful analytic solution, known as Gubser's flow, describes the axially symmetric collision of objects of a particular shape, with conformal equations of state. See the original papers or appendix B of [147].

16.5 The Little Bang versus the Big Bang: Sound Circles and Fluctuations

Let me emphasize several amazing analogies between heavy-ion physics and the Big Bang cosmology.

First of all, the same hot/dense hadronic matter, which heavy-ion physics is trying to produce in the laboratory, was present at an appropriate time after the Big Bang. More specifically, in both cases the matter goes through the QCD phase transition, from the QGP. This occurred at about 5×10^{-5} after the Big Bang.)

The first obvious similarities between Little Bangs and the cosmological Big Bang is that both are violent explosions. The fireball created in the heavy-ion collision of course explodes, as its high pressure cannot be contained: this is the Little Bang referred to in the title of this section. The Big Bang is a cosmological explosion, creating the Universe, which proceeded against the pull of gravity. So, the Little Bang is a laboratory simulation of a particular stage of the Big Bang.

Expansion of the created hadronic fireball approximately follows the same Hubble law as its bigger relative, $v(r) = Hr$ with $H(t)$ being some time-dependent parameter, although the Little Bang has a rather anisotropic (tensorial) H. However, by the end of the expansion, the anisotropy is nearly absent, and local expansion at freezeout at RHIC is nearly Hubble-like.

The final velocities of collective motion in the Little Bang are measured in spectra of secondaries: the transverse velocities now are believed to be reasonably well known, and

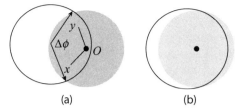

Figure 16.2. The perturbation is shown by the small filled circle at point O: its time evolution to points x and y is described by the Green function of linearized hydrodynamics, shown by two lines. The perturbed region—shown by the gray circle—is inside the sound horizon. The sound wave effect is maximal at the intersection points of this area with the fireball boundary: $\Delta\phi$ angle is the value at which the peak in two-body correlation function is to be found. Shifting the location of the perturbation, from (a) to (b), results in a rather small shift in $\Delta\phi$.

are not small, reaching about $0.7c$ at RHIC. For the Big Bang, there is no end and it proceeds even today: the current value of the Hubble constant, $H(t_{now})$ is, after significant controversies for years, reasonably well measured.

The next important phenomenon, present in both problems, is propagation of *collective excitations* on top of the smooth average geometry, which are *sound waves*. If the reader is not impressed by their presence in the Little Bang, let me point out that the system we are discussing have only $O(1,000)$ particles. If this sounds a lot, recall that we are in three dimensions and that $1000^{1/3} = 10$ particles in each dimension. And we still speak about sounds propagating inside this system!

Both in Little and Big Bangs, the sources of sounds are some fluctuations created when the system was initiated. In the former case we study harmonics of the azimuthal angle, deduced from two (and more) particle correlations

$$v_n = \langle \cos(n\phi) \rangle. \tag{16.20}$$

In the Big Bang there are angular harmonics C_{lm} of the temperature distribution over the sky.

To calculate perturbation evolution at later times, one needs to calculate the Green functions from the original location O to observation points x and y, as shown in Figure 16.2a. That was first done in [198] analytically for Gubser flow. One finds that the main contributions come from two points where the sound circle intersects the fireball boundary; see a sketch in Figure 16.2. In a single-body angular distribution, those two points correspond to two excesses of particles in the corresponding two directions.

(To explain what we are trying to do here, perhaps an analogy will help. Imagine an earthquake has happened, say, in Japan. We cannot for some reason see directly where and what has happened, but we can measure the amplitudes of the surface waves along the American Pacific coast. Correlation functions of these waves can reveal the location and timing of the earthquake.)

Extra particles will move predominantly along the radial flow at these two locations. The angle $\Delta\phi$ between them is about $120°$ or 2 rad. The correlation function, calculated in [198], is shown in Figure 16.3a. One of its feature is a peak at $\delta\phi = 0$: it is generated if both observed particles come from the same intersection point of the sound circle and the edge

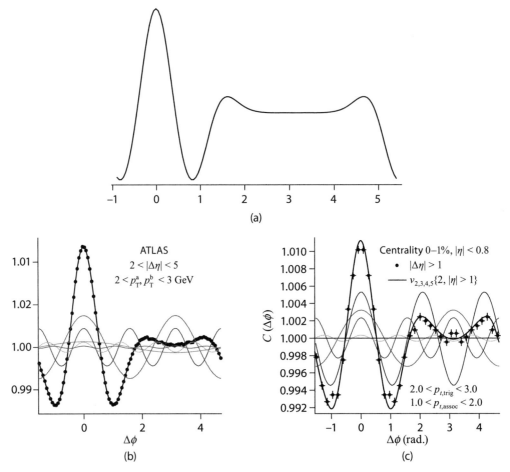

Figure 16.3. (a) Calculated two-pion distribution as a function of azimuthal angle difference $\Delta\phi$, for viscosity-to-entropy ratios $\eta/s = 0.134$. Adapted from Staig and Shuryak [198]. Experimental data for ultra-central collisions from LHC collaboration ATLAS and ALICE are shown in (b) and (c). Plots are from J. Jia (for ATLAS Collaboration), arXiv:1107.1468; and K. Aamodt et al. (ALICE Collaboration) Phys. Rev. Lett. 107, 032301 (2011), arXiv:1105.3865, respectively.

of the fireball. When two observed particles come from different points, one finds two peaks, at $\Delta\phi = \pm 2$ rad. This calculation had been presented on the first day of the Quark Matter conference at Annecy, France, *before* the experimental data were reported: these data from from LHC collaborations ATLAS and ALICE are shown in Figure 16.3bc. The agreement of the shape of the angular correlation is quite stunning, taking into account that it is just a single perturbation from a point source, and that in order to make the calculation analytic, certain approximations—a conformal QGP equation of state and a somewhat unrealistic nuclear shape—had to be made.

Extensive comparison of this expression with the heavy ion collision (AA) data, from central to peripheral, has been done in [199] from which we borrow Figure 16.4. Panel a in this figure shows the well known centrality dependence of the elliptic and triangular flows. v_2 is small for central collisions due to smallness of ϵ_2, and is also small for very peripheral

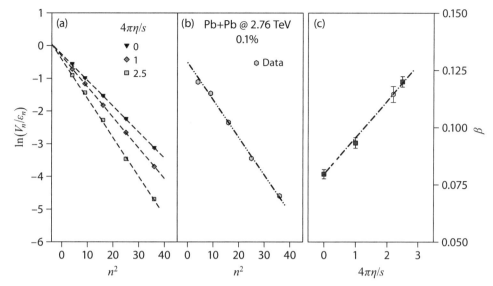

Figure 16.4. (a) ATLAS data for angular harmonics represented by $\ln(v_n/\epsilon_n)$ versus harmonic number squared n^2 from viscous hydrodynamical calculations for three values of specific shear viscosity as indicated. (b) $\ln(v_n/\epsilon_n)$ versus n^2 for Pb+Pb data. The $p_\perp$-integrated v_n results in panels a and b are from ATLAS 0.1% central Pb+Pb collisions at sNN = 2.76 TeV; the curves are linear fits. (c) exponent versus viscosity-to-entropy ratio $4\pi/s$ for curves shown in panels a and b. Adapted from Lacey et al. [199].

bins, because viscosity is large in small systems. Figure 16.4b shows the $\ln(v_n/\epsilon_n)$, which, according to the formula, is the exponent. As a function of the inverse system's size $1/R$, both elliptic and triangular flows show perfectly linear behavior. Further features—the n^2 dependence as well as the linear dependence of $\log(v_m/\epsilon_m)$ on viscosity value—are also very well reproduced; see Figure 16.4. Note that this expression works all the way to rather peripheral heavy ion collisions with $R \sim 1\,fm$ and multiplicities comparable to those in the highest pA binds. It also seems to work up to the largest n so far measured.

So, the acoustic damping provides correct systematics of the harmonic strength. This increases our confidence that—in spite of the somewhat different geometry—the perturbations observed are actually just a form of sound waves.

The observed hadrons (like microwave cosmic photons) are seen at the moment of their last interaction, or as it is termed technically, at their *freezeout stage*. The next comparison I make here deals with the issue of fluctuations. Very impressive measurements of the microwave background anisotropy made in the past decade have taught us a lot about cosmological parameters. First the dipole component was found (stemming from the motion of the Solar system relative to the microwave heat bath), and then the very small $(\delta T/T \sim 10^{-5})$ chaotic fluctuations of T originating from plasma oscillations at the photon freezeout. Recently it has been possible to measure some interesting structures in fluctuations of the cosmic microwave background, with angular momenta $l \sim 300$. The theoretical predictions for these are available, since they are related to primordial plasma-to-gas neutralization transition at temperature $T \sim 1/3\,$eV. Primordial fluctuations of

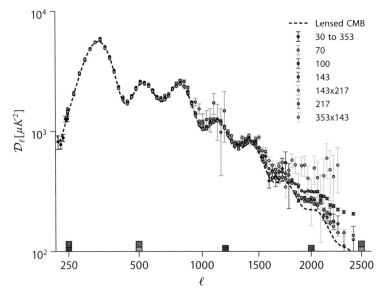

Figure 16.5. Power spectrum of cosmic microwave background radiation measured by the Planck collaboration. The plot is from the Planck Collaboration [200].

sufficiently long wavelength are attenuated by gravitational instability, until they reach the stabilization moment $\tau_{\text{stabilization}}$ when the instability changes to a regime called *Sakharov acoustic oscillations*. Hydrodynamics tells us that fluctuations disperse during this period with the (current) speed of sound, reaching the *sound horizon* scale

$$r_{\text{s.h.}} = \int_{\tau_{\text{stabilization}}}^{\tau_{\text{observation}}} c_s(t)\,dt. \tag{16.21}$$

This scale is the physical size of the spots observed and correspond to a peak observed on the sky; see Figure 16.5 from [200]. The so-called acoustic maxima and minima appear because each harmonic has its own oscillation frequency, but they all share the same freezeout time: therefore, the phase of the oscillation at which each was frozen changes.

For the same reason some calculations for the Little Bang, shown in Figure 16.6, show similar oscillations. Although the results are suggestive, so far we cannot say that we see acoustic maxima or minima in the Little Bang. What we do see is a decrease with the harmonic number, which provides the magnitude of the viscosity.

16.6 Phase Transitions in the Big Bang: Sound Cascades and Gravitational Waves

We think that our Universe was "boiling" at its early stages at least three times: (1) at the initial equilibration, when entropy was produced, and at the (2) electroweak and (3) QCD phase transitions. On general grounds, these events should have produced certain out-of-equilibrium effects. It remains a great challenge to us to observe their

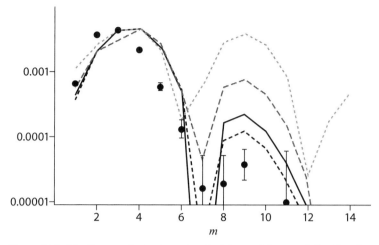

Figure 16.6. The lines are hydrodynamic calculations of the correlation function harmonics, v_m^2, based on a Green function from a point source for four values of viscosity $4\pi\eta/s = 0$, 1, 1.68, 2 (top to bottom at the right). The filled circles are the ATLAS data for the ultra-central bin. Reprinted from [198].

consequences experimentally, or at least evaluate their magnitude theoretically. The first question is: What observable do we have a chance of seeing through the subsequent evolution, back to the early moments of the Big Bang?

From the onset of QGP physics in heavy-ion collisions an especially important role has been attributed to the "penetrating probes": production of photons/dileptons. In this section (only) we will jump from our customary Little Bang to a discussion of the Big Bang. So it is quite logical to start with a question of whether the Big Bang also possesses some kind of a penetrating probe. The answer is quite clear: while the electromagnetic and weak interactions are in this case not weak enough, the gravitational one is. Thus the only penetrating probe of the Big Bang are *gravitational waves* GWs.

Thirty years ago Witten had discussed the cosmological QCD phase transition, assuming it to be of first order: he pointed out bubble production and coalescence, producing inhomogeneities in the energy distribution, and he mentioned production of GWs. Jumping many years to recent time, mention that Hindmarsh and colleagues recently found the hydrodynamical sound waves to be the dominant source of the GWs, while doing numerical simulations of (a variant of) the electroweak phase transition in the first-order transition setting.

Kalaydzyan and I [149] discussed sound-based GWs production and argued that generation of the cosmological GWs can be divided into three distinct stages, each with its own physics and scales. I list them starting from the UV end of the spectrum $k \sim T$ and ending at the IR end of the spectrum $k \sim 1/t_{life}$ cut off by the age of the Universe:

1. the production of the sounds,
2. the inverse cascade of the acoustic turbulence, moving the sound from UV to IR, and
3. the final transition from sounds to GWs.

Stage 1 remains highly nontrivial to analyze, associated with the dynamical details of the QCD (or electroweak) phase transitions. In contrast, stage 2 is in fact amenable to perturbative studies of the acoustic cascade, which is governed by the Boltzmann equation. It has been already rather well studied in the literature on turbulence, in which power attractor solutions have been identified. Application of this theory allows us to see how small-amplitude sounds can be amplified, as one goes to smaller k. Stage 3 can be treated via the standard loop diagram for sound+sound $\rightarrow$ GW transition.

Let us briefly recall the numbers associated with the QCD and electroweak transitions. Step one is to evaluate redshifts of the transitions, which can be done by comparing the transition temperatures $T_{QCD} = 170$ MeV and $T_{EW} \sim 100$ GeV with the temperature of the cosmic microwave background $T_{CMB} = 2.73$ K today. This leads to

$$z_{QCD} = 7.6 \times 10^{11}, \quad z_{EW} \sim 4 \times 10^{14}. \tag{16.22}$$

During the radiation-dominated era—to which both QCD and electroweak transitions belong—the solution to the Friedmann equation leads to the well-known relation between time and the temperature:[5]

$$t = \left(\frac{90}{32\pi^3 N_{DOF}(t)}\right)^{1/2} \frac{M_P}{T^2}, \tag{16.23}$$

where M_P is the Planck mass, and $N_{DOF}(t)$ is the effective number of bosonic degrees of freedom. Plugging in the corresponding T, we find the time of the QCD phase transition to be $t_{QCD} = 4 \times 10^{-5}$ s, and that for the electroweak transition $t_{EW} \sim 10^{-11}$ s. Multiplying these times by their respective redshift factors, we find that the t_{QCD} scale today corresponds to about 3×10^7 s ~ 1 year, and the electroweak to 5×10^4 s ~ 1 day.

GW from the electroweak era are expected to be searched for by future GW observatories in space, such as eLISA. The proposed observational tools for GW for the period scale of years are based on the long-term monitoring of the millisecond pulsar phases, with subsequent correlation between all of them. The basic idea is that when GW encounters the Earth and, say, stretches distances in a certain direction, then in the orthogonal direction, one expects distances to be reduced. The binary correlation function for the pulsar time delay is a known function of the angle θ between them on the sky. Existing collaborations—North American Nanohertz Observatory for Gravitational Radiation, European Pulsar Timing Array (EPTA), and Parkes Pulsar Timing Array—are actively pursuing both the search for new millisecond pulsars and collecting the timing data for some known pulsars. It is believed that about 200 known millisecond pulsars constitute only about 1 percent of the total number of them in our Galaxy. The current bound on the GW fraction of the energy density of the Universe is approximately

$$\Omega_{GW}(f \sim 10^{-8} Hz) < 10^{-9}. \tag{16.24}$$

---

[5] Note that we use not gravitational but particle physics units, in which $c = 1$ but Newton's constant $G_N = 1/M_p^2$.

Rapid progress in the field, including better pulsar timing and formation of global collaborations of observers, is expected to improve the sensitivity of the method, perhaps making it possible in a few years to detect GW radiation, either from the QCD Big Bang GW radiation we discuss, or that from colliding supermassive black holes.

The temperature T provides the micro (UV) scale of the problem: here the phase transition provides the sound source. The cosmological horizon is the IR cutoff on the gravitational radiation wavelength: in this case two sounds generate the GW. In between the UV and IR scales is the "dynamical range" of about 18 orders of magnitude! The challenge is to understand if and how the *inverse acoustic cascade* can be developed there and what n_k dependence is generated. It turns out that the answer crucially depends on the sign of the third derivative of the sound dispersion curve

$$\omega = c_s k + A k^3 + \cdots,$$ (16.25)

which remains unknown for both QGP and electroweak plasma.

If $A > 0$, sound decays 1 into 2 phonons is kinematically possible. If so, the turbulent cascade based on 3-wave interaction can only develop in the *direct* (that is, toward large k or UV) cascade, not the one we are interested in.

However, when the dispersive correction coefficient A is negative and there are no binary decays of the sounds, the answer is different. The turbulent cascade switches to higher order, of $2 \leftrightarrow 2$ scattering and $1 \leftrightarrow 3$ processes, generating an inverse cascade, with particle flow directed to IR. In this case of weak turbulence, the index of the density momentum distribution $n_k \sim k^{-s}$ is known to be

$$s_{\text{weak}} = 10/3.$$ (16.26)

Furthermore, the large value of the density at small k leads to the violation of the weak turbulence applicability condition, and the regime is known as "strong turbulence." We do not yet have a complete solution of this problem, only an estimate of the renormalized index, which is thought to be increasing to $s_{\text{strong}} \sim 5$. Given the huge (18 decades for QCD) dynamical range of the problem at hand, this will imply a significant increase of n_k at IR.

Another key result of our paper [149] is the calculation of the transition rate of the sound to GWs. We found it to be a rather simple process, an on-shell collision of the two sound waves.

With the historic detection in 2015 of GWs from merging black holes by the LIGO collaboration, it is a good time to work toward detection of the "sounds of cosmic storms" as well, namely, the QCD and electroweak phase transitions in the Big Bang.

17 | The QCD Vacuum I. Monopoles and Confinement

From this chapter to the end of the book we will be discussing nonperturbative phenomena in QCD and related gauge theories. The reader may thus wonder why these advanced topics, clearly belonging to a very specialized topic of QFT, are discussed in a book devoted to many-body theory. The reason is as follows: the most fascinating physics to be considered is not based on studies in terms of the perturbative Feynman diagrams and field harmonics (quarks and gluons) but in terms of certain topological solitons. The latter do not have infinitely many degrees of freedom, as do the QFTs themselves, but only a finite number of *collective coordinates*. Performing an integral over such collective coordinates is no different than integrating over positions of atoms in standard many-body calculations. Although it has not been done in many-body textbooks before, I think these calculations belong in such textbooks and include them here.

As the reader will soon see, many phenomena and ideas to be discussed below are just variants of those already discussed in previous chapters. For example, this particular section is a continuation of the chapters on quantum vortices, topological phases, and Bose-Einstein condensation. Chapter 18, on quark-antiquark pairing, is in essence just another incarnation of the BCS theory of superconductivity. The section on instanton-dyons discusses basically a four-dimensional plasma. Collectivization of a certain subset of topological quark eigenstates, breaking the chiral symmetry, and giving quarks (nucleons and ourselves) a mass are all quite similar to studying collectivized electrons in molten metals.

Let me now outline the content of the chapter in more detail. In Section 17.1 we discuss the *electric color confinement* phenomenon, relating it to BEC of colormagnetic monopoles in Section 17.2. The confining flux tubes are analogous (dual) to Abrikosov's magnetic flux tubes in superconductors. A more general picture of electric-magnetic duality, and the associated renormalization group (RG) flow of the coupling, is covered in Section 17.3. Unusual properties of QGP, especially near the phase transition, are explained

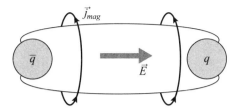

Figure 17.1. Schematic picture of the confining electric flux tube between quark and antiquark. The circular lines around it indicate presence of a magnetic current $\vec{j}_{\mathrm{mag}}$ needed to stabilize it.

in terms of the dual plasma, with both electrically and magnetically charged constituents, in Section 17.4.

17.1 Confinement and Flux Tubes

Permanent confinement of color-electric charges (or "confinement," for short) is the most famous nonperturbative feature of gauge theories. The Lagrangian of QCD-like gauge theories is similar to that of QED, with massless photons replaced by massless gluons, and massive electrons by (very light, or even massless) quarks. There are multiple definitions of the term itself (e.g., the statement that no object with a color charge can appear in a physical spectrum). There is a (so far) unclaimed million dollar prize offered for a mathematical proof that pure gauge theory has a finite mass gap.

Physicists are already sure that this conjecture is true beyond any reasonable doubt. Billions of observed high-energy collisions of hadrons and nuclei (as we have already briefly discussed) have produced large numbers of secondaries, and none of them has ever been a quark or a gluon. The formal limits on that event are so small that it makes no sense to even mention them.

We already discussed at the end of Chapter 14 the theoretical (although numerical) proof, the first demonstration[1] on the lattice, due to the pioneering work of Creutz [107]. It indicated that the confining string tension scales correctly, as the weak coupling RG scaling requires.

What was further observed on the lattice was that the electric flux from a color charge is not distributed radially outward, as in electrodynamics, but instead, being expelled from the QCD vacuum, is confined to a *flux tube* between the charges; see Figure 17.1. The energy per length of the tube is known as its tension

$$\sigma_T \approx (420\,\mathrm{MeV})^2 \approx 1\,\mathrm{GeV/fm}. \tag{17.1}$$

---

[1] Needless to say, no mathematician would take numerical evidence as a proof. In fact the very existence of any nontrivial QFT, such as QCD itself, is based on the path integral formulation, which includes taking a limit of vanishing lattice spacing $a \to 0$. The very existence of this limit has never been rigorously proven. All that was done numerically was to use a few values of decreasing a to see whether the results look convergent to a certain limit. Most physicists trust this procedure; mathematicians don't.

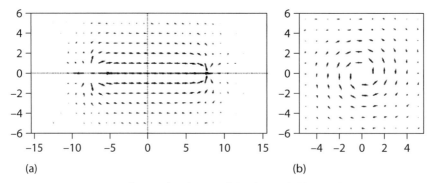

Figure 17.2. Lattice data on distribution of the electric field strength (a) and the magnetic current (b), for two static quark-antiquark external sources, from [170].

So a Colomb potential, $V(r) \sim 1/r$, at small r changes to a linear potential between charges, $V(r) = \sigma_T r$, at large r.

Exercise *Convert the tension σ into ordinary mechanical units. Can you pull this string with your two hands? If not, can you do it with the help of your car?*

Its most intuitive explanation was provided by the *dual superconductor model* by Nambu, 't Hooft, and Mandelstam [165]. The dual superconductor is a BEC of some magnetically charged objects—monopoles. The word "dual" means that while it is the magnetic field confined in flux tubes in a superconductor, made of electrically charged Cooper pairs, in the QCD vacuum we live in, it is the magnetic condensate that expels the electric fields into flux tubes. In other words, the QCD electric flux tubes are just a *dual copy* of the electric flux tubes in superconductors, already discussed in Chapter 11.

If so, the coil of the tube—the solenoidal-shape current around it, which keeps field inside and counterbalances its pressure—should also be dual (a "magnetic current" created by motion of *magnetic charges*). Let us postpone discussion of what those charges may be until the Section 17.2. We now address the subject more phenomenologically, using results of the lattice gauge theory simulations.

Figure 17.2 (displaying the result of lattice simulations summarized in [170]) shows distribution of the electric field (panel a) along the flux tube, and magnetic current (panel b) in a transverse plane. So, at least in numerical simulations, the flux tube is indeed clearly seen. At this point one would like to see whether its shape and other properties can be described by some generic effective model, such as Ginzburg-Landau theory.[2]

The profile of the electric field is shown by squares in Figure 17.3, where the lines are just fits.

Its curl, shown in Figure 17.3b, agrees well with the separately measured magnetic current k. The profile shown in Figure 17.3a is the resulting profile of the condensate.

---

[2] Recall that when the Ginzburg-Landau paper was written, the physical nature of the electric object that makes the condensate was also unknown: they argued for the form of effective action on general grounds, just as we do now.

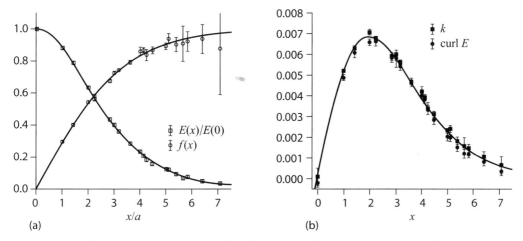

Figure 17.3. (a) Transverse profile of the electric field and the condensate. (b) The transverse distribution of the magnetic current. The points in both are lattice data, and the lines are fits using Abrikosov's vortex solution, from [170].

Note that, as in an Abrikosov vortex, it vanishes at the vortex center. The resulting parameters for two basic lengths, in "physical units" obtained by fixing the lattice scale to physical σ, are

$$\lambda = 0.15 \pm 0.02 \, \text{fm}, \quad \xi = 0.251 \pm 0.032 \, \text{fm}, \quad \kappa = \frac{\lambda}{\xi} = .59 \pm 0.14 < \frac{1}{\sqrt{2}}. \tag{17.2}$$

Their ratio (κ) is less than the critical Ginzburg-Landau value of $1/\sqrt{2}$. So, one can conclude that we all live inside a dual superconductor of type I. Those who are not convinced by the not-too-impressive accuracy of this important statement may wonder whether there are more direct manifestations of it. Indeed there are, but they remain controversial. Multiple flux tubes are routinely produced in high-energy collisions: for statistical reasons, the best-studied cases are LHC p-Pb collisions. When the impact parameter changes from peripheral to central collisions, the number of flux tubes is believed to change from about a dozen to 40 or so. Kalaydzhyan and I [171] suggested that not only is the long-distance interaction of strings attractive, as it should be in the type-I case, but at large enough densities of the flux tubes, it also produces a collective implosion. This leads to the formation of QGP and a subsequent explosion, of which indeed there are observed consequences: collective flows.

17.2 Monopoles in Non-Abelian Theories

Let us start by discussing the concept of electric-magnetic duality for pure Maxwellian electromagnetic fields. In this case it is next to trivial: introducing the usual vector potential A_μ, it has been decided to define the electric field to be its gradient, while the magnetic field is defined to be its curl. Introducing the *dual potential* C_μ, one can define the fields differently, $\vec{E}$ to be a curl, and $\vec{B}$ to be a gradient. As far as these correspond to the same fields, one definition is not better than the other.

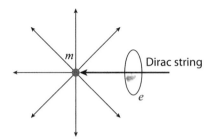

Figure 17.4. Magnetic monopole with the Dirac string (thick line) carrying the ingoing magnetic flux, equal to the total flux carried out by Coulomb-like field (thin lines with arrows). The circle around the Dirac string is a path of an electron e moving around it.

The duality seems to be lost when one introduce sources for those fields. QED has electrons, muons, and many other electrically charged particles, yet one does not observe any particles with magnetic charges. As explained in elementary physics textbooks, when cutting magnetic dipoles (e.g., a piece of magnetized iron) into two parts, one always get two dipoles and never north and south poles separately!

Yet, as famously shown by Dirac [133], monopoles in principle may exist, provided that the electric and magnetic charges $g_{\text{electric}}, g_{\text{magnetic}}$ satisfy the *Dirac condition*

$$\frac{g_{\text{electric}} g_{\text{magnetic}}}{2\pi} = \hbar n, \tag{17.3}$$

where n is an integer.

The existence of magnetic monopoles would not contradict Maxwell's equation $\text{div}\vec{B} = 0$ if there exists a singular magnetic flux—known as the Dirac string—that can carry the needed flux to the monopole location. See a sketch of it in Figure 17.4.

Exercise *Here is an example of a potential that corresponds to a monopole and a Dirac string:*

$$\vec{A} = g \left(\sin(\phi) \frac{1+\cos(\theta)}{r\sin(\theta)}, -\cos(\phi) \frac{1+\cos(\theta)}{r\sin(\theta)}, 0 \right), \tag{17.4}$$

where r, ϕ, θ are spherical coordinates. Using curl *in such coordinates, calculate the corresponding magnetic field. Check that the singularity of the vector potential does carry the needed magnetic flux.*

The wavefunction of an electron going around the Dirac strings (a circle in Figure 17.4) would have a certain phase ϕ. If the Dirac strings were visible, and had finite energy, they would produce a confining potential. However, Dirac argued, if the phase $\phi = 2\pi n$ (which happens when his condition (17.3) is satisfied), then the Dirac strings are invisible and thus unphysical. Note that this argument has a dual: if one would prefer to use the other potential C_μ, magnetic objects would not have Dirac strings, but the electric ones would. All such singularities, either in the vector potential A_μ or in C_μ, become unphysical pure gauge artifacts. Dirac emphasized that his relation is the *only known reason that electric charge should be quantized* and expressed hope that Nature would make use of it.

Needless to say, magnetic monopoles were looked for in all kind of substances, taken from the bottom of the oceans and from the surface of the Moon, and from secondaries produced at all accelerators. But in vain: QED magnetic monopoles were never found.

The surprising find was made not by experimentalists but by two distinguished theorists, G. 't Hooft [166] and A. Polyakov [167], so to say, "by the tip of their pens." The monopole solutions have been found for non-Abelian gauge theories interacting with scalar fields having color in the adjoint color representation.[3] As a detailed pedagogical introduction to monopole solutions, the reader may consult the book by Shnir [168].

Such fields and monopoles need to appear in theories with extended supersymmetry, such as the Seiberg-Witten theory [169]. Another set of theories with monopoles are models aimed at Grand Unification of the electroweak sector with QCD. All of this has resulted in large literature on the physics of monopoles in non-Abelian theories.[4]

Unfortunately, there are no scalar fields in QCD, which is the main focus of this and subsequent chapters. Thus it is unclear how to use the 't Hooft-Polyakov monopole solution in its framework. At finite temperatures, Lorentz invariance is anyway broken by the matter rest frame, and the fourth component of the gauge field A_4 (which is of course an adjoint color representation) can be used as a substitute for the adjoint scalar. We will discuss this option in Section 20.4, where we examine an explicit solution and its properties.

Here is the Lagrangian of the Georgi-Glashow model:

$$L = -\frac{1}{4}(G_{\mu\nu}^a)^2 + \frac{1}{2}(D_\mu\phi)^2 - \frac{\lambda}{4}(\phi^2 - v^2)^2. \tag{17.5}$$

It is a direct descendant of Ginzburg-Landau free energy of a superconductor. It differs from the electroweak sector of the Standard Model (the Weinberg-Salam model), precisely because ϕ has adjoint color representation. Nonzero vacuum expectation value (VEV), shown as v, provides different Higgsing of the gauge fields: their mass is proportional to the commutator of their color generator with that of the VEV. In this model the $SU(N_c)$ color group is broken into its $N_c - 1$ diagonal subgroups.

The simplest case is $N_c = 2$, in which there are three generators τ^a (three Pauli matrices) with $a = 1, 2, 3$. Let us take the nonzero VEV to be along the diagonal, so $\langle\phi^3\rangle = v$. Then two gauge bosons ($W^{+,-}$) get nonzero masses, while the boson number 3 (the neutral "photon") remains massless. (The Georgi-Glashow model was designed to avoid the existence of the Z boson, then unknown.)

---

[3] A model of this kind was previously suggested as a candidate for the theory of electroweak interactions by Georgi and Glashow. However, subsequent discoveries pointed to another option, scalar fields with the fundamental color representation, as suggested by Weinberg and Salam.

[4] If there are monopoles at the Grand Unification scale of energies, they must be stable and thus generated in the Big Bang. But we do not find them around us today, which is a problem. One possible solution to it is the invention of a "cosmic inflation stage" with a huge exponential expansion of the Universe. Today there is other cosmological evidence for inflation, and so now the idea can stand on its own legs. The fact that inflation solves the monopole problem is now a footnote in the history of physics.

Since there is such a drastic difference between these components, let us introduce a special notation for the Abelian-projected fields (without color indices):

$$A_\mu = A_\mu^a \hat{\phi}^a, \quad \hat{\phi}^a = \frac{\phi^a}{|\phi^a|}, \tag{17.6}$$

where I have introduced a unit vector (indicated by a hat). To define also the Abelian field strength, the field definition should not be just the usual Abelian expression based on A_μ, because it must be supplemented by a term canceling possible derivatives of the Higgs color direction. The definition is

$$F_{\mu\nu} = \partial_\mu A_\nu - \partial_\nu A_\mu - \frac{1}{e} \epsilon_{abc} \hat{\phi}^a \partial_\mu \hat{\phi}^b \partial_\nu \hat{\phi}^c. \tag{17.7}$$

The last term is of course zero for a constant Higgs field. In fact, it vanishes for all topologically trivial configurations of $\phi^a(x)$: it will be nonzero only for topologically nontrivial ones, and the possibility of having a gauge in which there is no Dirac string is based on this observation.

The magnetic current can now be defined from the definitions (17.6) and (17.7), it is a divergence of a dual field $\tilde{F}_{\mu\nu} = \frac{1}{2} \epsilon_{\mu\nu\alpha\beta} F^{\alpha\beta}$

$$k_\mu = \partial_\nu \tilde{F}_{\mu\nu} = \epsilon^{\mu\nu\rho\sigma} \epsilon_{abc} \partial_\nu \phi^a \partial_\rho \phi^b \partial_\sigma \phi^c \left(\frac{1}{2v^3 e} \right). \tag{17.8}$$

Unlike the usual Nether currents, this magnetic current k^μ is conserved by definition, without any underlying symmetry. The integral of its density is known in mathematics as the *Brouwer degree*. As for any topological quantity, for the appropriate normalization it yields an integer, which defines topologically distinct Higgs fields.

How this happens is made clear by an example of a "hedgehog"-like field[5]

$$\phi^a(r \to \infty) \to v \frac{r^a}{r}, \tag{17.9}$$

in which the "needles" go radially: the magnetic charge for it is

$$g = \int d^3 x k_0 = \frac{4\pi}{e}. \tag{17.10}$$

Let us now look for a solution consistent with that asymptotical trend, in terms of two spherically symmetric functions

$$\phi^a = \frac{r^a}{er^2} H(\text{ver}); \quad A_n^a = \epsilon_{amn} \frac{r^a}{er^2} [1 - K(\text{ver})]; \quad A_0^a = 0. \tag{17.11}$$

When these functions are plugged back into the expression for the Hamiltonian, we find the following expression for the monopole mass:

$$E = \frac{4\pi v}{e} \int_0^\infty \frac{d\xi}{\xi^2} \left[\xi^2 \dot{K}^2 + (1/2)(\xi \dot{H} - H)^2 + (1/2)(K^2 - 1)^2 + K^2 H^2 + \frac{\lambda}{4e^2} (H^2 - \xi^2)^2 \right]. \tag{17.12}$$

Here I have rescaled the radial coordinate $\xi = evr$, and its derivative is denoted by a dot over it. This expression can be used as an effective action corresponding to classical

---

[5] The relation between such a field configuration and this cute animal appeared for the first time in Polyakov's paper.

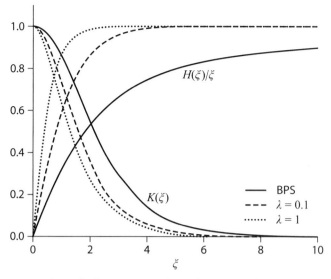

Figure 17.5. Classical solutions $K(\xi)$ and $H(\xi)$ for the 't Hooft-Polyakov monopole at $\lambda = 0$ (the BPS limit) and $\lambda = 1$. The graph is redrafted from Shnir [168].

motion in the K-H plane of a particle, with ξ being the time. Standard equations of motion are obtained from its variation.

Exercise *Since a monopole is a static classical solution, its energy is minimal for all functions K, H satisfying the following boundary conditions. The boundary conditions at small r correspond to $H \to 0$ and $K \to 1$, as then the Higgs field is smooth in spite of the hedgehog direction. At large distances, $H \to \xi$ and Higgs becomes of magnitude v, while $K \to 0$. Using variation of the Hamiltonian, derive the equations of motion for K, H. Solve them numerically using Maple or Mathematica.*

At large distances it is a Dirac monopole with a magnetic charge

$$g = \frac{4\pi}{e},$$ (17.13)

which is consistent with the quantization condition. In fact I prefer to write the result more symmetrically using electric and magnetic fine structure constants,

$$\frac{g^2}{4\pi} \frac{e^2}{4\pi} = \alpha_e \alpha_m = 1,$$ (17.14)

which are then inverse to each other.

The solution for the equation of motion can generally be obtained numerically. Two of them are shown in Figure 17.5 taken from Shnir [168]. On general grounds, the monopole mass can be written as

$$M = \frac{4\pi v}{e} f\left(\frac{\lambda}{e^2}\right)$$ (17.15)

with a smooth function f. Some of its values are $f(0) = 1$ and $f(1) = 1.787$.

The case $\lambda = 0$ is known as the Bogomolny-Prasad-Sommerfeld (BPS) limit, in which there is no scalar potential at all. In this limit both functions are known analytically:

$$K = \frac{\xi}{\sinh(\xi)}, \quad H = \frac{\xi}{\tanh(\xi)} - 1. \tag{17.16}$$

Exercise *Using these values for K and H, calculate the Abelian magnetic field for the BPS monopole. The Abelian magnetic field is defined as $\tilde{F}_{\alpha\beta} = \frac{1}{2}\epsilon_{\alpha\beta\gamma\delta} F_{\gamma\delta}$ with Abelian F defined in (17.7). Substitute it into the Hamitonian and plot the energy profile of the monopole.*

17.3 Electric-Magnetic Duality and RG Flow

Let us follow qualitatively the dynamics along the RG trajectory, starting from large momenta (UV) and ending at low ones (IR). In particular, this can be seen as the temperature dependence (from high to low T) or alternatively, as chemical potential dependence (from high to low μ).

In UV we start in a weak coupling regime $g_{electric} \ll 1$ because of asymptotic freedom. Monopoles, and any other solitons, must have strong fields, $A \sim 1/g_{electric}$, for the nonlinear commutator term in the field to be comparable to the derivatives. Therefore the mass

$$M \sim \frac{1}{g_{electric}^2} \tag{17.17}$$

is large compared to the temperature (whichever is defining the overall scale), and the monopole density

$$n \sim \exp(-M/T) \sim \exp\left(-\frac{const}{g_{electric}^2}\right) \tag{17.18}$$

is exponentially small. Thus at high T, monopoles (and other similar nonperturbative phenomena) are unimportant. Yet from the Dirac condition, we see that their magnetic coupling is large, and so they interact strongly.

Now, decreasing T, we try to go toward the IR. The electric coupling grows, the soliton mass M/T decreases, and their density grows. At some point the density of magnetic monopoles and electric charges (quarks and gluons) become comparable: this state was called the electric-magnetic equilibrium in [134]. In QCD-like theories, at this point a number of phase transitions happen: due to confinement and chiral symmetry breaking, the low-T phase of QCD becomes a hadronic phase (in which we live). How the transitions happen we will discuss in the next section.

But before that, let me describe what happens in some extended supersymmetric theories. Supersymmetry is a powerful Bose-Fermi symmetry, which may be do not exist in nature, but which simplifies problems a lot, sometimes to the degree that they can be solved. The word "extended" means that the theory has more than one supersymmetry. The simplest of those, called $\mathcal{N} = 2$, has two fermionic partners of a gluon; let us call them gluinos of types ψ and χ. Yet this is still an incomplete set (multiplet) of the extended supersymmetry group, since gluons have two polarizations, and so does each Majorana fermion, with $2 \times 2 = 4$ fermionic degrees of freedom. To balance the number of fermions and bosons, we need one more *complex* scalar ϕ, also with two degrees of freedom. (Times for all of them $N_c^2 - 1$ color components of the adjoint representation of $SU(N_c)$.)

The existence of scalars is crucial because, by Lorentz symmetry of the vacuum state, only scalars are allowed to have VEV. So this theory has a complex parameter $v = \langle \phi \rangle$ characterizing a whole plane of possible vacua. None of them is better than the other, since supersymmetry requires the vacuum energy of any of them to be zero. They are nevertheless all different since v defines a scale, and a running coupling $g(v)$ depends on it. In particular, far from the center of the v plane, where VEV is large, the coupling is small (due again to the asymptotic freedom).

Naively, one might suppose that the opposite strong-coupling regime would be at the center, where $v \rightarrow 0$. As shown by Seiberg and Witten [169], this assumption is wrong and in fact there are two singular points on the v plane. In one of them the monopole ($Q_{\text{electric}} = 0$, $Q_{\text{magnetic}} = 1$) becomes massless, in another the dyon ($Q_{\text{electric}} = 1$, $Q_{\text{magnetic}} = 1$) does. Both are perfect examples of what we are looking for: the ultimate IR end of RG flow. Indeed, in those vacua

$$g_{\text{electric}} \rightarrow \infty, \quad g_{\text{magnetic}} \rightarrow 0, \tag{17.19}$$

which is the opposite of the electric weak coupling in UV! In the vicinity of these points, the low-energy effective theory is the magnetic theory of light and weakly coupled monopoles (or dyons). Seiberg and Witten were able to show that the beta function of this theory and the effective theory just mentioned indeed match exactly.

What was observed in $\mathcal{N} = 2$ Super-Yang-Mills theory (SYM) is quite remarkable: a non-Abelian theory with asymptotic freedom in UV turns out to be continuously connected, by the RG flow, with a completely different magnetic theory, which is Abelian and has a beta function of the opposite sign! It turns out that it is in fact one and the same theory on the whole v plane, and whether we use the electric or magnetic formulation of the theory as input is just a matter of convenience: both are true!

What is remarkably different from QCD for $\mathcal{N} = 2$ SYM is that in this theory, the weak coupling and the strong coupling regimes are not separated by any phase transitions. (This happens due to supersymmetry; it is beyond the scope of this book to explain why.)

Returning from the hypothetical supersymmetric theory back to our real world, where we so far lack this powerful tool, we are back to QCD-like theories of quarks and gluons. There are no scalar fields in these theories at all, and thus one should (1) give up on monopoles entirely; (2) create some composite scalar, perhaps a gauge-dependent one; or (3) shift to the finite-T theory, which lacks Lorentz invariance and allows for nonzero VEV of A_4, as a substitute scalar.

Option (2) has been intensely studied on the lattice. Furthermore, these monopoles (still defined in certain gauges) do indeed form BECs precisely at $T < T_c$ in the confined phase. This is observed by following their Bose clusters, similar to our discussion of liquid He (see, e.g., [99]). Spatial correlations between lattice monopoles have been measured: they corresponds well to those in Coulomb plasmas.

More recent developments of confinement-phase transition are related to option (3). Using A_4 instead of a scalar field, one finds well-defined topological solitons called *instanton-dyons*, which are self-dual and have both electric and magnetic charges. For recent studies of their ensembles, see [201, 202].

Instead of going into details, let us outline a wider picture of electric-magnetic duality. An important motive of this book has been "running coupling constants" described by the RG. One may ask whether the naive Dirac condition just given above is in fact in agreement with such "running" (scale dependence). Remarkably, it is!

The usual gauge coupling $g_{\text{electric}}(k)$ becomes weak at large momenta k, but is growing as the momentum scale is reduced. The density of topological objects, such as monopoles, at temperature T

$$n(T) \sim \exp\left(-\frac{\text{const}}{g_{\text{electric}}^2(T)}\right) \tag{17.20}$$

grows as T goes down. At $T \sim T_c$ their density becomes comparable to that of electric charges like quarks. In some theories, such as $\mathcal{N} = 2$ supersymmetric gauge theory, one can go all the way to the dual phase, in which magnetic objects are weakly coupled and electric ones strongly suppressed. In nonsupersymmetric theories, we can only rely on lattice data, but we can still see evidence [135] of the fact that magnetic charges run in the opposite direction, in perfect agreement with the Dirac condition.

I end with a general point, perhaps surpassing the whole subject of this chapter in importance—the QCD vacuum problem. Quite different theories (like electric gauge theory weakly coupled in the UV; and the magnetic theory, weakly coupled in the IR) may in fact be just the limits of a much larger overarching theory. By following continuously the RG flow, it turns out to be possible to connect QFTs, which were studied separately and considered to be completely different in their properties. In particular, it is possible to connect a QCD-like electric theory with a QED-like magnetic one. Note that the Dirac condition keeps constant the product of two couplings, and thus *forbids both to be weakly coupled*. Therefore, it is very nontrivial to test the RG relation, as one needs to pass through the region where both couplings are strong. And yet, such a connection can be traced nonperturbatively, or at least can be indicated by the dualities!

17.4 Strongly Coupled QGP as a "Dual" Plasma

Discovery that the QGP is a perfect liquid with an extremely small viscosity resulted in significant interest in strongly coupled systems in general. (It is even now referred to as strong QGP, or sQGP for short, in literature.)

Two research directions that developed as a result curiously follow ideas of two dualities:

1. duality of conformal strongly coupled gauge theory (CFT) to weakly coupled string (gravity) theory in a special five-dimensional geometry known as anti-de-Sitter (AdS_5) space (AdS-CFT correspondence)
2. electric-magnetic duality, emphasizing the RG flow and transition from weakly coupled electric theory in UV to mixed strongly coupled electric-magnetic sectors around T_c.

The first direction now can be characterized as a mainstream, but (sorry to say) it does not belong to this book, as it requires too much background knowledge. The second one, on the contrary, is rather accessible, so I include it here.

In very general terms, we are going to discuss systems containing both electrically and magnetically charged particles. For brevity, let us call them "dual plasmas." We will start with classical behavior, first for a few particles, and then proceed to many-body theory. We will discuss quantum theory next: so far, it has been developed at the level of binary scattering only, not yet as a many-body theory.

Unexpected things appear along the way, starting from the very first step. Already in 1904, J. J. Thomson[6] observed that even two static charges, electric e and magnetic g, have a nonzero angular momentum J carried by a rotating electromagnetic field. Indeed, two Coulomb fields from these charges at an arbitrary point have nonzero Poynting vector $[\vec{E} \times \vec{B}] \neq 0$, making circles around the line connecting the two charges. So, the field is rotating even when both charges are not moving!

Exercise *Calculate Thomson's angular momentum of the field mentioned above, and prove that in quantum theory its quantization to $\hbar n$ produces another method of obtaining the same Dirac quantization condition (17.13) we discussed above.*

This has implications for cases when those charges start moving. Let us start with a magnetic charge moving in an electric Coulomb field of a static charge.[7] In fact even earlier than Thomson, in 1896, H. Poincaré [132] had already thought about this problem. The only force acting is the (dual) Lorentz force

$$m\ddot{\vec{x}} = g[\vec{E} \times \dot{\vec{x}}] \tag{17.21}$$

proportional to the velocity of the monopole. From this he observed that the motion takes place on the surface of a cone, now known as the *Poincaré cone*. Indeed, since the radius of a circle in a field gets smaller as the field gets stronger near the charge, $E \sim 1/r^2$, this focuses the path toward the cone's top ($r \to 0$), and thus a collision of both charges is inevitable.

Exercise *By solving equation of motion (17.21) numerically in Maple or Mathematics, in a field of a Coulomb charge, reproduce trajectories like those in Figure 17.6, which elucidate the parameters of the Poincaré cone. Continue to study monopole motion in the field of an electric dipole, or other arrangements of a few electric charges.*

Recall that a typical two-body potential results in classical motion in a collision plane: the motion is forced to be planar by the conservation of the angular momentum, normal

---

[6] Recipient of the 1906 Nobel Prize in Physics, J. J. Thomson was the first to measure the charge-to-mass ratio of the electron and thus discovered the first elementary particle. His other high distinction was that seven (!) of his students, including his son, also became Nobel Prize winners.

[7] More generally, it is a motion in relative coordinates, using the so-called reduced mass.

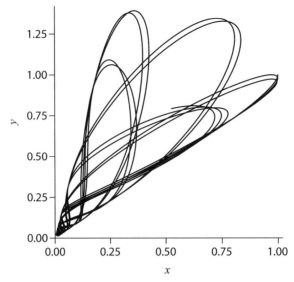

Figure 17.6. Trajectory of a monopole in a static electric Coulomb field. Adapted from Liao and Shuryak [134].

to this plane. What is different in the Poincaré problem at hand is precisely Thomson's effect: not all the angular momentum is due to the moving particle; part of it is due to the rotating field!

Jinfeng Liao (a Stony Brook grad student at a time) and I wrote a paper [134] proposing the so-called "magnetic scenario" for sQGP, in which it is considered to be a dual plasma made of electrically and magnetically charged particles.

We started by investigating several curious single-body motions in different settings. The Poincaré problem, in a field of one static charge, was the first one. The second one was two static electric charges, e and $-e$, with a moving monopole. We found that the monopole can be trapped between the two charges, bouncing[8] from one to the other on a surface consisting of two connected Poincaré cones (see Figure 17.7), colliding with the charges at the tops of the cones.

Our next configuration (not shown) was "a grain of salt," consisting of eight electric charges located at the corners of a three-dimensional cube, with alternating plus and minus charges. It was found that a monopole can get out of such a cage, but only with great difficulty, suffering hundreds of collision with the corners in the process. The mechanism is again due to focusing, as just described.[9]

Finally, we took few hundred charges, both electric and magnetic, both plus and minus to keep matter neutral, and did molecular dynamics simulations, numerically solving the equations of motion for all particles. Details are in [134]: the output consists of calculations of the diffusion constant and viscosity, from the corresponding relations

---

[8] We called it "two charges play ping-pong with a monopole, without even moving."

[9] We compared the monopole's behavior to that of a proverbial drunkard, who cannot go home from the city square, because there are lamp posts located at the corners, with which he keeps colliding.

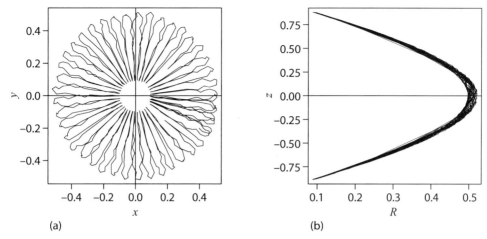

Figure 17.7. Trajectory of monopole motion in a static electric dipole field (with charges at $\pm 1\hat{z}$) as (a) projected on x-y plane and (b) projected on R-z plane ($R^2 = x^2 + y^2$). From [134].

known as Kubo formulas. The qualitative conclusion is that each electric charge is trapped by a cell of magnetic ones around it, and each magnetic charge by electric ones, as described above. The Lorentz nature of the forces between them does decrease particle (diffusion) and momentum (viscosity) transport by a lot!

The history of the *quantum monopole-charge scattering problem* is also very interesting and is not widely known, although it clearly belongs to any good quantum mechanics textbook, in my opinion.

There are indications that Dirac, after his paper on monopoles in 1931, tried to solve it, but did not succeed. It was a long time before the scattering problem was solved, a year after non-Abelian monopoles were discovered, by two distinguished teams [136,137].

To begin, let us think about the input parameters of the problem. Quantum mechanical scattering problem results are usually expressed in terms of scattering amplitudes. These depend on the scattering phase shifts $\delta_l(k)$ as a function of the angular momentum channel l and the collision momentum k. Note however, that the Coulombic field of a charge is produced by the dimensionless $e^2/(\hbar c)$ coupling. A background pointlike charge therefore cannot provide any length or momentum scale.[10]

Already our conclusion that the (pointlike) charge-monopole scattering has phase shifts independent of scattering momentum k has an important implication. The general theory of nonideal gases connects the contribution to the partition function with the number of states of a gas. Since scattering phases shift the phase of the wavefunction, quantization of states in a box are modified, and extra states appear. Their contribution to the partition

---

[10] The usual charge-charge scattering does not have well-defined phase shifts at all, due to the long-range nature of the Coulomb potential. However, in the charge-monopole scattering problem, the force is the dual Lorentz force, which is not long range; thus the scattering phases are well defined.

function is given by the Beth-Uhlebeck formula

$$\Delta Z = \frac{1}{\pi} \sum_l (2l+1) \int_0^\infty dp \frac{d\delta_l(p)}{dp} \exp\left(-\frac{p^2}{mT}\right), \tag{17.22}$$

where the integration is done over momentum, and summation over all angular momentum values l. The $\delta_l(p)$ is the scattering phase in channel l, a phase shift between the incoming and outgoing waves. The Boltzmann exponent does not have 2 in the denominator because for the scattering of two mass-m particles, the relative motion has effective mass $m/2$.

Exercise *Using the Beth-Uhlebeck formula, prove that narrow Breit-Wigner resonances in the binary scattering contribute to the partition function a sum that is identical to the contribution of stable particles with resonance masses.*

Since we concluded that $d\delta_l(p)/dp = 0$ for the charge-monopole problem under consideration, it follows that for it, no new states are created and thus the thermodynamics of the dual plasmas is that of an ideal gas.

The scattering phases do depend on the couplings, specifically, on the product of electric and magnetic charges. Yet according to the famous Dirac condition discussed above, this product must be an integer n (for example, just 1). This integer n is the only input to the charge-monopole scattering problem on which the scattering phase can depend!

A specific feature of the problem is that, after separation of the angular θ, ϕ, and the radial r coordinates, the angular part happens to be the most complicated. To explain why, recall that classically the motion takes place on the Poincaré cone rather than on a plane. The usual angular basis of spherical harmonics, $Y_{l,m}(\theta, \phi)$, at large quantum numbers $l, m \gg 1$ is in fact planar. So, one has to select more complicated angular functions that are conical in the classical limit.

Omitting technical details, let me just give the answer for the scattering phases. Indeed, they depend on the integer total angular momentum l and do not depend on energy: the expression for a phase is

$$\delta(l') = \frac{\pi}{2}l', \quad l'(l'+1) = l(l+1) - n^2, \tag{17.23}$$

where l' is to be found from this quadratic equation.[11] Note that while the total angular momentum l and n (the integer from the Dirac condition) are both integers, their combination l'—apparently the angular momentum of a particle—curiously is *not* an integer! That is the only reason that the rather nontrivial angular distribution of scattering appears!

---

[11] It is not hard to solve equation (17.23), but I think this expression is more instructive. It implies that the particle orbital momentum comes from the total momentum squared minus that contained in the rotation of the field.

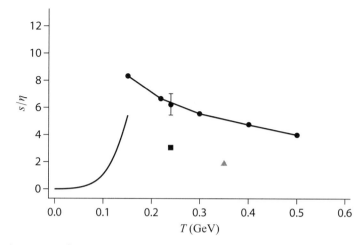

Figure 17.8. The entropy density to shear viscosity ratio s/η versus the temperature T (GeV). The upper range of the plot, $s/\eta = 4\pi$, corresponds to the value in an infinitely strongly coupled $\mathcal{N} = 4$ plasma [139]. The curve without points on the left corresponds to hadronic/pion rescattering according to chiral perturbation theory [140]. The single triangle corresponds to a molecular dynamics study of a classical strongly coupled colored plasma [142]; the single square corresponds to numerical evaluation on the lattice [143]. The single point with the error bar corresponds to the phenomenological value extracted from the data (see text). The series of points connected by a line on the right side correspond to gluon-monopole scattering [138].

In QGP with monopoles, one needs to calculate the scattering amplitudes of quarks and gluons colliding with a monopole. The latter problem is especially tedious, because the orbital angular momentum, spin, and color-spin of the gluon are all mixed in a set of coupled channels. It has been solved by C. Ratti and me [138]; the resulting angular distribution and its contribution to the transport cross-section can be found in this paper.

The results, shown in Figure 17.8 (together with those from other approaches and from phenomenology) do indeed indicate that gluon-monopole scattering in sQGP dominates its kinetic properties near the QCD phase transition temperature T_c.

To explain why this is the case, recall that the kinetic theory (valid for weak coupling) would interpret this dimensionless ratio of the entropy density s to that of the shear viscosity η as the product of the particle density, transport cross section, and inverse power of the mean velocity:

$$\frac{s}{\eta} \sim \frac{n\sigma_{transport}}{T\bar{v}} \tag{17.24}$$

or, a bit simpler, as the ratio of interparticle separation to the mean free path. Note that the densities of quarks and gluons go to zero at T_c. In contrast, the density of (noncondensate) monopoles, does not vanish—confinement is electric, not magnetic—and even has a peak there.

The peak in inverse shear viscosity correlates with similar peaks claimed for two more kinetic parameters: the heavy quark diffusion constant and the jet-quenching parameter $\hat{q}$. Those are perhaps also due to gluon-monopole scatterings, see [144, 145].

18 | The QCD Vacuum II. Chiral Symmetries and Their Breaking

In this chapter we focus on another symmetry aspect of QCD and strongly interacting matter. The *chiral symmetries*—from the Greek word meaning "a hand"—are symmetries existing separately for left- and right-handed components of the massless (light) fermion fields. These symmetries are exact if the fermions are massless and approximate if the masses are small.

Our primary interest in this chapter is focused on QCD, which has three light flavors of quarks u, d, and s (called "up," "down," and "strange"; see the discussion in the next section). Other applications in which fermion chirality matters are electrons in Dirac and Weyl semi-metals already discussed in Chapter 19.

In the next section we will introduce two types of chiral symmetries in QCD, known as $U(1)_a$ and $SU(N_f)_a$. The subscript a here stands for "axial," as the corresponding currents have the form of axial vectors $\bar{q}\gamma_\mu\gamma_5 q$, and N_f is the number of relevant quark flavors, mostly two or three.

The $U(1)_a$ chiral symmetry holds at the classical level only, because the *axial anomaly*, the subject of the Section 18.2, violates it, so that it basically is not a symmetry at all. The divergence of the corresponding axial vector current is nonzero but is a certain combination of gauge fields, with direct connection to gauge field topology.

The $SU(N_f)_a$ chiral symmetry holds both at the classical and quantum levels, and the divergence of the corresponding axial vector current does vanish. However, it is violated spontaneously in the QCD vacuum and at finite temperatures below some critical value $T < T_c$. This phenomenon is related to superconductivity, but is due to quark-antiquark pairing, rather than the quark-quark one. The first model of it, known as the Nambu-Jona-Lasinio model, was indeed built by a direct analogy to the BCS theory; see Section 18.3. The modern view on its mechanism relates these phenomena of the chiral symmetry breaking to very specific multifermion interactions induced by the spacetime topological solitons called *instantons*; see Section 18.4. So, the phase we live in, the QCD vacuum, is not only a superphase with a condensate of quark pairs but is also a form of *topological matter.* What is special is that instantons are four-dimensional objects, localized in space and time.

In this chapter we will show that the QCD vacuum and hadronic matter even have an analog of the Fermi surface: it is the "zero-mode zone" of quark eigenstates of the Dirac operator. A very small fraction of all quark states, located in this zone, in fact significantly affect the hadronic masses and many other properties of the world.

18.1 Light Quarks and Chiral Symmetries

When we discussed QGP in Chapters 15–17, we focused mostly on the gauge fields and related diagrams, as they are the most important (but also the most tricky) ones. Now we turn our attention to the real-world QCD, with its quark degrees of freedom and corresponding flavor symmetries.

The Standard Model of particle physics—including (nearly) all known fundamental building blocks and interactions (except gravity)—has six quark types collected into three generations of "up"-"down" doublets of the electroweak $SU(2)$ group. However, for our applications it is useful to group them into three light ones (u, d, s) and three heavy ones (c, b, t). This grouping refers to comparison of their masses to the typical scale of nonperturbative QCD

$$\Lambda_{\text{QCD}} \sim 200\,\text{MeV} \sim 1/\text{fm},$$

at which the coupling gets strong.

All heavy quarks have masses larger than the scale interval indicated, so they are suppressed in the vacuum, and we can ignore them in this chapter. The lightest of them, $m_c \sim 1.5\,\text{GeV}$, is at the edge above which the coupling can be considered weak, $\alpha_s(m_c) \ll 1$, and perturbative diagrams dominate the charm quark physics. In particular, $\bar{c}c$ bound states, known as charmonium, are approximately similar to positronium.

Masses of the three light ones are approximately equal to[1]

$$m_{u,d,s} \approx 4, 7, 120\,\text{MeV}.$$

Because these masses are too small to be important, there is approximate $SU(3)_f$ symmetry[2] between hadrons made of light quarks. This symmetry leads to grouping of mesons to an adjoint octet, baryons to another octet, and decuplet representations of the $SU(3)$ group. When all the members of these representations were subsequently discovered experimentally, completed with Ω^- or the sss baryon, the reality of the $SU(3)_f$ symmetry and the "eightfold way" led to the 1969 Nobel Prize for M. Gell-Mann. The

---

[1] The exact definition is required if one wants to be precise: in particular, there are powers of the logs of the normalization scale, etc., which we ignore for now. The more pedantic reader who wants a more precise definition should look, e.g., at the quark part of the Particle Data Group web page.

[2] Not to be confused with the color $SU(3)_c$ symmetry, which is exact: it has nothing to do with quark masses but with their color charges.

notion of quarks has since entered the physics vocabulary, although at the time their exact properties were far from clear.

To simplify our discussion further, let us start with the *chiral limit* in which all light quark masses vanish, $m_i = 0, i = u, d, s$, and discuss the massless QCD. We will also ignore electrodynamical effects, which also violate the symmetry between quarks due to different electric charge, $2/3$ for u and $-1/3$ for d, s quarks.

The mass term in the Lagrangian connects the left- and right-handed chirality components of the quark fields. If it is absent, one can perform independent global[3] rotations, belonging to the $SU(3)_L$ and $SU(3)_R$ groups, for left and right components.[4] The flavor symmetry group is now doubled.

(By rotations, we mean unitary transformation from the original quark basis $\psi_f = (u, d, s)$ into another linear combination of them:[5]

$$\psi'_f = \sum_{f'} \left(e^{in^a \lambda^a/2}\right)_{ff'} \psi_{f'}$$

using $SU(3)$ generators, the Gell-Mann matrices λ^a, $a = 1, 2, ..., 8$ and some arbitrary vector n^a. If all quarks are massless and the action only has flavorless terms like $\sum_f \psi_f^+ \psi_f$, then the conjugate will rotate with Hermitian conjugate matrix—which is the same as the inverse for a unitary group—and will cancel out.)

This new higher symmetry of the theory with massless quarks is the *chiral* symmetry. A massless quark field can be permanently split into two parts, the left and right handed components, denoted by

$$\psi_{L,R} = P_{L,R}\psi = \frac{1 \mp \gamma_5}{2}\psi.$$

Here we have introduced projection operators P_L, P_R into two chiral components. (Chirality of the antiquark is defined as $\bar{\psi}_L = \psi^\dagger P_L \gamma_0$, where the dagger is Hermitian conjugation)

Now, one can make two different flavor rotations, A, B

$$\psi_{L,f} \to A_{fk}\psi_{L,k}, \quad \psi_{R,f} \to B_{fk}\psi_{R,k},$$

doubling the flavor symmetry to $SU(3)_L \times SU(3)_R$ with 16 parameters. This can be rewritten as an ordinary vector rotation and a (parity-nonconserving) chiral or axial rotation, in which L and R are rotated with the opposite sign.

So far we have not used the full unitary groups $U(3)_L$, $U(3)_R$. There are two more $U(1)$ rotations with a phase factors $\exp(i\alpha_L)$ and $\exp(i\alpha_R)$ (times the unit flavor matrix),

---

[3] Unlike color rotations, which can be done independently at each spacetime point, flavor rotations are the same for all points.

[4] Chirality is the spin projection on the momentum. For a massive particle the direction of the momentum can be changed by going to another frame, for example, moving along the original momentum but with velocity greater than that of the particle: in this frame, chirality flips to its opposite sign. But for massless particles, which can only move with the speed of light, one cannot do the trick just described and move "ahead of them." This L, R chirality is a permanent label valid in all frames.

[5] The sum over f' is shown explicitly here but will be dropped below: henceforth we generally use Einstein's notation, with a twice-repeated index always summed.

with two independent phases. These two symmetries can be redefined as vector and axial vector, $U(1)_V$ and $U(1)_A$, ones. These symmetries have quite different fates, to be discussed below.

Existence of the chiral symmetry has multiple consequences. In particular, since the axial rotations can mix states with opposite parity, they should have the same masses. This statement appears to be in terrible disagreement with the real world: for example, we have hadrons like the ρ meson that have quite definite parity $P = -1$ and are not at all degenerate with particles possessing the opposite parity (e.g., the A_1 meson with $P = 1$).

After our discussion of several examples above, the reader should recognize that the presence of some symmetry of the action does not guarantee the symmetry of the ground state. For example, the Lagrangian of complex scalar fields has a symmetry of global phase rotations $\psi \to e^{i\phi}\psi$, but the BEC phenomenon in vacuum of this theory breaks it. Similar *spontaneous breaking of gauge symmetry* takes place in a superconductor. So, it should not be shocking for the reader to learn that the $SU(3)_a$ axial chiral symmetry is also spontaneously broken. The ground state (known as the QCD vacuum) is not a singlet under its transformation. Like the phase of a condensate, a particular direction in flavor space is spontaneously selected.

An important trace of the spontaneous symmetry breaking should remain the so-called Goldstone modes. Suppose that a condensate has a certain phase. But long-wavelength oscillations of the phase are approximately its global rotation and therefore, due to symmetry, should have (almost) no energy! In terms of excitations, we have discovered that the system has to have a massless excitation. For the complex field mentioned above, the Nambu-Goldstone mode is the ordinary density oscillations, the phonon. For chiral symmetry, the Goldstone mode needs to be a *pseudoscalar* field, violating parity.

How many such modes should one expect? Writing the symmetry transformation (with one extra parameter, the chiral phase)

$$\psi \to (1 + i\beta\gamma_5 + i\beta^a\lambda^a\gamma_5)\psi$$

and imagining small variation of the nine parameters, one expects nine pseudoscalar mesons. And experiment indeed has provided them, known as π(iso − triplet), K(two iso − doublets), η, η', with $3 + 4 + 1 + 1 = 9$ states.

Are we sure that those are indeed Nambu-Goldstone modes? In the chiral limit these particles are predicted to be massless. This fact helps explain apparent contradiction between the theorem about the presence of degenerate states of opposite parity and hadronic spectra in which particles of different parity have different masses. The theorem is correct: the state of opposite parity degenerate with any hadron is this hadron plus one massless pion at zero momentum.

If quark masses are nonzero, they will violate the symmetry, but only slightly. In a magnetized spin system it would mean the presence of a weak (but nonzero) external field in the Hamiltonian, a term like $\vec{S}\vec{H}$. If the angle between these two vectors is ϕ, we get $\cos\phi = 1 - \phi^2/2$, so that the magnon mass squared is proportional to the field H.

Analogous expressions in QCD take the form

$$m_\pi^2 = K(m_u + m_d)$$

$$m_K^2 = K(m_u + m_s),$$

where we have introduced some coefficient K. Its value is easy to obtain from the second expression, since m_s (known more or less from strange baryons) dominates: $K \approx 1.7\,\text{GeV}$. Putting this value into the first expression, we get the correct pion mass.

Continuing to the other member of the meson octet $\eta \sim (\bar{u} + \bar{d}d - 2\bar{s}s)$, one finds the following relation (for simplicity, we keep only the large m_s term):

$$m_\eta = m_K(4/3)^{1/2} = 566\,\text{MeV} \quad (\text{experiment: } 549\,\text{MeV}).$$

The agreement is good enough, which confirms that eight pseudoscalar mesons—pions, kaons, and η—are indeed Nambu-Goldstone modes.

But the ninth meson—the $SU(3)$ singlet $\eta' \sim (\bar{u}u + \bar{d}d + \bar{s}s)/\sqrt{3}$—brings in the following prediction:

$$m_{\eta'} = m_K(2/3)^{1/2} \sim 400\,\text{MeV},$$

which is a complete failure, because its experimental mass is 958 MeV! This is known as $U(1)_a$ problem: this symmetry of the QCD Lagrangian is apparently not broken spontaneously!

Understanding why it happens took some time. Formal explanation of this phenomenon is provided by the so-called *axial anomaly*. Its discussion goes beyond this introductory book: let me just state that "anomaly" refers to the situation in which a symmetry of the action is not a symmetry of the path integral, because it is violated *in its measure*. The practical explanation of how the anomaly works in QCD involues topological solitons (the instanton and its relatives; for a review, see [160]). We will discuss some elements of this theory in this chapter.

18.2 The (Nonexistent) $U_a(1)$ Chiral Symmetry

18.2.1 The Chiral Anomaly Near the UV Cutoff

The term "anomaly" does not sound very attractive, reflecting historical attitudes toward this phenomenon. Indeed, it was found unexpectedly, and, in spite of significant efforts by many theorists to eliminate it by various tricks, has survived and was eventually explained. Now its nature is well understood and the term "anomaly" no longer creates any negative feelings. In fact it leads to various new chiral effects which we discussed in Chapter 19.

One learns in lectures on classical mechanics that when the Lagrangian is invariant under certain transformations, the corresponding Neuter currents are classically conserved by the equations of motion. In particular, the electromagnetic current is conserved in Maxwell's electrodynamics.

Quantum physics is represented by path integrals, which include two factors: (1) the $\exp(-S)$ weight and (2) the measure of the integration over fields. Symmetries of the Lagrangian keep the former factor invariant, but they may or may not be the symmetries of the path integrals, depending on the second factor. In addition, the continuum $a \to 0$ limit is rather nontrivial: in perturbation theory there are divergent terms that are removed by certain prescriptions. So in principle, one may encounter surprises along the way, and this indeed was what happened to some classical symmetries. Although we will not study it explicitly, I have to mention the modern view on anomalies, pointed out by Fujikawa [178]: it is the measure of the integration over fermion fields that spoils the invariance.

Let us instead follow the path of historical development. The chiral anomaly had been discovered in the perturbative framework: people just did explicit calculations of Feynman diagrams, from the simplest to more complex ones, and at some point ran into trouble.

Recall that for the electrodynamics of an electron gas we discussed in Section 7.5 the Lindhard function, and for chromodynamics of QGP a similar polarization tensor in Section 15.3. Both come from the two-current one-loop diagram (Figure 18.1a) for the correlator of two vector currents,

$$\Pi_{\mu\nu}(x, y) = \langle j_\mu(x) j_\nu(y) \rangle. \tag{18.1}$$

Earlier we did not present the solution for $\Pi_{\mu\nu}(x, y)$ in its full glory, focusing instead on the resulting physics rather than the formalism. Yet, if one has such an expression at hand, one should check whether the result is in agreement with current conservation. Its mathematical expression, $\partial_\mu j_\mu = 0$, is trivially satisfied by free fermions. To see whether it is also true for virtual fermions in a loop, we need to check whether the divergence of the polarization tensor is zero:

$$\frac{\partial}{\partial x_\mu} \Pi_{\mu\nu}(x, y) \stackrel{?}{=} 0. \tag{18.2}$$

Without showing such calculations explicitly (it could be another exercise), let me just state that it is in fact always zero. So, to second order, the current is conserved. Changing from vector current to axial—that is, inserting two γ_5 matrices in two corners—does not change anything for massless fermions.

However when researchers proceeded to the triangular diagrams with three currents (see Figure 18.1b), they encountered a problem. Specifically, the offending diagram[6] has one axial and two vector currents $\langle j_\mu(x) j_\nu(y) j_\sigma^5(z) \rangle$, and the conservation of the axial current (even in the chiral limit $m_q \to 0$) failed to work. The divergence was not zero!

---

[6] The diagram first came out in 1949 in Steinberger's paper [172], in which he considered $\pi^0 \to 2\gamma$ decay (to which we turn below). The divergence of the axial current was not zero. Note that in 1949, of course the fermions under consideration were not quarks but "nucleons." The framework can be considered suspect from the modern perspective, but the calculation itself was correct! Still, Steinberger was perhaps so disappointed by the absence of clarity on the issue that he soon switched to experiment. Fortunately he later won the Nobel Prize for discovering two types of neutrinos.

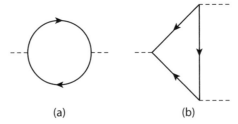

(a)   (b)

Figure 18.1. One-loop diagrams with (a) two and (b) three currents. The solid lines are quarks, while the definitions of vertices and the dashed lines depend on the particular currents. Vector and axial currents correspond to γ_μ and $\gamma_\mu\gamma_5$, for gluons there are also color matrices $t^a/2$, for photons and W there appear quark charges (or flavor matrices).

There are two ways of looking at this triangular diagram: using standard Feynman rules and momentum representation and using the Schwinger background field method. Let me briefly discuss both of them.

It is easy to construct the corresponding expression for such diagrams using standard Feynman rules. The loop has one d^4k and three $\sim 1/k$ propagators, so it must be linearly divergent. If we have a divergence in the axial current, it has one more power of momentum, so in total it is a quadratically divergent expression. The indices of two vector currents are arbitrary and may be multiplied by some polarization vectors of the two phonons. The gauge invariance of QED requires that photons be represented by two field strengths, not just the potentials, so one must obtain an expression containing

$$(k_\alpha^{(1)}\epsilon_\beta^{(1)} - k_\beta^{(1)}\epsilon_\alpha^{(1)})(k_\gamma^{(2)}\epsilon_\delta^{(2)} - k_\delta^{(2)}\epsilon_\gamma^{(2)})$$

with two photon momenta appearing explicitly. The coefficient has no dimension and can in principle be a convergent $O(1)$ integral . What structures in the four indices can it possibly be? There are two possible candidates: $g_{\alpha\gamma}g_{\beta\delta}$ and $\epsilon_{\alpha\beta\gamma\delta}$, and the calculation has shown it is the second one.

A useful observation is that for massless fermions one can consider triangular diagrams with the left- and the right-handed fermions handled separately, since neither γ_μ nor $\gamma_\mu\gamma_5$ (in three corners of the loop) change the chirality of the fermion. As was shown by Bell and Jackiw [175] and especially Adler [174], these triangular graphs cannot be regularized in such a way that the L or R contributions vanish. Furthermore, if these contributions cancel each other for a vector current, they must double for the axial current, since γ_5 reverses the sign.

The pion decay from the triangular diagram was considered by Schwinger [173].[7] He used the background field method and considered motion of the electron in some external electromagnetic field A_μ. The important point was that the currents were given the gauge-invariant (and thus nontrivial) definition

$$j_\mu(x) = \lim_{\epsilon \to 0} \mathrm{Tr}\left[\gamma_\mu S_F(x, x+\epsilon)\right], \tag{18.3}$$

---

[7] It is the same famous paper in which he also calculated the anomalous magnetic moment of the electron.

where S_F is the quark propagator in the external field. One might think that the r.h.s. simply follows from the convolution of two fermion operators, but the need to define the limit at small ϵ makes this point very nontrivial and fruitful. In particular, the current on the l.h.s. is a physical local quantity, so we want it to be gauge invariant. The r.h.s. includes a bi-local quantity. The propagator and the vertex themselves are not gauge dependent.

Schwinger added "by hand" a new factor to the r.h.s.:

$$\exp\left(ie\int_x^{x+\epsilon} dx_\mu A_\mu\right),$$

which makes it gauge invariant as well! Now, let us try to expand the propagator in powers of ϵ. The free propagator is highly singular, $\sim \hat{\epsilon}/\epsilon^4$, but this certainly should be subtracted out. There are no photons, so it is a purely vacuum loop. There are two other dangerous terms: (1) the one with singularity $O(1/\epsilon^2)$, and (2) terms like $O((\epsilon \cdot x)^2/\epsilon^2)$, which are formally of zeroth order but depend on the orientation of the epsilon vector. However, if one adopts the Schwinger "splitting prescription," both disappear and the finite answer in the $\epsilon \to 0$ limit is obtained

$$\partial_\mu \text{Tr}\left[\gamma_5 \gamma_\mu G(x,x)\right] = \frac{2\alpha}{\pi}(\vec{E}\vec{H}). \tag{18.4}$$

This relation, discussed and better understood later, is now known as the Adler-Bardeen-Jackiw anomaly. We will return to it in Section 18.2.2.

As already mentioned, the famous application of this formula was related to π^0 meson decay. In the effective nucleon theory of the 1950s, with nucleon lines instead of quarks, one had the same answer for the triangular diagram except for the factor N_c, the number of colors. So the QCD calculation has an extra N_c^2 in the pion lifetime, and the experiments precisely confirmed this factor. Ironically, the historically tricky anomaly relation turned into a triumph, one of the most convincing early arguments that QCD quarks have three colors!

Another set of applications we will discuss in Section 18.2.2 arises when one photon is a background constant magnetic field and the second is a vector current: but the axial vertex is traded into a chiral imbalance or chiral chemical potential μ_5 (which can be considered to be the zeroth component of the axial gauge field). This fixes the coefficient in the chiral magnetic effect (CME).

Now we finally turn to the main topic of this section, namely, the nonexistence of $U_a(1)$ chiral symmetry. If two vector currents are interaction with colored gluons, we just need to insert the proper color matrices into the same triangular diagram. This only changes the coefficient, and so we get the following result for the η' axial current

$$j_\mu^{\eta'} = \left(\frac{1}{\sqrt{3}}\right)(\bar{u}\gamma_\mu\gamma_5 u + \bar{d}\gamma_\mu\gamma_5 d + \bar{s}\gamma_\mu\gamma_5 s) \tag{18.5}$$

$$\partial_\mu j_\mu^{\eta'} = \sqrt{3}\left(\frac{1}{32\pi^2}\right)\epsilon^{\alpha\beta\gamma\delta} G_{\alpha\beta}^a G_{\gamma\delta}^a. \tag{18.6}$$

So this current has a nonzero divergence and thus it is not conserved. Note that the r.h.s. in the field notation is also the scalar product of the electric and magnetic fields, which here are of course the colored gluon fields of QCD. We will study the r.h.s. of this relation further in Section 18.4. This correlates with the phenomenological observation we made in the previous section: the $SU(3)$ singlet η' meson is not like all other pseudoscalars (e.g., pions and kaons). Even in the chiral limit $m \to 0$ when all pions and kaons (the Goldstone modes) must be massless, the η' meson is massive. In fact, it is surprisingly heavy, with a mass about equal to the mass of the nucleon! It is not a Goldstone mode simply because there is no $U(1)$ axial symmetry to be broken!

In summary, the QCD does not have the $U(1)_A$ symmetry. Of course, the actual violation only happens if the r.h.s. of (18.6) is nonzero, that is, in the presence of parallel chromo-electric and chromo-magnetic fields. We will discuss in Section 18.4 when precisely this situation takes place in the QCD vacuum.

For clarity, let us note that the nonsinglet axial currents of the $SU(3)_a$ chiral symmetry (containing not only the γ_5 but also some flavor matrices) is still conserved. Indeed, putting such a flavor matrix as the $SU(3)$ matrix $\lambda^a_{\text{flavor}}$ in the axial corner of the triangular diagram, one only needs to use $\text{Tr}\lambda_{\text{flavor}} = 0$ to see that the diagram is zero. For example, one may follow it comparing the η' current (with nonzero anomaly) to the π^0 current. It has the flavor matrix $\lambda^3_{\text{flavor}}$, or explicitly,

$$j^{\pi^0}_\mu = \frac{1}{\sqrt{2}}(\bar{u}\gamma_\mu\gamma_5 u - \bar{d}\gamma_\mu\gamma_5 d).$$

In this case the triangular diagrams with u and d quarks cancel each other.[8]

Let us take a theoretical digression on the UV cutoff and anomaly. The observation that the triangular diagram produces a finite answer for the divergence, which is independent of the fermion mass, may appear to be very troubling. Should one include in the loop all known heavy quarks c, b, t or perhaps all unknown ones that are so heavy that they have not yet even been discovered?

What one normally does, in (renormalized) field theories, is to define a cutoff at some UV scale, needed to make all divergent diagrams meaningful. Of course, we do not know what happens at arbitrarily large masses or how many fermion fields can exist, so a practical cutoff scale should be not higher than a TeV or so. For example, the method of Pauli and Villars introduces heavy "regulator fermions" for each species, which should be included in any loop with the opposite sign. As a result, for heavy quarks c, b, t the triangle diagram gives its contribution, at respective scales m_c, m_b, m_t, which is then canceled by the triangular diagram with the regulator mass. So, we can forget about the heavy quark contribution. For the light quarks the situation is as follows. The triangular diagram is converging at scale $\sim 1/m$, where m is the quark mass. However, the inverse of the light quark mass is much larger than a femtometer, and at such large distances any perturbative calculations are completely meaningless because of confinement. The only meaningful contribution of the light quark loops is in fact the contribution coming from its regulator.

---

[8] The case with two gluons we now consider should not be confused with the case of two photons discussed above, for the electromagnetic π^0 decay. In the latter case the u and d quark loops include the squares of their electric charges, $(2/3)^2$ and $(-1/3)^2$, respectively, and therefore do not cancel each other.

18.2.2 Chiral Anomalies, the IR Approach

Another view of the same phenomenon can be provided from the opposite end of the energy spectrum, which we discuss in this section. We will use here constant Abelian $\vec{E}$ and $\vec{B}$ fields as the simplest example of the kind, following Ambjorn et al. [176], which in turn is inspired by the Nielsen-Ninomiya theorem [177]. The final demystification of the phenomenon in its practically important form will be made in Section 18.4 in a discussion of fermionic zero modes of instantons.

We will do this in two steps. First, let us switch on only a constant magnetic Abelian field $\vec{B}$, directed along the z-axis. The relativistic Dirac equation for massless fermions is solved in the same way as is done nonrelativistically in quantum mechanics textbooks. The Landau levels are at energies

$$E_n = \sqrt{k_z^2 + e\,B(2n+1) \pm e\,B}, \tag{18.7}$$

where $\pm$ stands for spin projections. We see that there is a $k_z = 0, n = 0$ state with a negative spin term that has exactly zero energy.

Note that the second Landau term is positive—this is the so-called localization energy, present for spinless (scalar) particles as well. The $\pm$ term is the Pauli term due to interaction of the magnetic moment with the field. The zero-energy state exists only for a particular—minus—sign. This is very important. Another zero-energy state is formed for antiparticle $e \to -e$ with the opposite chirality.

As the second step, we introduce a weak electric field E_z in the same direction. This implies the gauge potential $A_0 = zE_z$, which means that the original levels are now slightly tilted along the z direction. The slightly negative energy states now start to move along the field direction z, accelerated by it to positive kinetic energy. Antiparticles or holes have the opposite charge, so they move in the opposite direction. This phenomenon does not produce total charge, but it does produce a nonzero *chiral charge*, because the moving particles and holes have opposite chiralities.

If one calculates correctly the density of the Landau levels, the total rate of chirality production rate is actually

$$\frac{d\,Q_5}{d^4x} = \frac{e^2}{8\pi^2}\vec{E}\vec{B}. \tag{18.8}$$

It is remarkable that (18.8) is exactly the same result as follows from the UV-based derivation of the anomaly discussed in Section 18.2.1.

We may further generalize the argument. The surface of the Dirac sea—the states with the zero energy—is not really special. In an external field with a nonzero $(\vec{E} \cdot \vec{B})$, there occurs a continuous *spectral flow* of states. It must happen for all energies, ranging from $E = -\infty$ to ∞, since the spectrum itself remains the same. Only the occupation numbers change: for one chirality they move upward in the spectrum, for the other they move downward.

The same phenomenon happens near the UV regulator cutoff, namely, states of one chirality sink below it and disappear from our view; the other chirality states appear from below the cutoff.

Let me now deviate from the QED and QCD we have discussed, to generalize the argument to the weak interaction sector of the Standard Model. The setting is the same, with photons/gluons in the vector corners of the triangular diagram replaced by $W^+ W^-$ fields. Since those interact with left-handed fermions only, there is no difference between the vector and axial currents: the anomaly diagrams give nonzero divergence for all of them. As the vector singlet current for quarks corresponds to the baryon number, the anomaly implies that the baryon charge can be changed by weak interactions! The divergence is proportional to $(\vec{E}_W \cdot \vec{H}_W)$, where the index W means it is the weak gauge field. Therefore, if one manages to create sufficiently strong W fields that make the r.h.s. of the anomaly equation integrated over spacetime equal to 1, one may indeed see the appearance/disappearance of the proton! One can literally create matter out of (weak) gauge fields.

The *baryonic asymmetry of the Universe*—the presence of matter and absence of antimatter—needs for its explanation three pre requisites, as pointed out by Sakharov. They are: (1) deviation from equilibrium, (2) baryon-number changing processes, and (3) violation of the charge-parity CP Symmetry. The Standard Model has them all, and yet known mechanisms can only generate the baryonic asymmetry orders of magnitude smaller than observed. This asymmetry remains one of the outstanding problems of physics today.

18.3 The Nambu-Jona-Lasinio Model

For the systems we have discussed so far, we have always started with the zeroth-order ground state, and we will do so now as well. The vacuum of QCD with (nearly) massless noninteracting quarks can be viewed as the *Dirac sea*, in which all positive-energy quark states are empty, while all negative-energy states are occupied. The excitation from negative to positive energy state leads to particle-hole pairs. Holes in the Dirac sea are the antiquarks. The gap is the quark mass. If it is zero, there is no gap between the quark and antiquark excitations.

Now let us switch on some interaction between the quarks. Experience with super-conductors tells us that even if the interaction is very weak at some scales, but is logarithmically increasing in the IR direction, it must get strong at some scale. This perhaps creates dynamical pairing of quarks near the surface of the Dirac sea. Quarks and antiquarks may form some bosonic pairs, and their Bose condensate may be created, spontaneously violating the symmetries of the Lagrangian. Furthermore, what if it creates a gap near zero energy? This gap would mean spontaneous dynamical generation of the effective quark mass.

These ideas were proposed very early, in 1961 [161, 162], obviously inspired by the BCS theory of pairing of 1957 and long before the discovery of QCD. Yoichiro Nambu was awarded the Nobel Prize in Physics in 2008 for this discovery. The model, proposed by Nambu and Jona-Lasinio, introduced four-fermion interactions with exact chiral symmetry, which gets broken spontaneously in the ground state. For a good pedagogical review, see [163].

(A skeptical reader may ask whether all the arguments above make any sense at all, since we know that color confinement occurs in QCD, due to which there are no colored states in the spectrum, with or without gaps. The first answer to this objection is that the reasoning I describe above follows the historical thinking of 1961: there were no notions of colored quarks or confinement then. The fermions in those pioneering papers were in fact assumed to be nucleons, and the gap represented their mass.)

As we have done many times now, let us start with some preliminary discussion of possible bosonic condensates. We will (somewhat illogically) use the language of the naive (i.e., nonrelativistic) quark model. Assuming that the vacuum must be Lorentz invariant and have the unbroken color group, we further assume that the condensate should have the mesonic $\bar{q}q$ form. So, unlike in superconductors, a Cooper pair is not made of two particles but of a particle-antiparticle pair. Unbroken parity relates to orbital momentum by $P = (-1)^{L+1} = 1$ (here, the +1 comes from the intrinsic fermion-antifermion parity; see QFT textbooks for an explanation). Thus the orbital momentum $L = 1$, while the total angular momentum of the pair $\vec{J} = \vec{L} + \vec{S}$ should remain zero. The only solution then is $S = L = 1$.[9] The charge parity of the condensates is then $C = (-1)^{L+S} = 1$.

The NJL model [161] was the first microscopic model that attempted to derive dynamically the properties of chiral symmetry breaking and pions, starting from some *hypothetical four-fermion interaction*:

$$L_{\text{NJL}} = \bar{\psi} i \not{\partial} \psi + G[(\bar{\psi}\psi)^2 + (\bar{\psi} i \gamma_5 \vec{\tau} \psi)^2], \tag{18.9}$$

where G is some unknown coupling constant added to the usual free-fermion Lagrangian.

Recall that a Lagrangian is kinetic minus potential energy, so the four-fermion term corresponds to attractive interaction. In quantum mechanics courses, one inevitably starts with the attractive potential well,

$$V(r) = -V_0 \text{ for } r < R,$$

present at distances less than some R and zero otherwise. The NJL model is a similar example in QFT, except that it acts in momentum space, for momentum modulus $k < \Lambda$. Small momenta correspond to large distances, and NJL interaction switches off at small $r < 1/\Lambda$.

Exercise *Calculate the effective NJL potential in coordinate space by doing a three-dimensional Fourier transformation from momentum to coordinate representation. Plot the resulting function. (Hint: Integrate over $z = \cos\theta$, the angle between $\vec{k}$ and $\vec{r}$, first.)*

The NJL interaction is a sum of squares of the scalar and pseudoscalar currents, referred to below as σ and $\vec{\pi}$, respectively. The vector τ in the latter stands for the three Pauli matrices, the isospin $SU(2)$ generators. The $SU(2)_a$ chiral symmetry we discussed before mixes these currents, but one can show that the sum of their squares remains invariant under the exact chiral symmetry.

---

[9] This is reminiscent of the superfluid B phase of ^{3}He.

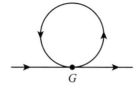

Figure 18.2. The mass operator for the NJL model.

This interaction is a pointlike local attraction between quarks. If this Lagrangian is understood literally, true for all momenta scales, it will generate bad nonrenormalizable QFT with uncontrolled power divergences. But NJL defined it as some effective Lagrangian, assuming that it disappears above a certain cutoff, at $|p| > \Lambda$. Thus the model has another parameter, Λ. The main idea is that if G is strong enough, it can indeed make the normal perturbative vacuum unstable and create the quark-antiquark condensate.

We have already discussed spontaneous breaking of symmetry for complex bosonic fields, when the symmetry is $U(1)$, an arbitrary phase. Degenerate vacua then constituted a one-dimensional circle. Now we are dealing with $SU(2)_a$ chiral symmetry breaking: the corresponding space of possible vacua should be three-dimensional, since this is the number of group generators. It is indeed a 3-sphere $O(3)$ because four fields, $\sigma, \vec{\pi}$, have a relation $\sigma^2 + \vec{\pi}^2 = const$. The presence of small quark mass adds a term $m\sigma$, tilting the effective potential slightly in favor of the σ field, which therefore gets a nonzero VEV. In terms of the original quark fields it is $\langle \bar{\psi}\psi \rangle \neq 0$, so it is called the *quark condensate*. Its numerical value is known empirically and is roughly

$$-\langle \bar{\psi}\psi \rangle \approx (240\,\text{MeV})^3 \approx 1.8\text{fm}^{-3}.$$

This means a couple of quark-antiquark bound pairs per fm^3, for each of the quark three flavors u, d, s separately. So the vacuum condensate is rather dense, an order of magnitude denser than the valence quark density in nuclear matter of $3 \times 0.16\,\text{fm}^{-3}$.

We will discuss its physical meaning and nature later, noting now only that scalar $\bar{\psi}\psi$ can be written as $\bar{\psi}_L\psi_R + \bar{\psi}_R\psi_L$, and therefore it really couples left and right components of the quark fields. This VEV is changed under axial transformation, which rotates R,L components into different directions. So the condensate, as expected, does break the chiral symmetry.

The simplest diagram is the mass operator shown in Figure 18.2. The condensate is related to the mass by a simple relation $m = -2G\langle \bar{\psi}\psi \rangle$, so we will use the mass or the VEV interchangeably in the calculation below.

In the lowest order in NJL-coupling G, the only diagram is the two-loop diagram with two massless propagators. Since the momentum integral is supposed to be done only for $|p| < \Lambda$, there are no divergences.

In subsequent orders there are many other diagrams. They are divided into a set preserving two-loop structure with some mass operator, and all others. In the mean field approximation, the contribution to the vacuum energy is represented by diagrams of the first kind only: one just uses some nonzero mass in the propagators. The expression for the (free) energy takes the obvious kinetic+potential energy form

$$F = (\text{kinetic one loop}) + G(\text{loop})^2, \tag{18.10}$$

where each loop is, as usual, a Matsubara sum times the momentum integral:

$$\text{loop} = \sum_n \int^{\Lambda} \frac{d^3 p}{(2\pi)^3} \text{Tr}\left(\frac{1}{(\omega_n - \mu)\gamma_4 + \vec{p}\vec{\gamma} + m}\right), \tag{18.11}$$

in which, the γ_μ are (Euclidean) gamma matrices. For generality, we consider here a nonzero temperature T and chemical potential case.

Before calculating any sums/integrals, let me point out the main logical assumption that the mean field approximation uses: the value of the mass m should be such as to minimize[10] the (free) energy. Differentiating the energy over m and putting it to zero, we derive the *gap equation* for the NJL model, which, after canceling common loops and taking the Matsubara sum via standard Sommerfeld-Watson method, has the form

$$m = 4 N_c N_f m \int^{\Lambda} \frac{d^3 p}{(2\pi)^3} \frac{G}{E_p} \left[1 - f^+(\vec{p}, \mu) - f^-(\vec{p}, \mu)\right] \tag{18.12}$$

with the virtual part plus particles and antiparticles and Fermi distribution functions $f_p^\pm \equiv 1/[e^{(E_p \pm \mu)/T} + 1] = f^\pm(\vec{p}, \mu)$. The quasiparticle energy here is that for a massive quark, $E_p^2 = p^2 + m^2$. In vacuum the expression is the same with only 1 in the brackets, without the f terms. We will focus on it first.

The gap equation is used to calculate m. Note first that there is always a trivial solution $m = 0$. If $m \neq 0$ however, we can cancel it out. Then the r.h.s. has the mass m only inside E_p, and so we have to calculate the integral and find its m-dependence: when the r.h.s. is 1, this is *a* solution. To know whether it is *the* solution we need to calculate the thermodynamical potential (or just energy of the vacuum, if $T = 0$, $\mu = 0$) and show that this state is lower than the one with zero m.

The gap equation in vacuum ($T = \mu = 0$) has a simpler form:

$$\frac{\pi^2}{2 G N_c N_f} = \int_0^{\Lambda} dp \frac{p^2}{\sqrt{p^2 + m^2}}. \tag{18.13}$$

Exercise *Use Maple or Mathematica for the r.h.s. integral, plot it as a function of m/Λ, and solve numerically the gap equation.*

With the model parameters mentioned, the gap, or the *constituent quark mass* turns out to be

$$m \approx 350 \, \text{MeV}. \tag{18.14}$$

---

[10] The reader may be amused to learn that minimizing the ground-state energy by the values of the gap was in fact used in the very first BCS paper on the superconductor: NJL simply follows that paper's lead. One may thus think of this procedure as a variational one-parameter set of wavefunctions, out of which "the best" one, with the lowest energy, is then chosen.

So what comes from this exercise are, in particular, two important statements:

1. The expected scale for normal (i.e., non-Goldstone) $l = 0$ two-quark mesons is

$$M_{\text{mesons}} \approx 2m \approx 700 \, \text{MeV},$$

to be compared with spin-0 $m_\sigma \approx 550 \, \text{MeV}$ and spin-1 $m_\rho \approx 770 \, \text{MeV}$.

2. The expected mass scale for baryons is

$$M_{\text{baryons}} \approx 3m \approx 1000 \, \text{MeV},$$

to be compared with spin-1/2 $m_N \approx 938 \, \text{MeV}$ and spin-3/2 $m_\Delta \approx 1200 \, \text{MeV}$.

Not bad, for a very schematic model!

Exercise *Show that the nonzero solution of the gap equation for m appears only above a critical value G_c and evaluate its value. (Note that it is different than for superconductivity, in which case the coupling runs logarithmically and eventually becomes large enough, however weak its initial value!)*

Exercise *Calculate the integral in the gap equation and fix the two parameters, G, Λ, of the model, which need to be fixed by some experimental inputs. Those are $\langle \bar{u}u \rangle = \langle \bar{d}d \rangle = (-0.25 \, \text{GeV})^3$ and $f_\pi = 93 \, \text{MeV}$. This gives*

$$G = 5 \, \text{GeV}^{-2}, \quad \Lambda = 0.65 \, \text{GeV}. \tag{18.15}$$

The details can be found in [163].

The model can also tell us what happens at nonzero temperatures and densities. The expression for the thermodynamical potential at nonzero T can be put in the form

$$\Omega(m) = \frac{m^2}{4G} - 2 N_c N_f \int \frac{d^3 p}{(2\pi)^3} E_p - 2 N_c N_f T \int \frac{d^3 p}{(2\pi)^3} \log \left\{ [1 + e^{-\beta(E_p + \mu)}][1 + e^{-\beta(E_p - \mu)}] \right\},$$

where $\beta = 1/T$ and $\gamma = 2 N_c N_f$ is the degeneracy factor. The first term is the "potential energy" (rewritten using the gap equation), while the latter two terms are just the sum of all energies of all fermionic levels, in the negative energy states of the Dirac sea (the second term) and in the positive energy ones, the heat bath (the third).

The calculations at finite temperature and density (omitted) lead to the phase diagram. For 40 years, from the original NJL paper to the end of the century, the NJL phase diagram calculated looked like that shown in Figure 18.3.

After the papers on color superconductivity peaked around 1998, it was realized that the good old NJL model should have actually predicted color superconductivity as well! We will discuss this interesting subject in Chapter 19: let me here just say that color superconductivity refers to Cooper pairing of two quarks. It is "color," because two quarks are never color neutral and carry a charge. For completeness, I mention the corrected phase diagram [164], from which I have borrowed Figure 18.4. (The lines to

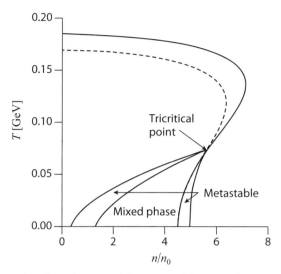

Figure 18.3. The phase diagram of the NJL model, in the density-temperature plane. Adapted from Schwarz et al. [164].

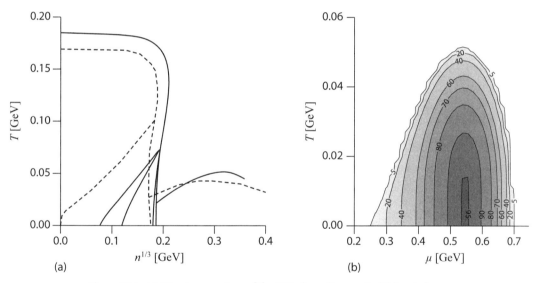

Figure 18.4. (a) Direct comparison of the NJL phase diagram (solid lines) shown as a function of T and $n^{1/3}$. (Ignore the dashed lines corresponding to an older paper.) (b) The diquark gap parameter Δ, shown as a function of μ and T. The contour values are given in MeV. Graphs are redrawn from Schwarz et al. [164].

the right of $n^{1/3} > 0.19$ are calculated in [164] and correspond to the superconducting phase transition; the rest of the diagram is older.)

Finally, few parting comments about the NJL model.

1. It was the first bridge between the BCS theory of superconductivity and QFT, leading the way to the Standard Model. It first showed that the vacuum can be truly nontrivial, a superconductor of a kind, with the mass gap $\Delta = 330-400$ MeV, known as the constituent quark mass.

2. The symmetries of the NJL model are not quite right; there is no $U(1)_A$ breaking, and η (the analog of η' for the $N_f = 2$ version we are considering) is as massless as the pion. It should not be, and we have already noted that experimentally η' is a very heavy meson, unlike its eight Goldstone relatives.

3. The NJL model has two parameters: the strength of its four-fermion interaction G and the cutoff $\Lambda \sim 0.8-1$ GeV. The latter regulates the loops (the model is nonrenormalizable, which is OK for an effective theory) and is known as the "chiral scale." For example, the size of the constituent quark is of this scale. We will return to the question of whether the NJL-type interaction exists in QCD and why its parameters have such values in Section 18.4. We will relate Λ to the typical instanton size ρ, and G to a combination $n\rho^2$ of the size and density of instantons.

4. One may wonder whether the model can be refined so as to be able to calculate the constituent quark mass without using the variational principle. Also one may wonder whether one can include many higher-order diagrams that can predict specific values for meson and baryon masses. Perhaps it could have been done, but in practice there were mostly conflicting papers on these issues. The truth is that since the NJL model is nonrenormalizable, higher-order diagrams are not well defined. The results depend on the precise shape of the momentum cutoff, and with extra parameters, the model quickly looses its predictive power. A more general lesson: trying to improve schematic models rarely pays off.

18.4 Gauge Topology and Chiral Symmetry Breaking

Topology is a mathematical concept, which can be crudely defined as defining properties of manifolds that are dependent on their shape but not on their metric.

Gauge field configurations $A_\mu^a(x)$ are, mathematically speaking, some maps from the internal (color, spin) space to the external space x. Any map can be characterized by certain topological classes. The exact form of these classes crucially depends on details, such as the dimension of the space discussed. The simplest case is a one-dimensional boundary between two phases, known also as a "kink." It physically occurs in systems with discrete symmetries, as *domain walls*, say, between spin-up and spin-down phases of the Ising model. Both states correspond to (degenerate) minima of the free energy, and the kink profile interpolates between them. We will not further discuss the one-dimensional case.

In two dimensions the representative example is a quantized vortex, which we have already discussed. Let me review this case, emphasizing the topology involved. The space boundary, at large distances from the vortex, is a large circle on a plane, the points on which can be labeled by one parameter, say, the azimuthal angle ϕ. The internal space of the field—say, of one complex scalar $\phi(x)$—has the modulus and the phase we will call $\alpha(x)$. Maps from the latter to the former ($\alpha \to \phi$) can be done only with some integer n, known as the winding number (the number of times one circle traverses the other). For example, if $\phi_i \sim x_i$, $i = 1, 2$, it is a single winding, so $n = 1$.

In three dimensions we deal with the monopoles discussed in Chapter 17. The three-dimensional topology manifests itself via a hedgehog map, from the inner space of adjoint Higgs field ϕ^a, $a = 1, 2, 3$, to the radial direction of the large $O(2)$ sphere r^a, $a = 1, 2, 3$. It is the single winding $n = 1$, as in the previous example.

Now comes new topological objects in four dimensions,[11] the gauge field instantons. As you may by now have guessed, its internal space can also be mapped onto a large $O(3)$ sphere in some interesting way.

Before giving the solution itself, let me introduce certain notations important for four-dimensional solitons. The first point is that in four dimensions, the symmetry group is $O(4)$, which has six generators, which can be divided into 3+3 of two $SU(2) = O(3)$ subgroups. The subgroups can be shown to act separately on self-dual and anti-self-dual gauge fields:

$$G_{\alpha\beta} = \pm\tilde{G}_{\alpha\beta} \tag{18.16}$$

$$\tilde{G}_{\alpha\beta} \equiv \frac{1}{2}\epsilon_{\alpha\beta\gamma\delta}G_{\gamma\delta}.$$

Note that the duality in question is also an interchange of electric and magnetic fields, so it is related to the electric-magnetic duality we discussed before.

Furthermore, the gauge action can be identically rewritten as

$$S = \frac{1}{4g^2}\int d^4x\, G^a_{\mu\nu}G^a_{\mu\nu} = \frac{1}{4g^2}\int d^4x\left[\pm G^a_{\mu\nu}\tilde{G}^a_{\mu\nu} + \frac{1}{2}\left(G^a_{\mu\nu}\mp\tilde{G}^a_{\mu\nu}\right)^2\right]. \tag{18.17}$$

The first term is the topological charge: its definition in four dimensions[12] is

$$Q = \frac{1}{32\pi^2}\int d^4x\left(G_{\alpha\beta}\tilde{G}_{\gamma\delta}\right). \tag{18.18}$$

This combination of field strengths can also be written simply as the product $(\vec{E}\vec{B})$.

The second term in (18.17) is a square and so it is always positive: thus the action in proper units is larger than the topological charge. If the field is self-dual (anti-self-dual, for the lower sign), the second term is zero, so in this case the inequality becomes equality:

$$S = \frac{8\pi^2}{g^2}|Q|. \tag{18.19}$$

This relation has very important consequences. The r.h.s. is topological, which makes it insensitive to any perturbations smooth enough to not change the topological class. The l.h.s. is the action, defining the weight in the path integral. Therefore, we find that the weights of the topological objects in question are insensitive to their perturbations.

---

[11] The fourth dimension is the Euclidean time.

[12] As a side remark, note that it can be generalized to even higher dimensions by convoluting the appropriate $2n$-index epsilon with n field strengths. In odd dimensions, however, one has to use at least one potential A_μ: the issue of gauge invariance of the corresponding topological invariant, known as the Chern-Simons number, is too involved to be discussed here.

Projection to self-dual (anti-self-dual) fields can be done with the 't Hooft symbol $\eta_{a\mu\nu}$. It is defined by

$$
\eta_{a\mu\nu} = \begin{cases} \epsilon_{a\mu\nu} & \mu,\ \nu = 1,\ 2,\ 3, \\ \delta_{a\mu} & \nu = 4, \\ -\delta_{a\nu} & \mu = 4. \end{cases}
\tag{18.20}
$$

We can also define $\bar{\eta}_{a\mu\nu}$ by changing the signs of the last two equations in (18.20).

We can look for a solution by using the $O(4)$ symmetric ansatz

$$
A_\mu^a = \frac{2}{g}\eta_{a\mu\nu}x_\nu f(x^2)/x^2,
\tag{18.21}
$$

where the unknown function $f(x^2)$ has to satisfy the radial equations. The boundary condition for is $f \to 1$ as $x^2 \to \infty$; otherwise the commutator is not canceled by the derivative, and the field strength becomes nonzero. Substituting (18.21) into the definition of the field we get, after some algebra

$$
G_{\mu\nu}^a = -\frac{4}{g}\left[\eta_{a\mu\nu}\frac{f(1-f)}{x^2} + \frac{f(1-f)-x^2 f'}{x^4}(x_\mu\eta_{a\nu\gamma}x_\gamma - x_\nu\eta_{a\mu\gamma}x_\gamma)\right],
\tag{18.22}
$$

and the action is proportional to its square. With a Lagrangian, depending on functions f, f', one can obtain the equations of motion in the standard way.

Exercise *Derive equation (18.22) in detail, and derive the action in terms of the function f and its derivative. Find a mechanical Lagrangian to which it corresponds, and solve its equation of motion using energy conservation. (Hint: Define $\eta_{\alpha\mu\nu}$ in Maple or Mathematica, and make it perform all summations of indices explicitly.)*

Instead of doing so we will use a shortcut following the original Belavin et al. (BPST) work [179] using self-duality of the presumed solution. Indeed, the expression for the dual field is

$$
\tilde{G}_{\mu\nu}^a = -\frac{4}{g}\left[\eta_{a\mu\nu}f' + \frac{f(1-f)-x^2 f'}{x^4}(x_\mu\eta_{a\nu\gamma}x_\gamma - x_\nu\eta_{a\mu\gamma}x_\gamma)\right],
\tag{18.23}
$$

and from self-duality we get a simpler first-order differential equation:

$$
f(1-f) - x^2 f' = 0.
\tag{18.24}
$$

This equation and the boundary condition result in $f = x^2/(x^2 + \rho^2)$, which gives the famous BPST instanton solution [179]:

$$
A_\mu^a(x) = \frac{2}{g}\frac{\eta_{a\mu\nu}x_\nu}{x^2 + \rho^2}.
\tag{18.25}
$$

Here ρ is an arbitrary parameter characterizing the size of the instanton. As for the potential we discussed in Section 18.3, its appearance is dictated by the scale invariance

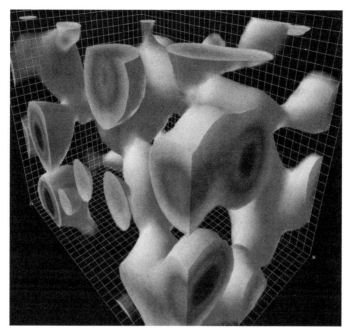

Figure 18.5. The distribution of a topological charge in a QCD vacuum. (From the Adelaide group, 2000), credited to Derek Leinweber, www.physics.adelaide. edu.au/theory/staff/leinweber.

of the classical Yang-Mills equations. The action density is highly localized:

$$(G^a_{\mu\nu})^2 = \frac{192\rho^4}{(x^2+\rho^2)^4}, \tag{18.26}$$

while the gauge potential is long range, $A_\mu \sim 1/x$. The invariance of the Yang-Mills equations under coordinate inversion implies that the singularity of the potential can be shifted from infinity to the origin by means of a (singular) gauge transformation $U = i\hat{x}_\mu \tau^+$. The gauge potential in singular gauge is given by

$$A^a_\mu(x) = \frac{2}{g} \frac{x_\nu}{x^2} \frac{\bar{\eta}_{a\mu\nu}\rho^2}{x^2+\rho^2}. \tag{18.27}$$

This singularity at the origin may look "unphysical," pure gauge, but this gauge is the only one in which many instantons can be superimposed.

Instanton ensembles in the QCD vacuum form the so-called instanton liquid, studied analytically and in separate numerical simulations (for a review, see [160]). Instead of describing this theory in detail, let me just show a snapshot of the typical topological charge distribution as seen in lattice numerical simulations (Figure 18.5).

Writing the gauge field as a sum of that of a classical soliton and quantum fluctuations

$$A^a_\mu = A^{a,\text{class}}_\mu + a^a_\mu, \tag{18.28}$$

we can put it into the action. Since the soliton is an extremum, there is no linear term, and expansion starts with quadratic terms $O((a^a_\mu)^2)$. Thus the corresponding integrals over fluctuations start with Gaussian path integrals, which we know can be reduced to the determinants of certain operators.

Higher-order corrections to topological solitons can be calculated using Feynman diagrams. They contain some nontrivial elements, and so far the corrections have been explicitly calculated only for quantum-mechanical instantons in two and three loops (see [182, 183]). Analogous QFT calculations are in progress.

Topological solitons have two types of zero modes: bosonic and fermionic. The former modes simply indicate that in some directions in functional Hilbert space, the action does not change. We know all those directions, since they simply represent classical symmetries of the solitons, such as displacements, scale change, and rotations. The bosonic determinant is in denominator, and a zero determinant produces infinity: this simply means that there exists not just one soliton, but perhaps many, with different locations, sizes, orientations and so forth.

Fermionic modes are solutions of the Dirac equation in the field of the topological soliton with some eigenvalues

$$i \, \displaystyle{\not}D \psi_\lambda = \lambda \psi_\lambda, \tag{18.29}$$

and integration over the fermions—always Gaussian—produces a factor

$$\det(i \, \displaystyle{\not}D) = \Pi_{\text{all modes}} \lambda_i \tag{18.30}$$

in the QCD partition function. So the existence of zero fermionic modes of the solitons implies, at first thought, that any configuration with a nonzero topology should be simply vetoed (meaning never included) in the ensemble of fields.

The Dirac equation (18.29) is for the massless quark. If the mass is nonzero, then it will appear in the equation and the determinant. The conclusion would then be that the contribution of instantons to the QCD partition function needs to be proportional to a product of all light quark masses, which is

$$n_{\text{inst}} \sim \frac{m_u m_d m_s}{\Lambda_{\text{QCD}}^3} \sim 3 \cdot 10^{-4}, \tag{18.31}$$

where for normalization we have $\Lambda_{\text{QCD}} = 200\,\text{MeV}$. This is what was concluded in 't Hooft's paper [180], in which those zero modes were discovered. If true, the instanton-induced effects would be negligibly small.

Fortunately, it is not true, as I now explain in several steps.

First, consider a configuration with an instanton and antiinstanton, say, separated by some distance exceeding their sizes so they are nonoverlapping. The total topological charge $Q = 0$, so no zero modes need to be present. A little thinking reveals that there would be two near-zero modes $\lambda_\pm = \pm |H|$, where H denotes the so called "hopping" (or nondiagonal) matrix element of the Dirac operator

$$H^{ij} = \langle \psi_0^{\text{antiinstanton } j} | \, \displaystyle{\not}D \, | \psi_0^{\text{instanton } i} \rangle \tag{18.32}$$

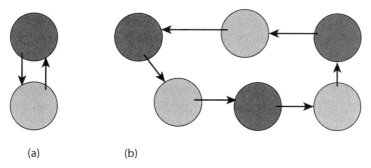

Figure 18.6. The Feynman closed quark loop diagrams involving (a) $N = 2$ and (b) $N = 6$ instantons. The fermionic determinant can be viewed as a sum of all closed loop diagrams.

exactly as in the problem with quantum states of, say, the H_2 molecule containing two identical atoms. Examples of the diagrams exemplifying such hopping are shown in Figure 18.6.

The lesson: topological solitons with fermionic zero modes actually become some new multifermion operators in the effective action, known as 't Hooft vertices. Such objects can exchange fermions among themselves, generating new Feynman diagrams.

Knowing typical parameters of the instantons, we can readily estimate the typical magnitude of this amplitude:

$$\langle H^{ij} \rangle \sim \frac{\rho_1 \rho_2}{R^3} \sim \frac{1}{1 \,\text{fm}} \left(\frac{0.3 \,\text{fm}}{1 \,\text{fm}} \right)^2 \sim 20 \,\text{MeV}. \tag{18.33}$$

Here $\rho_1, \rho_2 \sim 0.3$ fm are the typical sizes of each instantons, and $R \sim 1$ fm is the typical distance between them in the QCD vacuum. So, even though the hopping amplitude is small, it is still not as small as the light quark masses!

Now for the next step. If the previous argument works for one instanton-antiinstanton pair, it can equally well be applied to any number of them. For example, six such objects are shown in Figure 18.6b. Quarks can make longer paths: the only requirement (coming from chirality) is that they should alternate instantons with antiinstantons on their way. So, we recognize the possible appearance of clusters with arbitrary even number $2N$ solitons.

Now, let us ask: What should happen in the thermodynamic limit, when the volume of the QFT vacuum considered (say, the volume of the lattice used for its numerical simulations) goes to infinity $V \to \infty$? Does the cluster size N grow with V? The answer to this dynamical question can be positive or negative: it depends on the temperature, and on the number of flavors and colors in the theory. If the answer is positive, one find the "instanton liquid" phase [181], in which the chiral symmetry is broken, and there is nonzero VEV $\langle \bar{\psi} \psi \rangle \neq 0$. If the answer is negative (that is, N remains finite at $V \to \infty$), then the chiral symmetry remains unbroken. In the former case, which is realized in QCD-like theories with $N_f / N_c \sim 1$ and in real-world QCD, there are very long quark clusters with $N \sim V$.

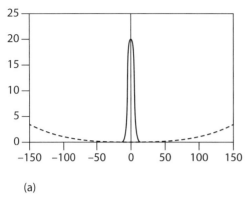
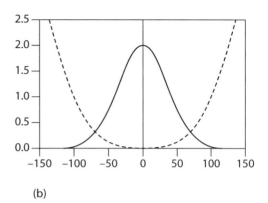

(a)   (b)

Figure 18.7. Schematic Dirac eigenvalue (in MeV) spectrum. (a) Spectrum for One instanton; the narrow peak is supposed to represent the delta function $\delta(\lambda)$. (b) A collectivized zero-mode zone (ZMZ) in the instanton ensemble. The dashed line is the contribution of the perturbative states.

Another way to discuss this issue is by considering the eigenvalue spectrum of the Dirac operator (see Figure 18.7). Figure 18.7a corresponds to one instanton, which has one zero mode (for ease of presentation, we draw the delta function with some small finite width). If there are only finite-N clusters, there would be finite eigenvalues, positive or negative, but none are close to zero. If $N \sim V \to \infty$, there appear eigenvalues $\sim 1/V \to 0$, and the spectrum has a finite density at zero (Figure 18.7b). Such collectivization of eigenvalues into a *zero-mode zone* (ZMZ) with a finite width is analogous to collectivization of electron wavefunctions in, say, a liquid metal.[13]

The number of quasizero quark modes on the lattice is given by the mean number of instantons in the lattice 4-volume, which is typically $O(10)$. This is a tiny number compared with the total number of fermionic modes, which is of the order 10^6 or so. And yet, those collectivized zero modes—the ZMZ states—play a very important role. This is nicely demonstrated by an experiment made by the Graz lattice group [184]: they simply remove a certain number k of the Dirac states with the lowest eigenvalues from the fermionic propagators, which are used to calculate the masses of mesons and baryons. The results of such "surgery" are shown in Figure 18.8, as hadronic masses versus the number of modes removed. As can be seen, by removing $O(10)$ ZMZ states (occupying a strip of width of only ~ 30 MeV around zero) out of $O(10^6)$, one can modify the masses substantially!

Some states—like the nucleon[14] shown by the lower line in Figure 18.8a—go up; most go down. What is also clear from these plots is that after this "surgery," the chiral symmetries are apparently restored, since baryons and mesons reduce to symmetric

---

[13] In a solid metal, electron wavefunctions are also collectivized, but in this case there is a crystalline background, with long-range order, which creates sharp allowed and forbidden zones. In a liquid metal, only local order prevails, so the zone boundaries are not sharp.

[14] Recall that our own weight is mostly due to the mass of the nucleon. So, it is interesting to learn that all of us seem to lose about 1/4 of our weight due to topological solitons in the vacuum!

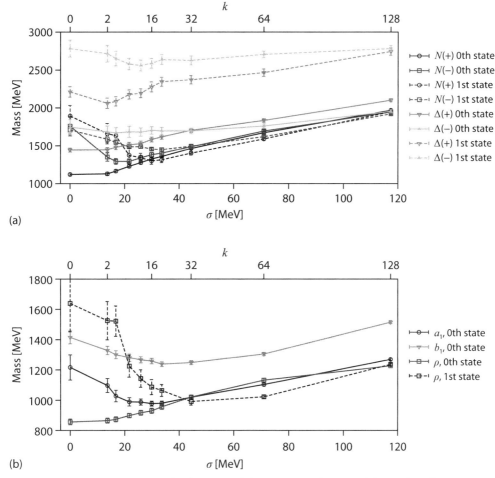

Figure 18.8. Masses of (a) baryons and (b) mesons as a function of the number of modes removed (upper scale) or as a function of the width of the strip σ in which the modes were removed (lower scale) from [184].

chiral multiplets. For example, the nucleon is united with its "chiral twin," the negative parity resonance, the vector meson ρ with axial vector meson A_1, and so forth.

This procedure is analogous to, say, defining the wavefunctions of all electrons of a normal metal and then removing from the calculation the contributions of all electrons located in a strip near the surface of the Fermi sphere. Such surgery, effectively making a normal metal into a superconductor, would obviously produce quite dramatic changes to its specific heat or conductivity.

Note that the ZMZ width revealed by this surgery is, as expected, of the order of the hopping amplitude estimated above. It is larger than m_u, m_d quark masses a of few MeV, but not by much. This apparent suppression of quark hopping tells us something very important about the QCD vacuum: the topological solitons in it are still well separated, forming a somewhat dilute system.

19 | Chiral Matter

The term "chiral matter" is rather recent, and we clearly need to start with its explanation.[1] What is meant by it is some form of matter in which the following two conditions hold:

1. a certain imbalance between the occupation of the left- and right-handed fermions is created; and
2. the lifetime of the chirality τ_5 is sufficiently long, compared to the timescale of the phenomenon considered, so that quantum effects induced by the chiral anomaly (discussed in Chapter 18) can be observed.

In Section 18.4 we discussed breaking of the chiral symmetry of QCD. We will see that in the QCD vacuum, $T = 0$ and, more generally, for subcritical temperature $T < T_c$, the nonzero quark condensate violates the chiral symmetry present in the Lagrangian. In contrast, the QGP at supercritical temperature $T > T_c$ keeps the chiral symmetry unbroken, and therefore, provided chiral imbalance is somehow created, it is our first example of "chiral matter." (Historically, it was also the first in which many effects to be discussed in this chapter were first considered.) Unfortunately, the study of QGP is quite expensive—one needs to have a relativistic collider handy to create it—and so one may ask whether there are other, cheaper alternatives.

The Dirac equation for a free massive fermion conserves the vector current (the fermion number),

$$\partial_\mu \left(\bar\psi \gamma_\mu \psi \right) = 0, \tag{19.1}$$

but not the axial current (the difference between the number of left- and right-chirality components):

$$\partial_\mu \left(\bar\psi \gamma_\mu \gamma_5 \psi \right) = 2m \left(\bar\psi \gamma_5 \psi \right), \tag{19.2}$$

because the mass term connects the left and right components of the fields.

---

[1] For a detailed discussion of the history of the chiral effects see, e.g. [126].

In QGP one finds two light quarks, called up and down or u, d. Their masses are of the magnitude of a few MeV, much smaller than any other relevant scales, and since they play no role, we can for simplicity consider them to be zero. Then the r.h.s. of equation (19.2) is also zero, and so the axial charge is also conserved. A simpler way to understand the situation is to state that the number of left- and right-polarized quarks are conserved independently. This is the *chiral symmetry*.

Chiral electron quasiparticles can also exist in the so-called *semi-metals*. The terminology is as follows. As the reader surely knows, the metals have the chemical potential inside the allowed zone of the electron state, and thus there are Fermi spheres with nonzero surface areas. Insulators have the chemical potential located inside the forbidden energy zone, and thus there is no Fermi sphere of excitations with small energies. Metals have free electrons to move in external fields. Insulators do not.

Semi-metals are in between, with pointlike contacts of the valence and conduction bands. We already mentioned graphene in Section 7.6, which has such an electron spectrum, but graphene is a two-dimensional material, while the anomaly phenomenon we need exists in $1+3$ dimensions.

The Dirac semi-metals have the so-called Dirac point, shared by "left" and "right" fermions, near which a linear relativistic-like dispersion relation is valid. There exist also Weyl semi-metals, for which the two chiralities have two separate touching points. The quasiparticle modes near those points have the following Hamiltonian:

$$H = \pm v_F \vec{\sigma} \cdot \vec{k}, \tag{19.3}$$

where $\vec{\sigma}$ is the Pauli spin matrices. It is of the kind originally suggested by Weyl for uncharged massless fermions. The first observation of a chiral magnetic effect (CME) in a condensed-matter setting [128] used the zirconium pentatelluride, $ZrTe_5$, a three-dimensional Dirac semimetal. We will discuss this experiment in Section 19.2. Before that, in Section 19.1, we consider general electrodynamics of chiral matter. In Sections 19.3 and 19.4 introduce the chiral vortical effect and the chiral wave phenomena.

19.1 Electrodynamics in CP-Violating Matter

Chirality—the product of spin and momentum $(\vec{s}\,\vec{p})$—for massless particles, fermions, or bosons (e.g., photons) has two values, commonly refered to as left- and right-handed polarizations. It is odd under P (parity) transformation $\vec{x} \rightarrow -\vec{x}$, since momentum changes sign and spin does not. Thus matter in which there is a left-right imbalance is P-odd (that is, not the same as its mirror image).

The subject of this section however goes well beyond violation of P parity, to that of CP-violating media. The physics of CP-violating media was originally discussed in the context of *axion* dynamics, which is still hypothetical and is well beyond the scope of this book. However it elucidated many interesting new effects that the existence of an effective pseudoscalar field $\theta(x)$ bestows. Our discussion of it, in the framework of modified Maxwell-Chern-Simons (MCS) electrodynamics, follows Kharzeev [130].

While Maxwell theory historically emerged, term by term, from certain ingenious experiments, in modern textbooks all of these terms are derived from a single *principle of gauge invariance*, requiring the change of all derivatives of the charged fields to their covariant form,

$$\partial_\mu \to D_\mu = \partial_\mu - ie\,A_\mu, \tag{19.4}$$

plus the statement that the gauge action must be given by the only four-dimensional gauge-invariant operator:

$$L_{\text{Maxwell}} = -\frac{1}{4} F^{\mu\nu} F_{\mu\nu}. \tag{19.5}$$

The last point is required for consistency with expected symmetries, such as CP invariance. The usual kinds of media modify Maxwellian theory only slightly, renormalizing the squares $(\vec{E})^2$, $(\vec{B})^2$ by coefficients known as electric and magnetic permitivities, ϵ and μ, respectively.

However, if the condition of CP invariance is not required, we have no reason not to add a term $(\vec{E}\vec{B})$ of the same dimension. We will use it in the form

$$L_{\text{MCS}} = -\frac{1}{4} F^{\mu\nu} F_{\mu\nu} - A_\mu J^\mu - \frac{c}{4}\theta\,\tilde{F}^{\mu\nu} F_{\mu\nu}, \tag{19.6}$$

where c is some coefficient, and θ will be treated as a time- and space-dependent field. If one added its kinetic and potential energies, this "axion" field θ could be upgraded to a separate dynamical entity: but we will not do so and will simply think of it as a matter-induced coefficient in a Lagrangian.

Now, if this field is just a constant, $\theta = const(t, \vec{x})$, we find that the last term in equation (19.6) does not change the equations of motion, because the added term is in fact a full divergence:

$$\tilde{F}^{\mu\nu} F_{\mu\nu} = \partial_\mu J_{\text{CS}}^\mu, \tag{19.7}$$

where the current is known as Chern-Simons current

$$J_{\text{CS}}^\mu = \epsilon^{\mu\nu\rho\sigma} A_\nu F_{\rho\sigma}. \tag{19.8}$$

The 4-volume integral of the full divergence can be rewritten as three-dimensional integral of the current flux over the volume boundary (that of a "large sphere"), which is usally considered to be zero, provided all fields vanish on this boundary.

However, if θ is time or space dependent, the derivative can be passed to it by integration by parts, so the last term in the action can also be written as $+(c/4)(\partial_\mu\theta)J_{\text{CS}}^\mu$. To make the presentation more familiar, let us write the equations of motion in nonrelativistic form, introducing the following notation for the vector and axial currents:

$$J_0 = \rho,\ \vec{J},\ M = \partial_0\theta,\ \vec{P} = \vec{\nabla}\theta. \tag{19.9}$$

Here are the resultant equations:

$$\vec{\nabla} \cdot \vec{E} = \rho + c\vec{P} \cdot \vec{B} \tag{19.10}$$

$$\vec{\nabla} \times \vec{E} + \frac{\partial \vec{B}}{\partial t} = 0 \tag{19.11}$$

$$\vec{\nabla} \cdot \vec{B} = 0 \tag{19.12}$$

$$\vec{\nabla} \times \vec{B} - \frac{\partial \vec{E}}{\partial t} = \vec{J} + c(M\vec{B} - \vec{P} \cdot \vec{E}). \tag{19.13}$$

One can see that both electric and magnetic fields have their sources—the r.h.s. of the first and last equations—modified. This leads to several new effects, some of them we will mention.

The Witten effect appears for nonzero $\vec{P}$. For example, consider a spherical defect, a region in which θ vanishes, while it is nonzero outside. (Such a defect can be a vortex or a magnetic monopole). Without new terms, those defects would support a magnetic field only, without $\vec{E}$, but at nonzero $\vec{P}$ the r.h.s. of equation (19.10) sources the electric field as well. Thus magnetic vortices or monopoles also become electric.

The electric charge separation in an external magnetic field also appears for nonzero $\vec{P}$: the r.h.s. of equation (19.10) may be zero when the two terms cancel each other.

If θ is time dependent and $M \neq 0$, one finds more unusual effects.

The chiral magnetic effect, which we will discuss more in Section 19.2, is one of them. It is a vector current along the magnetic field

$$\vec{J} = -c M\vec{B}, \tag{19.14}$$

which causes the r.h.s. of the last Maxwell-Chern-Simons equation (19.13) to vanish. One can put $\vec{E} = 0$ and take constant $\vec{B}$, which causes all other terms to vanish.

19.2 Chiral Magnetic Effect

We start this section with some general discussion of spacetime symmetries and currents in the medium. It will explain the required conditions under which the chiral magnetic effect (CME) may exist.

The expression we start with is the usual Ohmic current, induced by the electric field

$$\vec{J} = \sigma_{\text{Ohm}}\vec{E}. \tag{19.15}$$

If one watches this phenomenon in a mirror (which means performing the P-parity transformation $\vec{x} \to -\vec{x}$), both the l.h.s. and the r.h.s. vectors change sign, so the coefficient σ_{Ohm} remains unchanged.

Imagine now that one flips the sign of time, $t \to -t$, performing the T-parity transformation. The current is related to the velocity of the charge, and it changes sign. The electric field is only related to the positions of the charges that created it: therefore it is unchanged. The conclusion is that $\sigma_{\text{Ohm}} \to -\sigma_{\text{Ohm}}$. This is not surprising, since Ohmic current is dissipative, leading to increasing entropy of the media; and thus it should also be dissipative in an imagined world in which time goes backward.

In superconductors the expression for a current, proposed by London, is

$$\vec{J} = \sigma_{\text{London}} \vec{A}, \tag{19.16}$$

where the inducing field is the vector potential. Recalling that $\vec{E}$ contains $\partial_t \vec{A}$, it is clear that $\vec{A}$ must be T-odd. Therefore, σ_{London} should be the same in our world and in a "backward time" Universe.[2] It follows that it cannot lead to dissipation and entropy growth. And indeed, supercurrents are eternal and nondissipative!

The CME is the vector current along the magnetic field:

$$\vec{J} = \sigma_{\text{CME}} \vec{B}. \tag{19.17}$$

Transformation in a mirror of $\vec{J}$ is sign changing, while $\vec{B}$ remains unchanged. So, σ_{CME} must be P-odd. Thus we can consider weak interaction effects, as for neutrinos, since they violate P parity by involving left-handed fermions only.

Under the T-parity transformation both $\vec{J}$ and $\vec{B}$ change sign, as they are both related to the velocity of the charges. Thus σ_{CME} must be T-even, and therefore nondissipative(!), as the supercurrents are. This agrees with the idea that a magnetic field, although exerting a force on a moving charge, does not do any work on it. So, if one can find the conditions under which CME current can be produced, it will not dissipate. Unlike the supercurrent, it does not seem to require coherence and low temperatures.

We discuss the chiral symmetries and their violation in Chapter 18. Let me here just say that they lead to nonconservation of the axial current, or *chiral imbalace*, even when the fermions are massless. However, the fermions are not free but instead are under the influence of the gauge fields with particular properties: this is the chiral anomaly. Its consequence is not just the existence of CME in chiral matter, but also the universal coefficient. In its operator form, the expression is

$$\vec{J} = \frac{e^2}{2\pi^2} \mu_5 \vec{B}, \tag{19.18}$$

where μ_5 is the chemical potential for the chiral charge.[3]

Kharzeev and collaborators first suggested the use of QGP (a good chiral matter) and a superstrong magnetic field created by two positive colliding ions, directed normal to the collision plane. If a vector CME current appears, it will create an electric dipole along $\vec{B}$, which could be observed by the charge dependence of the elliptic flow of secondaries.

However, since strong interactions are CP-conserving, the chiral imbalance can only appear as a fluctuation. Thus we can only observe CP-even quadratic effects. Possible backgrounds of origins unrelated to CME, or even to $\vec{B}$, can influence observations. Therefore, strictly speaking, the observations of CME in heavy ion collisions are not yet

---

[2] According to a personal communication from Kharzeev, this argument has been made by V. I. Zakharov in a discussion.

[3] The first paper in which this formula was written was by Vilenkin [127]. The setting was P-violation due to left-handed fermions (neutrinos) of a weakly interacting sector near a rotating black hole. Existence of this current in equilibrium was questioned by C. N. Yang, and in his next paper, Vilenkin showed that the equilibrium current is indeed zero. The resolution lies in the realization that chiral matter is always metastable, not in equilibrium.

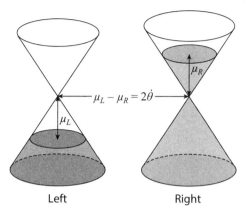

Figure 19.1. Dirac cones of the left and right fermions. In the presence of a changing chiral charge, there is an asymmetry between the Fermi energies of left and right fermions: $\mu_L - \mu_R = 2\mu_5$. Adapted from Kharzeev [130].

completely clear: a special run at RHIC with two nuclei, having the same number of nucleons but different charge Z, is planned, to at least separate $\vec{B}$-dependent effects from others.

It is easier to explain the CME experiment with the semi-metal. In the absence of fields, there is of course no chiral imbalance ($\mu_L = \mu_R$), and equal numbers of left and right fermions are present. But when parallel electric and magnetic fields are applied, the change in the chiral density ρ_5 appears, as illustrated in Figure 19.1 from [130]. The time evolution of the chiral imbalance can be written as

$$\frac{d\rho_5}{dt} = \frac{e^2}{4\pi^2}(\vec{E} \cdot \vec{B}) - \frac{\rho_5}{\tau_5}, \tag{19.19}$$

where the first term on the r.h.s. again stems from the chiral anomaly, and the second is the chiral density relaxation. The stationary condition is reached at late times then the l.h.s. is zero, because the gain and loss terms on the r.h.s. cancel each other. So we get

$$\rho_5(t \to \infty) = \frac{e^2}{4\pi^2}(\vec{E} \cdot \vec{B})\tau_5. \tag{19.20}$$

The chemical potential $\mu_5 \sim (\vec{E} \cdot \vec{B})$ as well, and putting it into expression (19.18), we obtain the current $\vec{J} \sim \vec{B}(\vec{B}\vec{E})$. This is indeed what was observed [128] in zirconium pentatelluride, $ZrTe_5$, and subsequently in other materials. Potentially, the CME current may be used in electronics, providing nondissipative currents at room temperature.

19.3 Chiral Vortical Effect

Can the magnetic field $\vec{B}$ be replaced by another quantity, possessing a similar P, T parity? for example, consider the *vorticity*, which we define in relativistic notation by

$$\omega_\mu = \left(\frac{1}{2}\right)\epsilon^{\mu\nu\lambda\rho}u_\nu\partial_\lambda u_\rho \tag{19.21}$$

with 4-velocity u_μ. The chiral vortical effect (CVE; introduced in [129, 130]) can be summarized by the relation

$$j_\mu = \sigma_{\text{CVE}} \omega_\mu, \tag{19.22}$$

with a vector current propagating along the vorticity. A similar relation—but of course with a different kinetic coefficient—can be written for the entropy current s_μ. Son and Surowka [129] argued that entropy production must be positive, $\partial_\mu s_\mu > 0$, but Kharzeev then argued that it should in fact be zero because of the nondissipative nature of the effect. These considerations lead to a specific expression for the coefficient σ_{CVE} in terms of the matter equation of state.

I am not aware of any specific applications of the CVE. In ultrarelativistic heavy ion collisions specifically, vorticity is not zero but is too small to be used for applying this phenomenon.

19.4 Chiral Waves

The CME expression (19.18) has an analog: changing the vector current to the axial one, and the axial chemical potential to the usual (vector) one μ, we get the following expression:

$$\vec{J}_A = \frac{e}{2\pi^2} \mu \vec{B}. \tag{19.23}$$

As argued in [131], combining the two equations (19.18) and (19.23), one finds a new oscillation mode called the *chiral magnetic wave*. Indeed, divergence of the currents can be replaced by time derivatives of the corresponding densities, and the two linear equations combine to produce one second-order equation.

So, starting with a certain baryon number density and μ, a heavy ion collision leads to a quadrupole excitation in which the density and the chiral imbalance should oscillate into each other. In Chapter 18 we saw how density oscillations—the sound modes—were observed. A search for the analogous chiral magnetic wave is in progress.

20 | Cold Quark Matter and Color Superconductivity

Cold quark matter may or may not be present at the centers of neutron stars: we do not yet know. Since gravitational waves from neutron star mergers were recently detected, supplemented by astronomical observations of the matter created, we will be able to learn more about these stars' equation of state.

Independent of that, one can think of cold quark matter as the ultimate limit of very dense cold matter, created by many orders of magnitude of compression over that of ordinary matter. Since quarks are fermions, they should form a Fermi gas with a Fermi surface. We have already discussed perturbative corrections to its properties.

The issue we discuss here is whether the Fermi surface would be a "normal" ungapped surface, a gapped one as a result of certain pairing phenomena, similar to other fermionic systems we have discussed.

Possible channels for forming quark Cooper pairs are discussed in Section 20.1. The topology-induced color superconductivity is discussed in Section 20.2 when two flavors of quarks are involved, and in Section 20.3 for three favors. Finally, in Section 20.4 we discuss a very interesting alternative magnetic pairing mechanism, working in asymptotically dense matter at weak coupling.

Let us start with a sketch of the phase diagram of hot/dense hadronic matter shown in Figure 20.1. We have already discussed the high-T QGP phase in Chapter 16, so let us now look at the two-dimensional plane, adding to the temperature T the second thermodynamic variable μ, the chemical potential conjugated to the baryon number.

(More generally, one can assign a separate chemical potential to each flavor of quark. There are works that focus on the nonzero strange chemical potential μ_s or the nonzero isospin chemical potential μ_I (or both). Furthermore, such work is necessary for quantitative treatment of neutron stars and low-energy heavy ion collisions. For simplicity we will not do so here and so we follow only one chemical potential.)

Before we proceed to discuss this phase diagram, a disclaimer is in order. Unlike other phase diagrams shown previously in the book, this one is not the result of

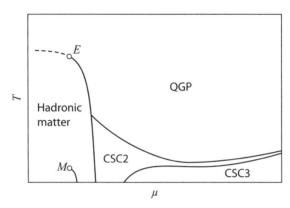

Figure 20.1. Schematic phase diagram of hot/dense matter on a plot of temperature T versus the baryonic chemical potential μ.

multiple experiments but is just a sketch of some theoretical ideas and still needs to be confirmed. Furthermore, many more phases have been proposed (e.g., crystals with various symmetries, with or without chiral symmetry breaking) that in principle may successfully compete with color superconductivity at certain places on the phase diagram. I have excluded these possibilities simply because in this book we do not have space for everything.

At small μ and high T—the upper-left part of Figure 20.1—is a dashed line, indicating that in QCD the deconfinement and chiral transitions seem to be crossovers, which are too close to each other to be separated. While current lattice simulations do not work at nonzero μ (because the integrand of the path integral is not positive definite), we can nevertheless study Taylor expansions in μ/T. Currently, terms including $(\mu/T)^4$ are now known.

At some chemical potential μ the endpoint of the first-order phase transition presumably exists (marked by E and the solid line in Figure 20.1). Following a suggestion in [146] to look for it via enhanced fluctuations, a set of RHIC runs at multiple energies— the Beam Energy Scan (BES) program—is being carried out. At the time of this writing, about half of the experimental runs are completed, and the data analysis is ongoing.

The point marked M is the critical endpoint of another first-order line, separating nuclear matter (a Fermi liquid) from a dilute gas of nucleons and nuclei fragments. This line is not hypothetical and was the subject of multiple experiments, in which the endpoint was located using the *multifragmentation* phenomenon, a production of nuclear fragments ranging in size from deuterons all the way to medium-sized nuclei. Its physics is similar to the *critical opalescence* observed near other second-order phase transitions. The exact location of (T_M, μ_M) is still somewhat debatable, but its order of magnitude is $T_M \sim 10\,\text{MeV}$. We will not discuss it further.

In this chapter we extend our discussion deeper into the r.h.s. region of the phase diagram shown in Figure 20.1. This diagram indicates approximate locations for the two color-superconductor phases, called the two-flavor-like (CSC2 or 2SC, for short) or three-flavor-like (called CSC3 or CFL) phases. We will discuss their dynamical origins and differences below. For those who would like a more technical review, see for example, [152,153].

20.1 Quark Cooper Pairs

Each system with the pairings we discussed above—electrons in metal, atoms of ^{3}He, protons and neutrons in nuclei, fermionic atoms in ultracold traps—have multiple attractive channels in which pairing can potentially occur. Before looking dynamically for the candidate that will eventually "win the race," we should always start by trying to get rid of the impossible pairs first.

As usual, let us look at general constraints following from the requirements of Fermi statistics. Now we go through the same routine for quarks, which is a bit more involved, because quarks have more quantum numbers. We will consider the cases[1] with two or three flavors (u, d or u, d, s); there are also three colors, spin 1/2, plus possibly a nonzero angular momentum l of the pair.

The basic choice for a quark Cooper pair is the flavor singlet (isospin-zero $I = 0$ for u, d antisymmetric in flavor indices), the color anti-triplet ($\epsilon_{ijk}q_jq_k$, antisymmetric in color indices as well), with spin zero (again, antisymmetric in spinor indices). So, for even angular momentum L (e.g., zero), one has the product $(-1)^3 = (-1)$. Therefore, the general Fermi statistics requirement is satisfied: the total wavefunction changes sign under two-quark permutation. We will call these particular pairs the scalar diquarks: they can be of ud, us, ds types.

Without involving a nonzero angular momentum, the only other possibility is the spin-1 and isospin-1 option: in this case only the color wavefunction is antisymmetric; the other two are symmetric under two-quark permutation. We will call these the vector diquarks. Note that in this case uu, dd, ss pairs are also included.

Now is the time to consider the dynamical details about interactions in those two competing channels. There is strong empirical evidence that the former (scalar) channel is much more attractive than the latter (vector) one. The former diquark appears inside the nucleon (or more generally, in the $SU(3)_f$ baryonic octet), while the latter vector diquark appears in the spin $- 3/2$, isospin $3/2$ Δ baryon (or in the $SU(3)_f$ baryonic decuplet representation). The $\Delta - N$ mass difference is positive and rather large $\sim 300\,\text{MeV}$, or 1/3 of the nucleon mass. There is evidence (from heavy-quark baryons, instanton models, and lattice simulations) that this difference is nearly completely due to the deeper binding of the scalar diquark.

Note how strong this binding is. Remember that the effective quark mass (due to chiral symmetry breaking) is about $M \sim 400\,\text{MeV}$. This is of the same order as the magnitude of the scalar ud diquark binding. This means we are now dealing with a completely relativistic Cooper pairing, putting a new twist on the BCS phenomenon.

---

[1] Just to be clear: I declared above that we will consider quark matter to be made of three quark flavors, u, d, s, and this remains true throughout the chapter. Here I mean something different: the two-flavor superconductor is the one in which only u, d quarks can be paired, but not s.

Any superconductivity needs to start with some attractive interaction between the fermions, which will grow in the vicinity of the Fermi sphere. Recall that for a metal having attraction between electrons was a nontrivial task, since naive expectations were that electrons obviously Coulomb repel each other. In QCD we do not have such a problem. In fact, there are three strong contenders for this role, which we will discuss subsequently:

1. the electric (Coulomb) interaction,
2. instanton-induced 't Hooft interaction in the qq channel, and
3. magnetic interaction.

The *Coulomb mechanism* is the most obvious one, and indeed that was the mechanism discussed in the earliest works on the subject. Unlike two electrons, two quarks can be in two color representations of $SU(3)$, the symmetric 6 and antisymmetric $\underline{3}$. In the latter they have negative scalar product of their relative color vectors,[2] which means their charges attract each other, as electrons and positrons do. The problem[3] with this mechanism is that the electric forces are strongly screened by the quark matter at the Debye screening length scale $R_D \sim 1/g\mu$. Thus, they can only produce a rather weak gap, $\Delta \sim O(1\,\text{MeV})$, of the same magnitude as in nuclear matter.

Historically, the second interaction discussed was the *topological or instanton-induced mechanism*, first in the context of the two-flavor problem [156, 154].[4] The forces induced by instantons are much stronger than the perturbative ones, accounting for the increase in the expected gap Δ (and T_c) from $O(1\,\text{MeV})$ to $O(100\,\text{MeV})$, with the corresponding $T_c \sim 50\,\text{MeV}$. Many people (including the referee of our paper) could not believe it.

Yet in hindsight it should not be surprising, since the same interaction in the $\bar{q}q$ channel is responsible for chiral symmetry breaking, producing a gap (the constituent quark mass) as large as 350–400 MeV.[5] The symmetries of the 2SC phase (two-flavor color superconductivity) are similar to the electroweak part of the Standard Model, with the condensed scalar isoscalar ud diquark operating as the Higgs. The chosen color direction of the condensate breaks the color group, making five out of eight gluons massive.

---

[2] Classically, color vectors of three quarks in a baryon can be represented by three equal-sized vectors added to zero. From symmetry it then follows that the angle between them is $2\pi/3$, and $\cos(2\pi/3) = -1/2$.

[3] Recall that in metals the Debye screening works in the opposite direction, reducing repulsion and thus helping weak attractive phonon exchange forces to act at large distances.

[4] These two rather similar papers were submitted to the preprint archive on the same day. That was no coincidence. Arriving at MIT for a sabbatical year, I discovered that our (Stony Brook) team was studying the same problem as the "Princeton group," represented at MIT by K. Rajagopal. The situation has been nicely resolved by an agreement between members of these two groups, which required all parties to refrain from discussing the subject until a predetermined day, on which both groups were expected to upload their papers to the preprint archive. After that day, the groups were expected to dissolve themselves, with full discussion and new collaborations possibly formed. It all took place like clockwork, as intended.

[5] Furthermore, in two-color QCD, there is the so-called Pauli-Gursey symmetry, which relates quarks with antiquarks and these two condensates. So at high density the chiral condensate $\langle \bar{q}q \rangle$ simply rotates into the superconductor one $\langle qq \rangle$, while the gap remains the same.

If in cold quark matter the only chemical potential is μ_B, the stable configuration should have the same[6] Fermi energies $\mu_B = \epsilon_f$, for all three quark species $f = u, d, s$. But equality of Fermi energies does not mean equality of the Fermi momenta: in fact, $p_F^s = \sqrt{\mu_B^2 - m_s^2} < p_F^{u,d}$. As a result, it is more difficult to combine a strange and a nonstrange quark into a Cooper pair. (Recall that the pair should have total momentum zero to be in the condensate!)

As a result, only at very high density can all three scalar diquarks du, ds, us be equally well paired. When it becomes possible, the condensation creates a new phase called the three-flavor color superconductor, CSC3. Its distinctive feature is *color-flavor locking* or CFL, a change to the higher symmetry-breaking pattern:

$$SU(3)_{\text{color}} \times SU(3)_{\text{flavor}} \rightarrow SU(3)_{\text{diagonal}}. \tag{20.1}$$

It also requires coexistence of both types of condensates, $\langle qq \rangle \neq 0$ and $\langle \bar{q}q \rangle \neq 0$ simultaneously. The CFL structure of the color superconductor was proposed in [155], based on one-gluon exchange interaction, but in fact it is favored by instantons as well [152]. It curiously combines features of the Higgs phase (eight massive gluons) and of the usual hadronic phase (eight massless pions).

Historically, the third mechanism of pairing, operating at asymptotically dense quark matter, is the *magnetic mechanism* pointed out by Son [157]. Indeed, at asymptotically high density, both the electric part of the one-gluon exchange and instantons are Debye screened.

The pairing due to magnetic one-gluon exchange forces requires densities that will perhaps be out of experimental reach forever, but they represent a theoretically interesting example. As we will see, at large μ the resulting gap is predicted to be of the order

$$\frac{\Delta}{\mu} \sim \exp\left(-\frac{3\pi^2}{\sqrt{2}g(\mu)}\right). \tag{20.2}$$

20.2 Topology-Induced Color Superconductivity: The 2SC Phase

As already discussed in Section 18.4, the topological solitons (e.g., instantons, or instanton-dyons) generate effective multifermion terms in the effective Lagrangian. These effective interactions between quarks may play the same role as four-fermion interactions of the BCS model, inducing Cooper pairing.

More specifically, this *effective t'Hooft interaction* is induced by fermionic zero modes of the topological configurations. In the simplest case of unit topological charge and two quark flavors, the vertex is schematically of the form

$$L_{\text{t'Hooft}} \sim \Pi_f(\bar{q}_f q_f) \sim (\bar{u}\Gamma^A u)(\bar{d}\Gamma^A d). \tag{20.3}$$

---

[6] Here we ignore the electron mass. A more accurate condition on the Fermi energies ϵ_u, ϵ_d is that weak decays like $d \rightarrow u + e^- + \bar{\nu}_e$ would not happen.

Therefore, it can only pair u and d quarks, not uu or dd ones. Fortunately, this flavor-asymmetric scalar diquark is the preferable channel. I do not present here the color-spin matrices Γ^A appearing in this effective Lagrangian, as those details would distract the reader from the main line of argument. They are complicated for an arbitrary number of colors and flavors: the instanton is an $SU(2)$ construction, embedded in an $SU(N_c)$ group, so the Lagrangian includes averaging over all possible embeddings. The number of fermions N_f gives the number of "quark legs" in the vertex, which are also involved in this averaging, since quarks have color indices.

Let me just mention that for the most interesting case, $N_c = 3$, $N_f = 2$, this four-fermion operator has the following structure:

$$L \sim \left[(\bar{\psi}\psi)^2 + (\bar{\psi}i\gamma_5\,\vec{\tau}\psi)^2 - (\bar{\psi}i\gamma_5\,\psi)^2 - (\bar{\psi}\vec{\tau}\psi)^2\right]. \tag{20.4}$$

The first two terms are the same as in the NJL model: recall that $SU(N_f)a$ chiral symmetry mixes these two currents into each other but preserves the sum of squares. It similarly mixes the second pair but preserves the sum of their squares. Yet the symmetry of this Lagrangian is different from the NJL one, because there is also $U(1)_a$ chiral symmetry, the one that has γ_5 but no flavor generators. It mixes the first with the third terms, and the second with the fourth. Now there is a minus sign between them, not a plus! Thus the $U(1)_a$ chiral symmetry is violated by the 't Hooft Lagrangian!

The coefficient (which we do not explicitly write here) contains the density of topological objects and a product of certain coupling constants per fermion. The analog of the cutoff parameter in the NJL model is given by the typical instanton size:

$$\Lambda \approx \frac{1}{\rho}.$$

Indeed, hard components of the quark fields with large momenta are exponentially cut off by the instanton formfactors, stemming from Fourier transform of the zero modes. Therefore, unlike in the NJL model, any higher-order diagram involving a 't Hooft vertex are UV-divergence free.

The form of the four-fermion Lagrangian given above, $(\bar{u}u)(\bar{d}d)$, is convenient for the discussion of the quark-antiquark channels and condensation, as done in Section 18.4 when discussing the spontaneous breaking of the chiral symmetry. Now we will apply the very same Lagrangian, emphasizing pairing in the scalar diquark channel. For this goal it is more convenient to identically rewrite the operator in a new form: $(ud)(ud)^+$. This step is technically called a Fierz transformation. I do not present any details of the calculation and only show schematic Gorkov's equations in Figure 20.2. Explicit calculations with the instanton-induced forces for two-flavor $N_f = 2$, $N_c = 3$ QCD have been made in two simultaneous papers [156,154]. Both suggested very robust Cooper pairs, with superconducting gaps on the order of $\Delta \sim 100$ MeV. After these two papers, the field was booming for a decade or so, resulting in hundreds of papers on the details of color superconductivity.

For the two-flavor pairing there is only one pair, ud. Its spin and isospin are both zero, $S = I = 0$, so the pair (and thus the condensate) has no indices other than the color one.

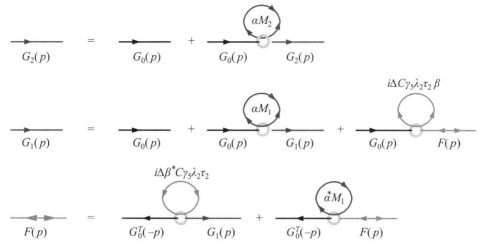

Figure 20.2. Schematic representation of the Gorkov-type equations for the propagators, both the usual and anomalous, for two-flavor theory with the four-quark operator. Note the direction of the arrows on each line. Reprinted from [152].

In $SU(3)c$ the diquarks are colored, and antisymmetric ϵ_{ijk}, $i, j, k = 1, 2, 3$, reduce two indices to one. In this case one has to select some random direction in the color space. The total number of Fermi spheres is 12—three colors times two flavors times two spins—and four of those, with the color along the random direction, are not gapped. This means that this 2SC phase is only partly super and partly normal phase.

Although we conclude that there is only one best diquark channel, we should not forget that there is also scalar quark-antiquark pairing. It is not possible to prefer one to another, especially because both presumably stem from the same effective 't Hooft Lagrangian. Solutions of the gap equations are extrema of the grand-canonical potential. Some of these solutions are minima, and only the absolute minimum represents a stable phase. There are four possible types of the solutions to the gap equations:

1. A chiral condensate, but no diquark one is present (i.e., M1 = M2 = M, $\Delta = 0$).
2. A diquark condensate, but no chiral one is present (i.e., M1 = M2 = 0, $\Delta \neq 0$).
3. Both condensates are present (i.e., all M1, M2, $\Delta \neq 0$).
4. No condensates occur at all (i.e., M1 = M2 = $\Delta = 0$).

(Here M1 = M2 are mass gaps on the Dirac sea surface (mass); Δ is a gap at the Fermi surface.) The last case corresponds to free quarks, and it is easy to see that, at $T = 0$, having at least one condensate is always more favorable. We shall call the first three types of solutions described above the phases 1, 2, and 3.

The resulting[7] grand canonical potential $\Omega = -Vp$, where p is pressure, is presented in Figure 20.3, as a function of μ. The winning phase at a given μ is the one with the smallest grand canonical potential (the largest p). The plot shows that the chiral symmetry

---

[7] The input includes the instanton size $\rho = 1/3$ fm and their density in vacuum $N/V = 0.94$ fm^{-4} from the Instanton Liquid Model proposed in 1982 and confirmed by lattice studies since mid-1990s.

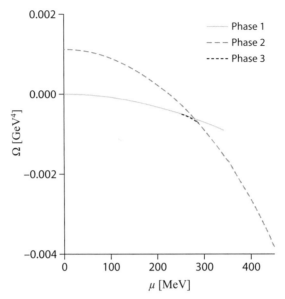

Figure 20.3. The three solutions to the Gorkov equations are shown as phases 1, 2, 3. Note that phase 3 occupies a very small domain between phases 1 and 2. Reprinted from [152].

breaking solution 1 with $\langle \bar{q}q \rangle \neq 0$ dominates at low $\mu < \mu_c \sim 300$ MeV, being then substituted by the color superconductor 2CS phase with $\langle ud \rangle \neq 0$ at high μ. However, a small (and hardly seen) window is still left, for phase 3, in which both manage to coexist!

20.3 Three-Flavor QCD: The Color-Flavor-Locked Phase

The much more symmetric *Color-flavor-locked* (CFL) phase has been discussed for the three-flavor QCD $N_f = N_c = 3$ in [155]. The authors gave up the topology-induced interaction in favor of a one-gluon-exchange interaction, to keep the four-fermion operators. They found that the winner is a very nice phase in which the color and the flavor indices are locked together, so that out of three $SU(3)$ groups (the color plus the L-handed and R-handed flavor ones) one residual $SU(3)$ managed to survive. In this phase all available Fermi spheres are gapped in a symmetric way.

A quite tricky feature of this particular approach was its somewhat unusual conclusion: there is zero quark-antiquark condensate, $\langle \bar{q}q \rangle = 0$, but the chiral symmetry is nevertheless broken.

The second detailed paper from our group solves the three-flavor instanton-induced model. It turned out that of all possible phases, the most symmetric CFL wins in this case as well. Furthermore, the chiral-symmetry-violating condensates are nonzero in this case, $\langle \bar{q}q \rangle \neq 0$. Let me explain why this happens. The "potential energy" in such approximations is the interaction Lagrangian convoluted with all possible condensates. Specifically, the instanton-induced vertex for $N_f = 3$ leads to the two types of diagrams shown in Figure 20.4, with $\langle \bar{q}q \rangle^3$ on the left and $\langle qq \rangle^2 \langle \bar{q}q \rangle$ on the right.

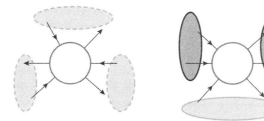

Figure 20.4. Two possible contributions to the potential energy for the three-flavor theory, with the six-fermion operator. The option shown on the left contains the cube of the chiral condensate $\langle \bar{q}q \rangle$, while the option on the right includes a combination of both type of condensates, namely, $|\langle qq \rangle|^2 \langle \bar{q}q \rangle$. Reprinted from [152].

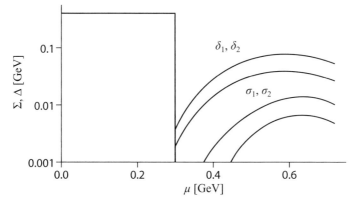

Figure 20.5. The $\langle qq \rangle$-induced gaps (marked by δ) and the $\langle \bar{q}q \rangle$-induced gaps (marked by σ), as functions of the global (baryon number) chemical potential per quark. There are two gaps of each kind, because nine Fermi spheres classify into an octet and a singlet, with different gaps. Below the critical value $\mu < \mu_c = .3$ GeV, only the usual chiral symmetry breaking takes place, and the gap $\sigma \approx 400$ MeV is just the constituent quark mass. Reprinted from [152].

All calculations follow the same traditional path as for the BCS or NJL papers: one minimizes the potential over all condensates and get multiple gap equations that are solved simultaneously. The algebra is involved, because the condensates are in fact not numbers but color-flavor matrices. The diquark condensate has the following structure:

$$\langle q_i^a C q_j^b \rangle = \bar{\Delta}_1 \delta_{ia} \delta_{bj} + \bar{\Delta}_2 \delta_{ib} \delta_{ja}, \tag{20.5}$$

where i, j are color and a, b flavor indices. It is very symmetric, reducing the original symmetry to the final one as $SU(3)_c SU(3)_f \rightarrow SU(3)_{\text{diagonal}}$. Gaps δ_i and masses σ_i (proportional to $\langle \bar{q}q \rangle$), following from instanton-based calculation, are shown as a function of μ in Figure 20.5.

The physics issues under discussion for the CFL phase include a fascinating idea of possible *hadron-quark continuity*. As proposed by Schäfer and Wilczek [158], the CFL phase not only has the same symmetries as hadronic matter (e.g., broken χ-symmetry), but also very similar excitations. Eight gluons become eight *massive* vector mesons,

$3 \times 3 = 9$ quarks become $8 + 1$ "baryons." The eight massless pions remain massless in the chiral limit. Furthermore, photon and gluons are combined into a massless γ_{inside}. Can these phases be distinguished, and should there be any phase transition in the $N_f = 3$ theory, separating it from nuclear matter? There is no need for it, at least from the symmetry point of view. Schäfer and Wilczek go as far as to suggest that the hadronic phase and this one can therefore not be separated by any phase transition.

20.4 Magnetic Pairing in Asymptotically Dense Matter

As already mentioned, we need to consider magnetic quark interactions, rather than electric ones, because the former do not suffer from static Debye screening by cold quark matter. In this section we follow closely the original paper by Son [157] (see also [159]), who very nicely found important modifications that appear in the case of one-gluon exchange in dense quark matter in the RG equation.

The first step is to note that the gluon propagator in the magnetic sector can be approximated by

$$D(q_0, q) = \frac{1}{q^2 + \frac{\pi}{2} m_D^2 \frac{|q_0|}{q}}, \tag{20.6}$$

if $q_0 \ll q \ll \mu$. The term $\frac{\pi}{2} m_D^2 \frac{|q_0|}{q}$ comes from the imaginary part of the ordinary loop-polarization operator. Recall that, in the static limit $q_0 = 0$, the magnetic field is not screened, but when $q_0 \neq 0$, the field is said to be "dynamically screened" on the scale $q \sim m_D^{2/3} q_0^{1/3}$.[8]

Now let us return to the RG formalism and apply it to the magnetic interaction mediated by one-gluon exchange. We will see immediately that we have serious trouble with very soft gluons. Indeed, on the Fermi surface, the tree-level small-angle ($\theta \ll 1$) scattering amplitude, due to one-gluon exchange, is

$$f_{\text{tree}}(\theta) = -\frac{2g^2}{3} \left(\frac{1}{\mu^2 \theta^2 + m_D^2} + \frac{1}{\mu^2 \theta^2} \right). \tag{20.7}$$

The two contributions on the r.h.s. come from the electric and the magnetic interaction, respectively (again, the factor $\frac{2}{3}$ comes from considering only the $\bar{3}$ channel.) All partial amplitudes diverge logarithmically. For example,

$$f_0 = \frac{1}{2} \int\limits_{q_{\min}/\mu}^{\pi} d\theta \, \sin\theta \, P_l(\cos\theta) \, f(\theta) \approx -\frac{g^2}{3} \log \frac{\mu}{q_{\min}}, \tag{20.8}$$

where $q_{\min}$ is the smallest allowed momentum exchange that one has to put in by hand to make f_0 finite.

But we still apply the RG procedure, starting the RG evolution at $t = 0$ with $\delta \sim m_D$. The evolution stops when δ is of the order of the gap, so typically $\delta \ll m_D$. At the tree level, the

---

[8] The statement is the same as derived previously for QGP. The only difference in cold quark matter is zero T, leading to the absence of nonperturbative magnetic mass.

fermions interact via one-gluon exchange, characterized by the momentum of the gluon $(q_0, \vec{q})$. Since all fermions have energy less than δ, the energy of the gluon q_0 is naturally of order or less than δ, while the momentum exchange q can be anywhere between 0 and 2μ.

Let us divide the four-fermion interaction that arises from the one-gluon exchange into instantaneous and noninstantaneous parts. The instantaneous interaction is mediated by the gluons that have momenta $q > q_\delta \equiv m_D^{2/3} \delta^{1/3}$. The Landau damping for these gluons is negligible, $m_D^2 \frac{|q_0|}{q} \lesssim q^2$. The gluon propagator, which is now simply q^{-2}, does not depend on q_0, which means that the four-fermion interaction they mediate can be considered as instantaneous. This part of the interaction is of the familiar type and will be treated in the conventional way. In particular, one can characterize this part by the partial-wave amplitudes f_l. For $q \lesssim q_\delta$, the Landau damping can no longer be neglected. This part of the interaction has a considerable temporal retardation and should be treated separately.

Now let us integrate out fermionic degrees of freedom in the strip of momenta, say, with energy between $e^{-1}\delta$ and δ. During this process the following will occur:

1. The partial-wave amplitudes f_l obtain the conventional renormalization, as in the BCS case.

2. One could ask whether the noninstantaneous coupling is renormalized during this integration. To answer this question, one should compute the correction to the noninstantaneous interaction that comes from integrating out the fermion degrees of freedom. In the appendix of [157], it is shown that the noninstantaneous part of the interaction does not get renormalized.

3. Most importantly, and what makes our RG distinctive, *part of the noninstantaneous interaction becomes instantaneous*. Specifically, the gluon exchange with q lying in the interval $(e^{-1/3}q_\delta, q_\delta)$, which was formerly treated as noninstantaneous, now becomes a part of the instantaneous interaction and contributes to f_l.

Small-angle scattering thus leads to a new term in the RG group equation for f_0:

$$\frac{d}{dt} f_0 = -\frac{g^2}{9\mu^2} - \frac{\mu^2}{2\pi^2} f_0^2. \tag{20.9}$$

The second term in the r.h.s. is the familiar term that gives rise to the BCS. What is new is the first term, which takes into account the fact that softer and softer gluon exchanges contribute to f_l. The noninstantaneous part of the interaction can be considered as an infinite pool, which continuously replenishes the instantaneous part during the RG evolution. Clearly, this should speed up the approach to the Landau pole.

To secure a solution, we also need to specify an initial condition on f_0. Recall that $t = 0$ corresponds to $\delta \sim m_D$, and from equation (20.7) we find, to the leading logarithm,

$$f_0(0) = -\frac{2g^2}{3\mu^2} \log \frac{1}{g}. \tag{20.10}$$

The solution to (20.9) with the initial condition (20.10) is

$$f_0(t) = -\frac{\sqrt{2}\pi g}{3\mu^2} \tan\left[\frac{g}{3\sqrt{2}\pi}\left(t + 6\log\frac{1}{g}\right)\right].$$

The coupling f_0 hits the Landau pole when the argument of the tangent is equal to $\pi/2$. This happens when

$$t = \frac{3\pi^2}{\sqrt{2}g} - 6\log\frac{1}{g}.$$

The Fermi liquid description thus breaks down at the energy scale

$$\Delta \sim m_D e^{-t} \sim \mu g^{-5}\exp\left(-\frac{3\pi^2}{\sqrt{2}g}\right), \tag{20.11}$$

which is the scale of the gap. Notice that $e^{-c/g}$ is parametrically larger than the naive estimate e^{-c/g^2} at small g. The reason for this enhancement is obviously the singularity of the magnetic interaction at small angles or large impact parameters.

21 | The QCD Vacuum III: Instanton-Dyons

All the material in this book so far has been well tested over time. But in this chapter we venture into rather recent advances, which perhaps need much more scrutiny. At the time of this writing this material can be considered to be a kind of current frontier. However, it definitely belongs to many-body theory, so it is tempting to give the reader some flavor of it.

Let me formulate its main ideas upfront. While the QFTs in general, and gauge theories in particular, deal with fields possessing infinitely many degrees of freedom, not all of them are equally important. Most of gauge fields are generalized gluons, plane waves perhaps modified by interaction and scattering. Some are nonperturbative solitons, with fields $A_\mu^a \sim O(1/g)$ making the nonlinear terms in the non-Abelian field strength as large as the terms derivatives with. And some of those solitons are topological solitons, possessing nontrivial topology: they are the focus of our attention in this chapter.

Topological properties are stable under any smooth deformation. No matter how much quantum fluctuations distort a topological soliton, it retains the same topology. As a result, in certain cases, the so-called *index theorems* apply, stating in particular that a Dirac equation for a quark in such a field background must have a certain number of zero modes (i.e., localized solutions with exact zero eigenvalues).

It seems like a miracle, but indeed, even in very complicated and chaotic gauge fields produced numerically in lattice simulations, it is possible to identify these zero modes. Furthermore, they resonate/interact mostly with zero modes of other solitons and create a special subset—known as the zero-mode zone (ZMZ)—which is small,[1] but as we will see plays a decisive role in gauge dynamics.

Section 21.1 describes the physics of the Polyakov line, or holonomy, at finite temperatures. This term refers to an integral over the Matsubara circle of the gauge field, which

---

[1] To give the reader an idea of the numbers involved, the number of quark states in modern lattice simulations can be of the order of 10^6, while the number of ZMZ states is 10–100.

(as for any integral over a closed loop) is gauge invariant. The existence of a nontrivial vacuum expectation value of the A_0 field implies a kind of "Higgsing" of the color group.

As a result, all classical solutions should be redefined, with asymptotic values of the fields at large distances tending not to zero but to the vacuum A_0 value. Applying this procedure to the instantons, it was found that they in fact split into their N_c constituents, called instanton-dyons or instanton-monopoles (see Section 20.2). We also discuss in this section relatively recent works on the instanton-dyon ensembles and show that they reproduce lattice data on properties of the deconfinement and chiral symmetry restoration phase transitions in QCD-like theories. In Section 21.3 we continue discussion of these phase transitions, but in modified (or deformed) QCD, with certain flavor-dependent periodicity phases (or imaginary chemical potentials). Their nontrivial dependence on those phases is used as a tool for revealing the properties of the fundamental topological solitons driving these transitions.

21.1 Nonzero Holonomy and Finite-Temperature "Higgsing"

In this section we will move from the discussion of the QCD vacuum to a version of it heated to some temperature T. As always, in path integrals with Euclidean time this simply means reducing the time period from infinitely large to finite $\beta = 1/T$.

One of the things associated with such change is the loss of four-dimensional rotational (Lorentz) symmetry: the fourth coordinate $x_4 = \tau$ is now different from the others. Thus the fourth component of the vector potential plays a special role. Furthermore, due to periodicity in τ, one can form the so-called Polyakov loop (or holonomy, as it is known in mathematics):

$$P = \frac{1}{N_c} \text{Tr Pexp} \left(i \int_{S^1} d\tau \hat{A}_4 \right), \tag{21.1}$$

in which the integral is done over the Matsubara circle, and the hat over A_4 reminds us that it is still a color matrix. "Pexp" means the product of small-step exponents over the loop: it forms some unitary matrix. While A_4 itself is gauge dependent, the integral over the circle, as over any closed loop, is in fact gauge invariant!

The vacuum expectation value $L = \langle P \rangle$ plays a very important role in finite-T QCD. Its temperature dependence is shown in Figure 21.1. As one can see, in the temperature range 100–400 MeV (roughly coincidental with the region studied experimentally at the Relativistic Heavy Ion Collider (RHIC) and the Large Hadron Collider (LHC), it changes from near zero to near one. A quark line making a similar circle over the Matsubara time feels the gauge field via this factor L, so at high T its role is nearly negligible, while at low T (where $L \approx 0$) the amplitude is near zero. What we just described is known as the *statistical confinement* phenomenon: the contribution of the quark-gluon plasma thermodynamics is effectively switched off at low T by the near-zero VEV of the Polyakov line.

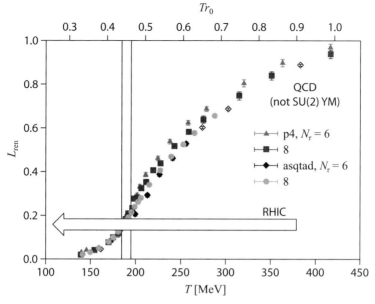

Figure 21.1. The temperature dependence of the expectation value of the renormalized Polyakov line. Different points correspond to different lattice actions and lattice spacings. The art is adapted from the Brookhaven National laboratory lattice group, Upton, NY.

To understand the phenomenon a little better, let us take a closer look at P. Suppose A_4 itself—the logarithm of the unitary matrix—has VEV.[2] Since its color orientation can be changed by a gauge transformation, we can think of it as a diagonal matrix. The simplest case of the $SU(2)$ color group, with three generators, is given by Pauli spin matrices τ^1, τ^2, τ^3. Only one of them (τ^3) is diagonal: so let one component have VEV, to be denoted by

$$\langle A_4^3 \rangle = 2\pi T v, \tag{21.2}$$

where v is called the holonomy parameter. Substituting it back into the definition of P, we find a simple expression for the Polyakov line:[3]

$$L = \langle P \rangle = \cos(\pi v). \tag{21.3}$$

So, $v \approx 0$ at high T and $v \approx 1/2$ at low T, in the confining phase.

The nonzero VEV of the field component $\langle A_4 \rangle \neq 0$ leads to a phenomenon known as "Higgsing." Since the QCD Lagrangian is the square of the field strength, it includes the square of the commutator $([A_m, A_4])^2$, $m = 1, 2, 3$ (here I use notation in which the fields are color matrices $A_\mu = A_\mu^a t^a$, including the group generators t^a, $a = N_c^2 - 1$). If

---

[2] Recall that the lower index of the gauge field is the Lorentz index $\mu = 1, \ldots 4$, and the upper one (if present) is the adjoint color index $a = 1, \ldots, N_c^2 - 1$. If there is no upper index, the object implies convolutio n with color generators and is therefore a color matrix by itself.

[3] Remember that P is associated with quarks, so the color generators in it are in *fundamental* or spinor representation. Therefore they are half of Pauli matrices, so $\hat{A}_4 = A_4^3 \tau^3 / 2$.

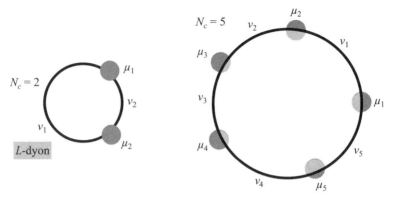

Figure 21.2. Explanation of the color holonomy notations, for theories with two and five colors.

the commutator with the nonzero VEV $\langle A_4 \rangle$ is nonzero, then the Lagrangian gets an extra term, which is quadratic in A_m. As we learned early in the book, this means that those fields, while initially massless, now become massive.

Selecting $\langle A_4 \rangle$ to be the diagonal matrix, we find that the commutator with it is zero only if A_m is also diagonal. Therefore, all the nondiagonal components of the gauge field become massive!

In QCD with three colors, the number of generators is $N_c^2 - 1 = 8$. Two of them are diagonal:

$$\lambda^3 = \begin{pmatrix} 1 & 0 & 0 \\ 0 & -1 & 0 \\ 0 & 0 & 0 \end{pmatrix}, \qquad \lambda^8 = \frac{1}{\sqrt{3}} \begin{pmatrix} 1 & 0 & 0 \\ 0 & 1 & 0 \\ 0 & 0 & -2 \end{pmatrix},$$

and the remaining six are nondiagonal. Therefore, six out of eight gluons become massive due to Higgsing, while the other two remain massless. General $SU(N)$ gauge theory gets Higgsed to its Cartan diagonal subgroup with $N_c - 1$ generators. The arrangement of holonomy phases for $N_c = 2, 5$ is shown in Figure 21.2.

To simplify this dicussion, we will focus instead on a simpler theory with two colors, $N_c = 2$. It has three generators, the fields A_m^1, A_m^2 become massive, and only one, A_m^3, remains massless. The $SU(2)$ color symmetry gets broken to one diagonal Abelian subgroup $U(1)$.

21.2 Instanton-Dyons and Their Ensembles

The Higgsing phenomenon of the gauge fields should also be reflected in the theory of topological solitons, such as instantons. Instead of classical solutions made of fields that vanish at large distances, we should instead look for those which go to the nonzero Higgs VEVs. For example, for the $SU(2)$ color $\langle A_4^3 \rangle \neq 0$, as explained above.

Instanton-like solutions with $Q = 1$ of this type has been derived in [185, 186]; for a review, see [187]. Unfortunately the method used and even the expressions for the solution are in general too complicated to be presented here. But we can still explore the dramatic discovery that those solutions reveal. It turns out that Higgsing of the gauge field splits instantons into N_c constituents, which can be separated. These constituents are known by several names—instanton-dyons, instanton-monopoles, or even instanton-quarks. In this chapter we will discuss whether these topological objects can be subject to many-body theory methods, and whether their ensembles in fact possess all the phase transitions, with the same properties, as QCD on the lattice.

In the case of the simplest $SU(2)$ gauge theory, the instanton splits into $N_c = 2$ instanton-dyons. The anti-instanton splits into two others, so in total there are four players in the ensembles we are going to study below. The diagonal massless gluons, often called the "photons," define their electric and magnetic charges. In total four dyons have all possible combinations of electric[4] and magnetic charges. By tradition, the time-independent selfdual ones are called M with charges $(e, m) = (+, +)$, and time-twisted L with charges $(e, m) = (-, -)$; the antiselfdual antidyons are called $\bar{M}$, $(e, m) = (+, -)$ and $\bar{L}$, $(e, m) = (-, +)$.

If these constituents are separated by large distances, the expression for them simplifies, as they become (in a sense) spherically symmetric. Moreover, they are already known to us as the BPS color-magnetic monopoles of 't Hooft and Polyakov, with the adjoint Higgs scalar being replaced by A_4. In the simplest "hedgehog," gauge,[5] the solutions take a very simple form:

$$A_4^a = \pm \hat{r}_a \left(\frac{1}{r} - v \coth(vr) \right),$$

$$A_i^a = \epsilon_{aij} \hat{r}_j \left(\frac{1}{r} - \frac{v}{\sinh(vr)} \right),$$

(21.4)

where $+$ corresponds to the M dyon and $-$ corresponds to the $\bar{M}$ dyon. Here r is the distance in the three-dimensional position space. The L and $\bar{L}$ dyons are obtained by the replacement $v \to 2\pi T - v$ and a certain time-dependent gauge change.

As promised, the large-r asymptotic behavior of A_4 is given by the second term, in which $\coth(vr) \to 1$ and $A_4 \to v$, the Higgs VEV. Massless components decrease as $1/r$, the Coulombic fields of a charge. (Note that dyons have both electric and magnetic charges). Massive components decrease exponentially $\sim \exp(-vr)$, as they should.

The reader is strongly encouraged to put (21.4) into Maple or Mathematica and do the following exercise, to get familiar with these objects.

Exercise *Substitute (21.4) into general formulas for non-Abelian electric and magnetic fields, and prove that (up to a sign) they are in fact equal to each other. Plot the field components and their colorless square as a function of distance.*

---

[4] Note that this electric charge is real as defined in Euclidean time formulation. If one attempts to continue this description in Minkowski time, the fourth component of the gauge field and the electric currents both become imaginary, and basically lose their meaning.

[5] In which the color direction of the Higgs field at large r is directed along the unit radial vector: $A_4^m \to v\hat{r}_m$.

Equation (21.4) describe M-dyons, which are time independent. Nevertheless, we are still in the finite-T setting, in which the bosonic fields such as A_μ need to be periodic in Matsubara time, with period $1/T$. The M-dyons just described are of course periodic, since they do not depend on time at all. The expression for "twisted" L-dyons I will not give here explicitly, but only note that its spatial part is like that for M-dyons, with substitution $v \to \bar{v} = 2\pi T - v$, and with the explicit exponential time dependence making them periodic.

Having formulated the objects we are going to study, let us now discuss a "plasma" made of them. It is important to point out that the instanton-dyons are not particles in the usual sense. They are four-dimensional solutions, constituents of the instantons, and thus they do not have paths. The instanton-dyon partition function is just an integral over all collective variables, which is basically just their positions. In this sense the problem resembles more the configuration integral of classical statistical mechanics rather than quantum path integrals we have discussed for other systems.

Unfortunately, the hedgehog gauge cannot be used in configurations with more than one such object. So, before putting them together, we need to "comb the hedgehogs" into a gauge in which there is one common Higgs direction. Such combing cannot in principle go smoothly: it creates singularities known as *Dirac strings*: we have already discussed these in connection to magnetic monopoles. These strings can either connect dyons with opposite magnetic charges or go to infinity. However, the Dirac strings are pure gauge artifacts and are actually invisible and have zero tension. Presence of the Dirac strings connecting the dyons inside the instanton explains why these objects may have fractional topological charge Q, avoiding conflict with topological classification of the fields.

Section 21.1 prepared us for the idea that, in order to understand the topological structure of the finite-T QCD, one needs to proceed from the instanton liquid to some kind of *triple plasma of these instanton-dyons*. We call it "triple" because of the three types of interactions: these objects have nonzero electric and magnetic charges (according to remaining massless $U(1)$s) plus the topological charges inducing the fermion exchanges. Omitting all the technicalities (for which one needs to look into the original papers) let me just emphasize that theory of the instanton-dyons was able for the first time to combine confinement and chiral symmetry breaking (see Figure 21.3). As in other chapters, we will not be able to do justice to the historical path of development and mention all significant papers. Instead we proceed to the methods that are the easiest to explain.

One such method is direct numerical simulation of the instanton-dyon ensembles, in which a certain number of dyons (typically 64 or 128) are placed in some three-dimensional compact space (in this case, a 3-sphere) and then integration over all collective variables (positions) of them is done using a standard Monte-Carlo algorithm (Metropolis; see Chapter 13). Standard methods are used to calculate free-energy dependence on input parameters of the model (holonomy value and fraction of dyons of various kinds), which is then minimized, and then properties of the optimal ensemble with their corresponding values is reported.

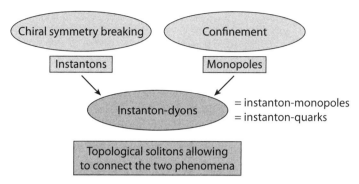

Figure 21.3. Schematic picture showing convergence of the theory of confinement based on monopoles, and chiral breaking based on instantons, to one common theory based on instanton-dyons.

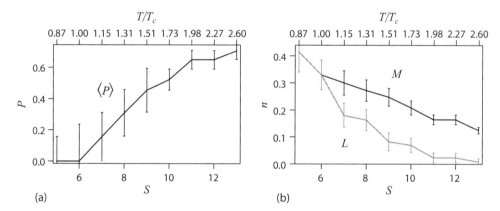

Figure 21.4. (a) The VEV of the Polyakov line is shown as a function of parameter S (the lower scale) or temperature T (the upper scale), corresponding to the minimum of the free energy of the instanton-dyon ensemble. (b) The densities of M- and L-type dyons. Both panels show a deconfinement phase transition at $S \approx 6$. Adapted from Larsen and Shuryak [202].

For pure gauge $SU(2)$ theory, some results of such simulations (from [202]) are shown in Figure 21.4. The left plot shows the dependence of the mean Polyakov line on the input parameter

$$S = \left(\frac{11}{3} N_c - \frac{2}{3} N_f \right) \log \left(\frac{T}{\Lambda} \right), \tag{21.5}$$

which is the action per instanton (a combination of the asymptotic freedom formula (15.10) and the instanton action (18.19)). This parameter S is eliminated in favor of the (approximately defined) critical temperature T_c on the upper scale of Figure 21.4.

The mean value of the Polyakov line is seen to go to 1 at high values of S and T (the right side of the plot), and to zero at low values (the left side of the plot). This is qualitatively the same behavior as shown in Figure 21.1, the lattice first-principle calculation. It has

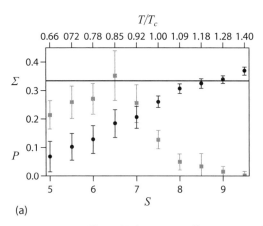
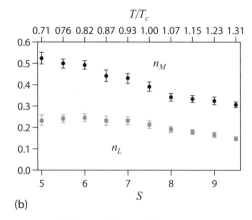

(a)   (b)

Figure 21.5. (a) Deconfinement transition represented by vanishing of the Polyakov line P (filled circles) and chiral symmetry restoration represented by chiral condensate Σ (squares) in the instanton-dyon model for two colors and two light quark flavors. (b) The corresponding densities of the M and L-type dyons. Adapted from Larsen and Shuryak [188].

the same meaning, namely: the ensemble of the dyons generates statistical confinement below a certain temperature T_c. The transition appears to be second order.

The main reason for that transition is back-reaction of the dyons to the holonomy potential. An important detail is revealed in the Figure 21.4b, in which the densities of two types of the dyons are shown: the low-T confining phase turns out to be a symmetric one, in which the density, masses, and other parameters of the dyons become the same. In the higher-T phase, this is not the case; L-dyons are heavier and more suppressed.

When two flavors of the light quarks are added to the $SU(2)$ gauge theory (via a fermionic determinant associated with quark hopping over the dyons, as described in Chapter 18), the results change substantially [188]. Figure 21.5a shows the temperature dependence of the mean Polyakov loop and the chiral condensate. One can see that the latter vanishes at high T—this is called chiral symmetry retoration. The dependence of bot is now very smooth: such transitions are known as *crossover transitions*, which are not really singularities of the thermodynamical quantities. The corresponding quasicritical temperatures are defined as the locations of inflection points on these curves (the maxima of second derivatives). While the statistical accuracy of these data are not sufficient to locate the inflection points precisely, it is clear from the plot that for both deconfinement and chiral restoration transitions, they should be about the same. The same conclusion has been previously reached from (much more extensive) first-principle simulations of lattice QCD with dynamical light quarks.

Figure 21.5b provides some explanation of why this is the case: it shows the corresponding dependence of the dyon densities. As one can see, L and M dyons are never present with equal densities, so the "democratic" symmetric properties of the confining phase seen above without fermions is no longer there. The reason for such disparity is that quarks interact via zero modes with the L dyons but not with the M ones.

21.3 Modifying Quark Periodicity in Euclidean Time

In the Euclidean time formalism used throughout this book, the boson and fermion fields are distinguished by their different periodicity condition over the Matsubara circle: boson fields are periodic, and the fermion fields are *antiperiodic*. The respective Matsubara frequencies are integer or semi-integer. As we have seen after using the Sommerfeld-Watson method of summation over corresponding frequencies of Section 3.2, this leads to Bose and Fermi distribution functions in the diagrams.

In this final section of the book let us dare to modify those fundamental rules by playing with some arbitrary phase in the periodicity condition of quarks. We will assume it has some phase

$$\psi(\beta) = e^{i\alpha_f}\psi(0),\tag{21.6}$$

which implies that our quarks will not in general be fermions but "anyons," with some crazy statistics. (Fermions are obtained when $\alpha_f = \pi$ and bosons when $\alpha_f = 0$.) This phase has index f, which indicates that in principle one can make it different for different quarks: often this phase is called *flavor holonomy*. Why would we want to do that?

Because in this way, we can reveal very nontrivial properties of the QCD vacuum and the instanton-dyon model. To explain requires clarifying the setting with color and flavor holonomies. In a general $SU(N_c)$ theory the Pexp in the Polyakov loop is a unitary $N_c \times N_c$ matrix, and so it has N_c eigenvalues denoted by $\exp(i\pi\mu_i)$. Tracelessness implies that $\sum_i \mu_i = 0$. More important than μ_i are their differences, defined so that

$$\nu_i = \mu_{i+1} - \mu_i, \quad \sum_i \nu_i = 1,\tag{21.7}$$

where by definition, $\mu_{N_c+1} = 1 + \mu_1$. The simple geometric meaning of the construction is shown in Figure 21.2, for two and five colors. In the former case $\mu_2 = -\mu_1$, so we call it simply μ. Also ν_1 was called simply ν and $\nu_2 = 1 - \nu$. The more generic case with five colors explains the general setting. What is most important is that the ν_i represent the lengths of the segments between μ_i phases, in units of the circle circumference: that is why all ν_i add to 1, the total circle. One of the direct applications of these fractions ν_i is that the actions of the instanton-dyons are proportional to them:

$$S_i = \frac{8\pi^2}{g^2}\nu_i,\tag{21.8}$$

adding them together. Note that one can use relation (21.7) and see that the sum is the instanton action $S = 8_n^2/g^2$.

Let us now add light fermions to the gauge theory. First, recall what we know about fermionic zero modes for the instanton. The number of quark (fundamental color representation) zero modes is equal to the topological charge. Since $Q = 1$ for an instanton, there is one zero mode for a quark and one for the antiquark. As a result, an instanton generates a multifermionic vertex with $2\,N_f$ legs.

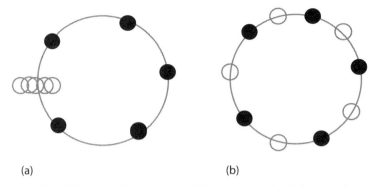

(a)   (b)

Figure 21.6. The gauge (filled circles) and flavor (open circles) holonomy phases, in the (a) usual QCD and the (b) Z_N-symmetric model. The picture shown corresponds to $N_c = N_f = 5$.

An instanton can be seen as all types of the dyons together, again with $Q = 1$ and one zero mode. Yet, given the fact that an instanton is split into N_c constituents, which may be far from each other, the question is: What does this zero mode look like? The answer has been worked out by Kraan and van Baal [185], who gave the general solution for the zero mode. Instead of presenting those lengthy expressions, I will just describe the answer in simple terms. It is illustrated graphically in Figure 21.6. The filled circles indicate, as before, the positions of Polyakov loop eigenvalues μ_i. The new open circles indicate flavor holonomies α_i. Figure 21.6a corresponds to the QCD in which all quarks are fermions, and thus for all flavors $\alpha_f = \pi$. Figure 21.6b corresponds to the Z_N-symmetric QCD invented in Japan [190]:[6] in it flavor holonomies are distributed "democratically" around the circle. The van Baal prescription for the fermionic zero mode tells us that in the former case, all quarks have zero modes with only one of the dyons, namely, the one to which segment ν_i (the open circles) belong. In the latter case the answer is different: each dyon type has one zero mode associated with it.

The question is whether these flavor holonomies do or do not matter. It can be answered either by lattice simulations in this setting [191] or by analogous simulations with the instanton-dyons [189]. The results of both simulations have been quite spectacular: both deconfinement and chiral symmetry phase transitions change dramatically: some of them are shown in Figure 21.7 from [189].

The expectation value of the Polyakov line changes from smooth crossover (in $N_c = N_f = 2$ QCD) to a strong first-order transition in Z_2-symmetric QCD. In contrast, the chiral restoration transition, seems to never happen at all. What is instead observed is a change from a symmetric confining phase,[7] with equal quark condensates $\langle \bar{u}u \rangle = \langle \bar{d}d \rangle$ at low T to an asymmetric phase, with nonequal condensates. And indeed, when L, M dyon

---

[6] It was invented in a context not related to instanton-dyons.

[7] The attentive reader may note that the exact transition point between the two phases seems to be different for the Polyakov line and the quark condensate. Since one is dealing with a first-order transition, associated with two competing maxima, simulations are difficult due to the metastability of those maxima. In other words, in the region in S between the two vertical lines in Figure 21.7b the simulations perhaps do not correpond to perfectly equilibrated ensembles.

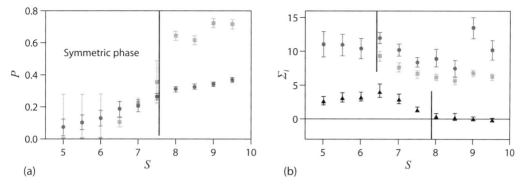

Figure 21.7. (a) The mean Polyakov loop P as a function of action parameter S, for Z_2-symmetric model (squares), compared to that for the $N_c = N_f = 2$ QCD with the usual antiperiodic quarks (filled circles). (b) Chiral condensate generated by u quarks and L dyons (squares) and d quarks interacting with M dyons (filled circles) as a function of action S, for the Z_2-symmetric model. For comparison the results are shown from II for the usual QCD-like model with $N_c = N_f = 2$ by triangles. Adapted from Larsen and Shuryak [189].

densities and properties are no longer the same, why should the corresponding quark condensates be equal?

The first lattice results [191] are consistent with these findings. These differences are so striking that I dare to say that if they survive further scrutiny, they will perhaps prove "beyond a reasonable doubt" that the instanton-dyon mechanism of the phase transition is indeed the correct explanation.

22 | Parting Comments

The reader who has come this far may be overwhelmed with the many physical settings and ideas presented in this book. Hence it makes sense to spend a little time here looking back and emphasizing the central ideas of this book—those that appear in a variety of applications. In these final comments, I also provide some hints on the further development of these ideas, which we were not able to cover. Perhaps they may inspire the reader to find out more about them in the literature.

22.1 Feynman Diagrams and Their Resummations

The first subject we discussed was Feynman's *path integral formulation for the density matrix*. We had seen how it can be used at zero and nonzero temperatures, as well as at finite chemical potential. This formulation can be used in three basic ways:

1. to develop Feynman perturbation theory, in weak coupling;
2. to develop semiclassical approximations and;
3. to evaluate directly Euclidean path integrals numerically.

Most of our time was devoted to the first approach, while the third was discussed in Section 12.3. The semiclassical theory (the second approach) was mentioned only in passing near the end of the book, when we discussed matter made of the topological solitons—monopoles, instantons, and their descendants. Let us recapitulate the main ideas underpinning various applications of perturbation theory that we have discussed.

The Feynman form of perturbation theory has many celebrated achievements. Some types of diagrams have predicted/described many phenomena. The single-end loop diagram corresponds to the matter density. The two-end diagram (the particle-hole polarization operators) gave us the mass operators for the interaction: the Coulomb or gluon force, the Lindhard function for nonrelativisitic Fermi systems, and a similar relativistic one for quark-gluon plasmas. They describe screening and plasmon waves. Another four-end one-loop diagram (the "fish") leads to pairing and superconductivity in the particle-particle channel.

The triangular diagram with three currents gave us the chiral anomaly and led to the realization—quite shocking at the time of its discovery—that not all symmetries of the Lagrangian survive quantization. We focused on $U(1)_a$ chiral symmetry, but there are other anomalies as well. For example, classical Yang-Mills or QCD with massless quarks are naively conformally (scale) invariant, as the Lagrangian contains no dimensional quantities. But quantum theory has no scale invariance because of the RG flow of the coupling. This is the so-called scale anomaly.

Moreover, the power of the Feynman form of perturbation theory goes well beyond particular diagrams. Unlike the versions of perturbation theory discussed in quantum mechanics textbooks, this one provides simple and clear Feynman rules, by which we can construct expressions for perturbations of any desired order. Therefore, using the Feynman rules, we can see that certain subsets of the diagrams lead to series that can be resummed, as the series are recognized as expansions of some known functions. A particularly clear and simple example of that is the exponentiation of the diagrams, originally derived for the partition function Z but then resummed into diagrams for its logarithm, the free energy. Recall for example how this resummation elegantly allows us to get rid of all disconnected diagrams.

Another major example is a resummation of the Green function corrections, forming (in frequency and momentum representation) a simple geometric series. Its result is simply the transfer of the mass operators in the denominator. The results of that are extremely important, because the poles of the Green functions correspond to quasiparticles, elementary excitations that can exist in the described system.

In QFT courses attention is focused on UV divergences and their removal. Since the finite-T, μ version of the same theory cannot have different UV divergences, we have basically ignored those issues in this book, focusing on the IR divergences instead. In this book we have discussed few examples in which resummation of IR-divergent diagrams leads to finite and physical results. The lesson is that such divergences does not constitute some fundamental deficiency of the theory; they simply reflect the fact that some quantities may depend on the coupling constant g by a function, not expandable in a good series in powers of g^2.

Finally, let us devote some time to a few more general questions, which were inevitably asked at lectures that I have given on this material. *Can all diagrams be summed up, at least for some theories? Are the perturbative series convergent or not? If not, do they at least specify some unique function of the coupling? And, more generally, is perturbation theory a promising way to provide a strict definition of what the given QFT is, outside of the weak coupling domain?* Unfortunately, the answer to all these question is "no."

It is easy to argue that the number of diagrams of order n is very large, $\sim n!$. So the first question one can ask here is whether they do or do not cancel each other. The main idea explaining why the perturbative series *must* diverge was suggested by Dyson [192] in 1952. The argument goes as follows. In QED one makes expansion in e^2, order by order.

Let us think a bit about what would happen if $e^2 < 0$. Then protons and electrons would no longer attract each other, so atoms would dissolve. In contrast, electrons would start to attract each other and congregate in large numbers, and so would protons or positrons. A bit of thought shows that the binding energy in this case can be made arbitrarily large; there would be complete collapse of the theory, with no stable ground state possible. Now, if the positive and negative vicinity of the $e^2 = 0$ point are that different, Dyson argued, the series expansion around this point obviously cannot be nice and convergent!

Similar arguments can be made for other theories. Furthermore, in gauge theories we know that there are nonperturbative phenomena that—even for physical value of the coupling $g^2 > 0$—lead to answers behaving like $\sim \exp(-1/g^2)$. Note that this function has a strong singularity at $g = 0$, but its Taylor expansion in g^2 consists of only zeros. Clearly, those effects are invisible for the perturbative expansion.

French mathematician J. Ecalle in the 1980s defined *transseries* as triple series including not only powers of the parameter (= coupling), as in perturbation theory, but also exponential and logarithmic functions:

$$f(g^2) = \sum_{p=0}^{\infty} \sum_{k=0}^{\infty} \sum_{l=1}^{k-1} c_{pkl} g^{2p} \left(\exp(-\frac{c}{g^2}) \right)^k \left(\ln \left(\pm \frac{1}{g^2} \right) \right)^l . \tag{22.1}$$

His argument was based on symmetries: such a set of functions is *closed* under manipulations including the so-called Borel transform plus analytic continuations.

Divergent perturbative series often lead to ambiguous results, and so do the exponential and exp-log terms. Ecalle's point is that the correctly formed transseries should have all such unphysical and ambiguous terms to cancel out. Examples studied in mathematics are numerous, but they were limited to some ordinary one-dimensional integrals depending on parameters and defining certain special functions. Quantum mechanical examples deal with path integrals, which are infinite-dimensional, but in this case one can also explicitly build the transseries and demonstrate that indeed some ambiguities do cancel. Furthermore, certain relations between the coefficients of transseries were found, known as *resurgence relations*. Work on building the transseries for QFTs remains in progress (see, e.g., [11] and references therein).

22.2 Bose-Einstein Condensation and the Reality of Cooper Pairs

The particles belonging to the condensate were represented in the Feynman rules by "anomalous" lines and Green functions (denoted by dashed lines) in Section 3.6, which is devoted to weakly coupled Bose gas. The resummation of the Green function, including the anomalous terms, leads to the so-called Belyaev quasiparticles. At large momenta k they are Bose particles, but at small k they become phonons, quanta of Bogolyubov sound. The diagrammatic approach allowed us to reproduce previously known results related to a weakly coupled Bose gas, and also to calculate new effects, such as the quasiparticle lifetime.

A technique similar to the anomalous Green functions was applied by Gorkov to superconductors. It reproduced results known from BCS theory and also clarified some time-dependent and kinetic properties of superconductors.

Let us now discuss the issue of the reality of Cooper pairs, comparing different settings discussed earlier in the book. In ordinary superconductors the Fermi energy is very large compared to superconductor's gap:

$$\epsilon_F \sim eV \sim 10^4 \, K \gg T_c \sim \Delta \sim 1 \, K.$$

Cooper pairs are very long-range correlations between two electrons, located (in momentum space) at the Fermi surface. But in coordinate space, these two electrons are far from each other, with billions of other electrons in between. The idea that they are "bound to each other" is true, but only in some narrow technical sense. It is not possible, say, to take a Cooper pair out of the superconductor and put it in your pocket. In fact, you cannot even take it out of the condensate! The same is true of other weakly coupled (in UV) systems we discussed, such as liquid ^{3}He and nuclear matter.

Color superconductivity is a somewhat intermediate case, in the following curious sense. Of course, you still cannot put that quark pair in your pocket and walk away: after all, this Cooper pair has a color charge, and confinement phenomenon will generate a flux tube if we were to try to separate it from the piece of quark matter. Yet in $SU(3)$, it has the quantum numbers of an antiquark, and so it can be pulled out together with another quark. So, one can pull out a nucleon, which, one may argue is made of a quark plus a Cooper pair (diquark). Of course, we and all matter around us *is* made of nucleons, and thus the nuelcon structure is of great interest—it has been and will be studied in great detail.

Other intermediate cases that we have not discussed in this book are cuprate superconductors. They are believed to have a phase in which there are Cooper pairs without the condensate, at some locations on the phase diagrams.

Curiously, it is indeed true that a quark-quark Cooper pair can exist outside the dense quark matter and its condensate: in fact the pair is well preserved inside the nucleon! It is in fact is made of one valence quark (u for a proton, or d for a neutron) plus a well-correlated scalar isoscalar diquark ud, the Cooper pair. Among several indications that this is the case is the relatively recent lattice studies that found that d-quark contribution to the proton spin is very small ($\sim 1\%$), within the errors bounds.

Finally, let me remind the reader that strongly coupled ultracold fermionic gases, which we discussed in Section 11.5, do produce really well-bound bosonic molecules, existing with and without the condensate, or even by themselves in vacuum.

22.3 Pairing Can Occur Even without the Fermi Surface!

Most of the examples of Cooper pairing we discussed, required, as a starting point, the existence of a well-defined Fermi sphere. Indeed, at its surface the pairing interaction becomes effectively one dimensional. This leads to the logarithmic enhancement of the initial weak coupling, eventually leading to the formation of Cooper pairs. It was all nicely explained by the BCS theory, especially in its more modern form using RG flow.

But what happens if the initial interaction is not weak? The ultracold fermionic gases at the Feshbach resonance—that is, at strong coupling—have no parameters to justify the assumption of $\Delta \ll \epsilon_F$. And indeed, experiments have found superfluidity but have not found any indications of the presence of the Fermi surface!

Even more unusual is the pairing in the QCD vacuum, which we discussed in Section 18.4. The Lagrangian includes near-massless quarks. If these quarks do not interact with the gauge fields, they would form the Dirac sea. All positive energy states with $E_p \approx p$ would be empty, and all negative energy states with $E_p \approx -p$ would be occupied. The density of states at the surface between the two, namely, at $p = 0$, is zero. Since this factor is crucial for the BCS theory, one may wonder whether any pairing may occur at all.[1]

The phenomenology tells us that pairing does occur, creating a significant gap—the constituent quark mass, $m \approx 400\,\text{MeV}$. It happens only because the effective four-fermion interaction between the quarks is very strong. As we discussed in Chapter 18 this interaction is due to topological solitons: instantons, instanton-dyons, or both.

22.4 Topological Matter

This book is on many-body theory, and in most cases the identity of those bodies clear. We discussed electrons in the electron gas, nucleons in nuclear matter, ^{3}He and ^{4}He atoms in liquid helium and so forth.

The resulting ensemble often can be understood not in terms of the original particles put together to make this matter, but in terms of some quasiparticles. Sometimes they carry the same quantum numbers as the original particles, like those in Landau Fermi liquids; sometimes they are collective excitations, like the plasmons, zero sounds, or the phonon-roton excitation mode in liquid helium.

Yet not all "bodies" we discuss have such obvious origins. As we discussed, many curious phenomena in liquid helium are related to matter made of vortices and vortex rings.

The quantized vortices in superfluids and superconductors can form what is called topological matter. Its simplest form is a two-dimensional Abrikosov crystalline lattice, but it can be a gas of bound pairs or a liquid without long-range order as well. Multiple phases are separated by phase transitions: for example, we have discussed in Section 11.1 the Berezinsky-Kosterlitz-Thouless transition between the phases of a random vortex plasma and the phase made of correlated pairs of vortices.

---

[1] Note also that in this application, we discuss quark-antiquark (or particle-hole) pairing. The quark-quark pairing still needs the Fermi surface and leads to color superconductivity, discussed in Section 20.1.

While in a two-dimensional setting the quantized vortices have only two collective coordinates (their two-dimensional location), in three dimensions the vortex rings may have nontrivial shapes, and therefore even their single-body dynamics may be far from simple. In some conditions, the vortex rings lead to very complex turbulent behavior.

Other examples of topological matter we discussed in Section 17.2 include more complicated quantized objects, the three-dimensional monopoles and even four-dimensional instantons and instanton-dyons in Section 18.4. Since the partition function includes an integration over their collective coordinates, studies of ensembles of such solitons provide the newest and rather complex applications of many-body theory. For example, the instanton-dyons possess electric and magnetic charges, so their ensemble is a kind of dual plasmas. Some of them also possess fermionic zero modes, which generate effective four-quark terms in the Lagrangian, which in turn lead to quark-antiquark or quark-quark pairings. So, there is enough work left to do for new generations of many-body theorists.

22.5 RG Flows and Dualities

It is fitting to end this book by praising one of our most precious tools, the renormalization group.

Its first application in this book, in Section 2.4, was to the theory of second-order phase transitions. Using RG flow, we learned about a huge number of very different systems that flow in the IR limit to the same universal fixed points.

Its next application—the (somewhat modified) BCS theory—has shown how weak coupling in UV may become strong in IR, producing such unusual phases of matter as superconductors. Of course, a very similar RG flow is the asymptotic freedom of QCD.

We also learned that in some cases (e.g., the $\mathcal{N} = 2$ SYM) the RG flow may take us from the electroweak-like weakly coupled theory, though the strongly coupled domain, all the way to the weakly coupled QED-like magnetic theory! So, some seemingly very distinct QFTs may in fact just be different limits of a larger overarching theory. These ideas in QCD inspired the magnetic scenario for QGP discussed in Section 17.4, as well as other "magnetic" or dual formulations of QCD that we have not discussed.

We have not discussed the AdS-CFT duality [193] in this book, but it is impossible not to mention it in this context. In this approach RG flow obtains its most explicit realization: the flow is just a motion along the additional fifth coordinate z. One limit (say, $z \to 0$) corresponds to the UV limit, and another ($z \to \infty$) to the IR.

For example, the complicated problem of thermal relaxation of various modes of QGPs becomes dual to the very first problem of physics, a stone falling under gravity to (not Earth but) a black hole; see [194]. While theorists propose different scenarios of equilibration, some propagating from UV to IR and some from IR to UV, the holographic duality makes the choice obvious: all objects fall in the direction prescribed by the gravitational force (in the fifth dimension).

The fields living in this five-dimensional holographic space are weakly coupled, and the omnipresent highest spin-2 field is gravity. If motion in z gets quantized, it generates a tower of states with quantized four-dimensional momentum squared, or the four-dimensional masses. Suppose that the quantized gravity field generates a tower of $J^{PC} = 2^{++}$ hadrons. One can read about applications of such models, collectively known as AdS/QCD, in [195, 196]. Holography is the most fascinating form of the duality, since it not only relates some weakly and strongly coupled theories, but also relates theories living in flat four-dimensional to curved five(+six)-dimensional spaces. Furthermore, it is based on string theory, which possesses multidimensional solitons, known as "branes." Combining branes of different dimensions, the holographic engineers create multiple brane constructions, each providing a kind of new duality between the five-dimensional bulk and four-dimensional boundary theory. While the number of branes involved often goes to infinity ($N \to \infty$), so far no multibrane construction has been subjected to quantum many-body theory. We shall see whether that happens in the future.

References

[1] A. Zee, *Quantum Field Theory in a Nutshell*, 2nd edition. Princeton, NJ: Princeton University Press, 2010.

[2] R. P. Feynman and H. R. Hibbs, *Quantum Mechanics and Path Integrals*. New York: McGraw-Hill, 1965, chapter 10.

[3] R. P. Feynman, *Statistical Mechanics*. Boston: W. A. Benjamin, 1972.

[4] A. A. Abrikosov, L. P. Gorkov, and I. E. Dzyaloshinski, *Methods of Quantum Field Theory in Statistical Physics*, English revised ed. New York: Dover, 1963.

[5] A. L. Fetter and J. D. Walecka, *Quantum Theory of Many-Particle Systems*. New York: McGraw-Hill, 1971.

[6] J. Negele and H. Orland, *Quantum Many-Body Systems*. New York: Perseus Books, 1972.

Euclidean Time and QFT Methods in Statistical Mechanics

[7] T. Matsubara, "A New Approach to Quantum-Statistical Mechanics," *Progress in Theoretical Physics* (Kyoto), 14, 351 (1954).

[8] E. S. Fradkin, "The Asymptote of Green's Function in Quantum Electrodynamics," *Soviet Physics JETP* 28, 750 (1955).

[9] A. A. Abrikosov, L. P.Gorkov, and I. E. Dzyaloshinski, "Superconducting Alloys at Finite Temperatures," *Soviet Physics JETP* 9, 220 (1959).

[10] M. A. Escobar-Ruiz, E. Shuryak, and A. V. Turbiner, "Quantum and Thermal Fluctuations in Quantum Mechanics and Field Theories from a New Version of Semiclassical Theory," *Physical Review* D 93, 105039 (2016) [arXiv:1601.03964 [hep-th]]; M. A. Escobar-Ruiz, E. Shuryak, and A. V. Turbiner, "Fluctuations in Quantum Mechanics and Field Theories from a New Version of Semiclassical Theory. II," *Physical Review* D 96, 045005 (2017) [arXiv:1705.06159 [hep-th]].

[11] G. V. Dunne and M. Unsal, "Uniform WKB, Multi-Instantons, and Resurgent Trans-Series," *Physical Review* D 89, 105009 (2014) [arXiv:1401.5202 [hep-th]].

The Real Scalar Fields, RG and the Second Order Phase Transitions

[12] K. G. Wilson and M. E. Fisher, "Critical Exponents in 3.99 Dimensions," *Physical Review Letters* 28, 240 (1972).

[13] K. G. Wilson and J. B. Kogut, "The Renormalization Group and the Epsilon Expansion," *Physics Reports* 12, 75 (1974).

[14] G. Parisi, *Statistical Field Theory*. Boston: Addison-Wesley, 1988.

Complex Scalar Fields and Bose Gas

[15] J. I. Kapusta, "Bose-Einstein Condensation, Spontaneous Symmetry Breaking, and Gauge Theories," *Physical Review* D 24, 426 (1981).

[16] N. N. Bogolyubov, "On the Theory of Superfluidity," *Journal of Physics* (USSR) 11, 23 (1947).

[17] T. D. Lee and C. N. Yang, "Many-Body Problem in Quantum Mechanics and Quantum Statistical Mechanics," *Physical Review* 105, 1119 (1957); T. D. Lee, K. Huang, and C. N. Yang, "Eigenvalues and Eigenfunctions of a Bose System of Hard Spheres and Its Low-Temperature Properties," *Physical Review* 106, 1135 (1957).

[18] S. T. Belyaev, "Energy-Spectrum of a Non-ideal Bose Gas," *Soviet Physics JETP* 7, 299 (1958).

[19] T. T. Wu, "Ground State of a Bose System of Hard Spheres," *Physical Review* 115, 1390 (1959); N. M. Hugenholtz and D. Pines, "Ground-State Energy and Excitation Spectrum of a System of Interacting Bosons," *Physical Review* 116, 489 (1959); K. Sawada, "Ground-State Energy of Bose-Einstein Gas with Repulsive Interaction," *Physical Review* 116, 1344 (1959).

[20] E. Braaten and A. Nieto, "Renormalization Effects in a Dilute Bose Gas," *Physical Review* B 55, 8090 (1997) [hep-th/9609047].

Hydrodynamics of Electron Currents

[21] R. N. Gurzhi, "Minimum of Resistance in Impurity-Free Conductors." *Soviet Physics JETP* 17, 2 (1963); R. N. Gurzhi, "Hydrodynamic Effects in Solids at Low temperature," *Soviet. Physics Uspekhi* 11, 255 (1968).

[22] L. Levitov and G. Falkovich, "Electron Viscosity, Current Vortices and Negative Nonlocal Resistance in Graphene," *Nature Physics* 12, 672 (2016) [arXiv:1508.00836 [cond-mat.mes-hall]].

[23] R. Krishna Kumar, D. A. Bandurin, F. M. D. Pellegrino, Y. Cao, A. Principi, H. Guo, G. H. Auton, M. Ben Shalom, L. A. Ponomarenko, G. Falkovich, K. Watanabe, T. Taniguchi, I. V. Grigorieva, L. S. Levitov, M. Polini, and A. K. Geim. "Superballistic Flow of Viscous Electron Fluid through Graphene Constrictions," *Nature Physics* 13, 1182 (2017) [arXiv:1703.06672 [cond-mat.mes-hall]].

Superfluid ⁴He and BEC of Trapped Gases

[24] D. R. Tilley and J. Tilley, *Superfluidity and Superconductivity*, 3rd ed. Philadelphia: IOP Publishing, 1990.

[25] I. M. Khalatnikov, *Introduction to the Theory of Superfluidity*, trans. P. C. Hohenberg. New York: W. A. Benjamin, 1965. (Note: this book includes reprints of two relevant Landau papers.)

[26] P. L. Kapitza, "Viscosity of Liquid Helium at Temperatures below Lambda Point," *Nature* 141, 74 (1938).

[27] J. F. Allen and A. D. Misener, "The Properties of Flow of Liquid He II," *Proceedings of the Royal Society of London* A 172, 467 (1939).

[28] W. H. Keesom and H. P. Keesom, "On the Heat Conductivity of Liquid Helium," *Physica* 3, 359 (1936).

[29] L. D. Landau, "The Theory of Superfluidity of Helium II," *Journal of Physics* (USSR) 5, 71 (1941).

[30] L. D. Landau, "On the Theory of Superfluidity of Helium II," *Journal of Physics* (USSR) 11, 91 (1947).

[31] E. L. Andronikashvili and Yu. G. Mamaladze, "Quantization of Macroscopic Motions and Hydrodynamics of Rotating Helium II," *Reviews of Modern Physics* 38, 567 (1966).

[32] H. Palevsky, K. Otnes, K. E. Larsson, and E. L. Andronikashvili, "Investigation of the Thermal Structure of Helium II," *Soviet Physics JETP* 16, 780 (1946); H. Palevsky, K. Otnes, and K. E. Larsson, "Excitation of Rotons in Helium II by Cold Neutrons," *Physical Review* 112, 11 (1958).

[33] R. P. Feynman and M. Cohen, "Energy Spectrum of the Excitations in Liquid Helium," *Physical Review* 102, 1189 (1956).

[34] G. Placzek and L. van Hove, "Crystal Dynamics and Inelastic Scattering of Neutrons," *Physical Review* 93, 1207 (1954).

[35] C. J. Pethick and H. Smith, *Bose Einstein Condensation in Dilute Gases*. Cambridge: Cambridge University Press, 2002.

[36] L. P. Pitaevskii and S. Stringari, *Bose Einstein Condensation*. Oxford: Clarendon Press, 2003.

[37] C. Raman, M. Khl, R. Onofrio, D. S. Durfee, C. E. Kuklewicz, Z. Hadzibabic, and W. Ketterle, "Evidence for a Critical Velocity in a Bose-Einstein Condensed Gas," *Physical Review Letters* 83, 2502 (1999).

[38] E. Shuryak, "Metastable Bose Condensate Made of Atoms with Attractive Interaction," *Physical Review* A 54, 3151 (1996) [cond-mat/9512008].

[39] A. Eleftheriou and K. Huang, "Instability of a Bose-Einstein Condensate with an Attractive Interaction," *Physical Review* A 61, 43601 (2000) [cond-mat/9908229].

[40] J. M. Gerton, D. Strelkov, I. Prodan, and R. G. Hulet, "Direct Observation of Growth and Collapse of a Bose-Einstein Condensate with Attractive Interactions," *Nature* 408 (2000).

Fermions: Electron Gas, Nuclear Matter, Liquid ^{3}He

[41] F. A. Berezin, *The Method of Second Quantization*. New York: Academic Press, 1966.

[42] D. J. Candlin, "On Sums over Trajectories for Systems with Fermi Statistics," *Nuovo Cimento* 4, 231 (1956).

[43] K. Huang and C. N. Yang, "Quantum-Mechanical Many-Body Problem with Hard-Sphere Interaction," *Physical Review* 105, 767 (1957).

[44] C. DeDominicis and P. C. Martin, "Saturation of Nuclear Forces," *Physical Review* 105, 1419 (1957).

[45] V. N. Efimov and M. Ya. Amusya, "Ground State of a Rarefied Fermi Gas of Rigid Spheres," *Soviet Physics JETP* 20, 388 (1965).

[46] V. M. Galitskii, "The Energy Spectrum of Non-ideal Fermi Gas," *Soviet Physics JETP* 34, 151 (1958).

[47] E. P. Wigner, "On the Interaction of Electrons in Metals," *Physical Review* 46, 1002 (1934).

[48] D. Pines, "A Collective Description of Electron Interactions: IV. Electron Interaction in Metals," *Physical Review* 92, 626 (1953).

[49] J. Lindhard, "On the properties of a gas of charged particles," *Danske Matematisk-fysiske Meddeleiser. Det Kongelige Danske Videnskabernes Selskab* 28 (8): 1 (1954).

[50] M. Gell-Mann and K. A. Brueckner, "Correlation Energy of an Electron Gas at High Density," *Physical Review* 106, 364 (1957).

[51] G. E. Brown, *Unified Theory of Nuclear Models and Forces*. Amsterdam: North-Holland, 1967.

[52] B. D. Serot and J. D. Walecka, "The relativistic Nuclear Many-Body Problem," *Advances in Nuclear Physics* 16 (1984); J. D. Walecka, *Theoretical Nuclear and Subnuclear Physics*, 2nd ed. Singapore: World Scientific, 2004.

[53] S. K. Bogner, R. J. Furnstahl, and A. Schwenk, "From Low-Momentum Interactions to Nuclear Structure," *Progress in Particle and Nuclear Physics* 65 (2010) [arXiv:0912.3688].

Cooper Pairing and the RG Flow of the Coupling

[54] M. Gell-Mann and F.E. Low, "Quantum Electrodynamics at Small Distances," *Physical Review* 95, 1300 (1954).

[55] T. Banks and A. Zaks, "Chiral Analog Gauge Theories on the Lattice," *Nuclear Physics* B 206, 23 (1982).

[56] V. L. Ginzburg and L. D. Landau, *Soviet Physics JETP* 20, 1064 (1950).

[57] R. Prozorov, "Equilibrium Topology of the Intermediate State in Type-I Superconductors of Different Shapes," *Physical Review Letters* 98, 257001 (2007).

[58] J. Bardeen, L. N. Cooper, and J. R. Schrieffer, "Microscopic Theory of Superconductivity," *Physical Review* 108, 1175 (1957).

[59] Y. Li, J. Hao, H. Liu, Y. Li, and Y. Ma, "The Metallization and Superconductivity of Dense Hydrogen Sulfide," *Journal of Chemical Physics* 140, 174712 (2014); A. P. Drozdov, M. I. Eremets, I. A. Troyan, V. Ksenofontov, and S. I. Shylin, "Conventional Superconductivity at 203 Kelvin at High Pressures in the Sulfur Hydride System," *Nature* 525 (2015).

[60] L. P. Gor'kov, "On the Energy Spectrum of Superconductors," *Soviet Physics JETP* 7, 505 (1958).

[61] A. A. Abrikosov, L. P. Gorkov, and I. E. Dzyaloshinski, "On the Application of Quantum-Field-Theory Methods to Problems of Quantum Statistics at Finite Temperatures," *Soviet Physics JETP* 9, 636 (1959).

[62] A. J. Leggett, "Interpretation of Recent Results on He^3 below 3 mK: A New Liquid Phase?" *Physical Review Letters* 29, 1229 (1972).

[63] L. D. Landau, "Theory of Fermi Liquid," *Soviet Physics JETP* 32, 59 (1957).

[64] G. Baym and C. Pethick, *Landau Fermi Liquid Theory: Concepts and Applications.* Hobokin, NJ: Wiley, 1991.

[65] J. Polchinski, "Effective Field Theory and the Fermi surface," in J. Harvey and J. Polchinski, eds., *Recent Directions in Particle Theory: From Superstrings and Black Holes to the Standard Model: Proceedings of the 1992 Theoretical Advanced Study Institute in Elementary Particle Physics.* Singapore: World Scientific, 1993. [hep-th/9210046].

[66] R. Rapp, E. V. Shuryak, and I. Zahed, "A Chiral Crystal in Cold QCD Matter at Intermediate Densities?" *Physical Review* D 63, 034008 (2001) [hep-ph/0008207].

[67] A. B. Migdal, "Interaction between Electrons and Lattice Vibrations in a Normal Metal," *Soviet Physics JETP* 7, 996 (1958).

[68] J. Hormuzdiar and S. D. H. Hsu, "Effective Field Theory of Neutron star Superfluidity," nucl-th/9811017.

Superconducting Vortices and Topological Matter

[69] A. A. Abrikosov. "On the Magnetic Properties of Superconductors of the Second Group," *Soviet Physics JETP,* 5, 1174 (1957).

[70] M. A. Alpar, P. W. Anderson, D. Pines, and J. Shaham, "Vortex creep and the Internal Temperature of Neutron Stars," *Astrophysical Journal* 276, 325 (1984).

[71] V. L. Berezinsky, "Destruction of Long-Range Order in One-Dimensional and Two-Dimensional Systems Having a Continuous Symmetry Group I. Classical Systems," *Soviet Physics JETP* 32, 3 (1972) 493500; "Destruction of Long-Range Order in One-Dimensional and Two-Dimensional Systems Having a Continuous Symmetry Group II. Quantum Systems," *Soviet Physics JETP* 34, 3 (1972) 610616.

[72] J. M. Kosterlitz and D. J. Thouless, "Ordering, Metastability and Phase Transitions in Two-Dimensional Systems," *Journal of Physics C: Solid State Physics* 6, 7 (1973) 11811203.

[73] J. B. Kogut, "An Introduction to Lattice Gauge Theory and Spin Systems," *Reviews of Modern Physics* 51, 659 (1979).

[74] L. Onsager, "Statistical Hydrodynamics," *Nuovo Cimento* Suppl. 6, 249, 279 (1949); G. L. Eyink and K. R. Sreenivasan, "Onsanger and the Theory of Hydrodynamic Turbulence," *Reviews of Modern Physics* 78, 87 (2006).

[75] P. Wiegmann and A. G. Abanov, "Anomalous Hydrodynamics of Two-Dimensional Vortex Fluid," *Physics Review Letters* 113, 034501 (2014) [arXiv:1311.4479].

Phase Transitions in Rotating Nuclei

[76] G. Scharff Goldhaber, C. B. Dover, and A. L. Goodman, "The Variable Moment of Inertia (VMI) Model and Theories of Nuclear Collective Motion," *Annual Review of Nuclear Science* 26 (1976).

[77] A. Bohr, B. R. Mottelson, and D. Pines, "Possible Analogy between the Excitation Spectra of Nuclei and Those of the Superconducting Metallic State," *Physical Review* 110, 936 (1958).

[78] A. B. Migdal, "Superfluidity and the Moments of Inertia of Nuclei," *Nuclear Physics* 13, 655 (1959).

[79] S. T. Belyaev, *Effect of pairing force correlations on nuclear properties*, "Kgl. Danske Videnskab Selskab," *Mat. Fys. Medd.* 31, 11 (1959).

[80] B. R. Mottelson and J. G. Valatin, "Effect of Nuclear Rotation on the Pairing Correlation," *Physical Review Letters* 5, 511 (1960).

[81] R. Budaca and A. A. Raduta, "Semi-Microscopic Description of the Double Back-Bending in Some Deformed Even-Even Rare Earth Nuclei," *Journal of Physics* 40, 025109 (2013) [arXiv:1301.6004 [nucl-th]].

[82] D. Pines and A. Alpar, "Superfluidity in Neutron Stars," *Nature* 316, 27 (1985).

[83] B. Haskell and A. Melatos, "Models of Pulsar Glitches," *International Journal of Modern Physics D* 24, 3 (2015) 1530008 [arXiv:1502.07062].

[84] Y. V. Stadnik and V. V. Flambaum, "Reply to Comment on 'Searching for Topological Defect Dark Matter via Nongravitational Signatures,'" *Physical Review Letters* 116, 16 (2016) 169002 [arXiv:1507.01375].

Strongly Coupled Fermionic Systems

[85] G. Bertsch, "Proceedings of the Tenth International Conference on Recent Progress in Many-Body Theories," in R. F. Bishop, Klaus A. Gernoth, Niels R. Walet, and Yang Xian, eds., *Recent Progress in Many-Body Theories*. Seattle: World Scientific, 2000.

[86] M. Randeria and E. Taylor, "BCS-BEC Crossover and the Unitary Fermi Gas," *Annual Review of Condensed Matter Physics* 5, 209 (2014) [arXiv:1306.5785].

[87] M. J. H. Ku, A. T. Sommer, L. W. Cheuk, and M. W. Zwierlein, "Revealing the Superfluid Lambda Transition in the Universal Thermodynamics of a Unitary Fermi Gas," *Science* 335, 6068 (2012) 563567.

[88] M. W. Zwierlein, J. R. Abo-Shaeer, A. Schirotzek, C. H. Schunck, and W. Ketterle, "Vortices and Superfluidity in a Strongly Interacting Fermi Gas," *Nature* 435 (2005) 10471051.

[89] B. A. Gelman, E. V. Shuryak, and I. Zahed, "Ultracold Strongly Coupled Gas: A Near-Ideal Liquid," *Physical Review A* 72, 043601 (2005) [nucl-th/0410067].

[90] Allan Adams, Lincoln D. Carr, Thomas Schaefer, Peter Steinberg, and John E. Thomas, "Strongly Correlated Quantum Fluids: Ultracold Quantum Gases, Quantum Chromodynamic Plasmas, and Holographic Duality," *New Journal of Physics* 14, 115009 (2012) [arXiv:1205.5180 [hep-th]].

[91] W. Ketterle and M. W. Zwierlein, "Making, Probing and Understanding Ultracold Fermi Gases," *Nuovo Cimento Rivista Serie* 31, 247 (2008) [arXiv:0801.2500].

Numerical Path Integrals for Particles

[92] N. Metropolis, A. W. Rosenbluth, M. N. Rosenbluth, A. H. Teller, and E. Teller, "Equation of State Calculations by Fast Computing Machines," *Journal of Chemical Physics* 21 (1953) 10871092.

[93] M. Creutz and B. Freedman, "A Statistical Approach to Quantum Mechanics," *Annals of Physics* 132, 427 (1981).

[94] M. Creutz, "Confinement and the Critical Dimensionality of Space-Time," *Physical Review Letters* 43, 533 (1979); M. Creutz, "Monte Carlo Study of Quantized SU(2) Gauge Theory," *Physical Review* D 21, 2308 (1980).

[95] E. V. Shuryak and O. V. Zhirov, "Testing Monte Carlo Methods for Path Integrals in Some Quantum Mechanical Problems," *Nuclear Physics* B 242, 393 (1984).

[96] D. M. Ceperley, "Path Integrals in the Theory of Condensed Helium," *Reviews of Modern Physics* 67, 279 (1995).

[97] R. P. Feynman, "The λ-Transition in Liquid Helium," *Physical Review* 90, 1116 (1953); R. P. Feynman, "Atomic Theory of the λ-Transition in Helium," *Physical Review* 91, 1291 (1953).

[98] M. Cristoforetti and E. Shuryak, "Bose-Einstein Condensation of Strongly Interacting Bosons: From Liquid He-4 to QCD Monopoles," *Physical Review* D 80, 054013 (2009) [arXiv:0906.2019].

[99] A. D'Alessandro, M. D'Elia, and E. V. Shuryak, "Thermal Monopole Condensation and Confinement in Finite Temperature Yang-Mills Theories," *Physical Review* D 81, 094501 (2010) [arXiv:1002.4161].

[100] A. Ramamurti and E. Shuryak, "Effective Model of QCD Magnetic Monopoles from Numerical Study of One- and Two-Component Coulomb Quantum Bose Gases," *Physical Review* D 95, 076019 (2017) [arXiv:1702.07723 [hep-ph]].

[101] O. O. Tursunov and O. V. Zhirov, "Numerical Simulation of Fermi Systems," *Physics Letters* B 222, 110 (1989).

[102] B. L. G. Bakker, M. I. Polikarpov, and A. I. Veselov, "Pauli-Potential and Green Function Monte-Carlo Method for Many-Fermion Systems," *Few Body Systems* 25, 101 (1998) [arXiv:quant-ph/951109].

[103] D. M. Ceperley, "Path Integral Calculations of Normal Liquid ^{3}He," *Physics Review Letters* 69, 331 (1992).

[104] M. Boninsegni, N. V. Prokofev, and B. V. Svistunov, "Worm Algorithm and Diagrammatic Monte Carlo: A New Approach to Continuous-Space Path Integral Monte Carlo Simulations," *Physical Review* E 74, 036701 (2006).

QCD and Weakly Coupled Quark-Gluon Plasma

[105] C. N. Yang and R. Mills, "Conservation of Isotopic Spin and Isotopic Gauge Invariance," *Physical Review* 96, 191 (1954).

[106] K. G. Wilson, "Confinement of Quarks," *Physical Review* D 10, 2445 (1974).

[107] M. Creutz, "Confinement and the Critical Dimensionality of Space-Time," *Physical Review Letters* 43, 553 (1979).

[108] E. V. Shuryak, "Theory of Hadron Plasma," *Soviet Physical JETP* 47, 2 (1978).

[109] J. Kapusta, "Quantum Chromodynamics at High Temperature," *Nuclear Physics* B 148, 461 (1979).

[110] E. V. Shuryak, "The QCD Vacuum, Hadrons and Superdense Matter," *Physics Reports* C 61, 71 (1980).

[111] A. Hietanen, K. Kajantie, M. Laine, K. Rummukainen, and Y. Schroder, "Three-Dimensional Physics and the Pressure of Hot QCD," *Physical Review* D 79, 045018 (2009) [arXiv:0811.4664].

[112] D. J. Gross, R. D. Pisarski, and L. G. Yaffe, "QCD and Instantons at Finite Temperature," *Reviews of Modern Physics* 53, 43 (1981).

[113] E. Braaten, "Solution to the Perturbative Infrared Catastrophe of Hot Gauge Theories," *Physical Review Letters* 74, 2164 (1995) [hep-ph/9409434]; E. Braaten and A. Nieto, "On the Convergence of Perturbative QCD at High Temperature," *Physical Review Letters* 76, 1417 (1996) [hep-ph/9508406]; E. Braaten, "Free Energy of QCD at High Temperature," *Physical Review* D 53, 3421 (1996) [hep-ph/9510408].

[114] K. Kajantie, M. Laine, K. Rummukainen, and Y. Schröder, "Resumming Long-Distance Contributions to the QCD Pressure," *Physical Review Letters* 86, 10 (2001) [hep-ph/0007109].

[115] E. Braaten and R. D. Pisarski, "Soft Amplitudes in Hot Gauge Theories: A General Analysis," *Nuclear Physics* B 337 (1990); J. Frenkel and J. C. Taylor, "High-Temperature Limit of Thermal QCD," *Nuclear Physics* B 334 (1990).

[116] J.-P. Blaizot and E. Iancu, "Kinetic Theory and Quantum Electrodynamics at High Temperature," *Nuclear Physics* B 390, 589 (1993); "Kinetic Equations for Long-Wavelength Excitations of the Quark-Gluon Plasma," *Physical Review Letters* 70, 3376 (1993).

[117] A. M. Polyakov, "String Representations and Hidden Symmetries for Gauge Fields," *Physics Letters* B 82, 247 (1979).

[118] A. D. Linde, "Infrared Problem in the Thermodynamics of the Yang-Mill Gas," *Physics Letters* B 96, 289 (1980).

[119] P. Arnold and C. Zhai, "Three-Loop Free Energy for Pure Gauge QCD," *Physical Review* D 50, 7603 (1994); P. Arnold and C. Zhai, "Three-Loop Free Energy for High-Temperature QED and QCD with Fermions," *Physical Review* 51, 1906 (1995); C. Zhai and B. Kastening, "Free Energy of Hot Gauge Theories with Fermions through G 5," *Physical Review* 52, 7232 (1995).

[120] E. Braaten and A. Nieto, "Free Energy of QCD at High Temperature," *Physical Review* D 53, 3421 (1996).

[121] A. Peshier, B. Kämpfer, O. P. Pavlenko, and G. Soff, "Massive Quasiparticle Model of the SU(3) Gluon Plasma," *Physical Review* D 54, 2399 (1996); A. Peshier, "Quasiparticle Description of Strongly Coupled Plasmas," (1998) [arXiv:hep-ph/9809379].

[122] J. O. Andersen, E. Braaten, and M. Strickland, "Hard-Thermal-Loop Resummation of the Free Energy of a Hot Gluon Plasma," *Physical Review Letters* 83, 2139 (1999); J. O. Andersen, E. Braaten, and M. Strickland, "Screened Perturbation Theory to Three Loops," *Physical Review* D 63, 105008 (2001).

[123] J. P. Blaizot, E. Iancu, and A. Rebhan, "Entropy of the QCD Plasma," *Physical Review Letters* 83, 2906 (1999); J. P. Blaizot, E. Iancu, and A. Rebhan, "Self-Consistent Hard-Thermal-Loop Thermodynamics for the Quark-Gluon Plasma," *Physics Letters* B 470, 181 (1999); J. P. Blaizot, E. Iancu, and A. Rebhan, "Approximately Self-Consistent Resummations for the Thermodynamics of the Quark-Gluon Plasma: Entropy and Density," *Physical Review* D 63, 065003 (2001).

[124] I. B. Khriplovich, "Green's Functions in Theories with Non-Abelian Gauge Group," *Soviet Journal of Nuclear Physics* 10, 235 (1969).

[125] R. A. Soltz, C. DeTar, F. Karsch, S. Mukherjee, and P. Vranas, "Lattice QCD Thermodynamics with Physical Quark Masses," *Annual Review of Nuclear and Particle Science* 65, 379 (2015) [arXiv:1502.02296 [hep-lat]].

[126] D. E. Kharzeev, "The Chiral Magnetic Effect and Anomaly-Induced Transport," *Progress in Particle and Nuclear Physics* 75, 133 (2014) [arXiv:1312.3348 [hep-ph]].

[127] A. Vilenkin, "Equilibrium Parity-Violating Current in a Magnetic Field," *Physical Review* D 22, 3080 (1980).

[128] Q. Li, D. E. Kharzeev, C. Zhang, Y. Huang, I. Pletikosić, et al., "Observation of the Chiral Magnetic Effect in ZrTe$_5$," *Nature Physics* 12, 550 (2016).

[129] D. T. Son and P. Surowka, "Hydrodynamics with Triangle Anomalies," *Physical Review Letters* 103, 191601 (2009) [arXiv:0906.5044 [hep-th]].

[130] D. E. Kharzeev, "Topologically Induced Local P and CP Violation in QCD x QED," *Annals of Physics* 325, 205 (2010) [arXiv:0911.3715 [hep-ph]].

[131] Y. Burnier, D. E. Kharzeev, J. Liao, and H. U. Yee, "Chiral Magnetic Wave at Finite Baryon Density and the Electric Quadrupole Moment of the Quark-Gluon Plasma," *Physical Review Letters* 107, 052303 (2011) [arXiv:1103.1307 [hep-ph]].

Magnetic Monopoles and Confinement

[132] H. Poincaré, "Remarques sur une expérience de M. Birkeland," *Comptes rendus de l'Académie des Sciences* B 123, 530 (1896).

[133] P. A. M. Dirac, "Quantised Singularities in the Electromagnetic Field," *Proceedings of the Royal Society* A 133, 60 (1931).

[134] J. Liao and E. Shuryak, "Strongly Coupled Plasma with Electric and Magnetic Charges," *Physical Review* C 75, 054907 (2007) [hep-ph/0611131].

[135] J. Liao and E. Shuryak, "Magnetic Component of Quark-Gluon Plasma Is Also a Liquid!," *Physical Review Letters* 101, 162302 (2008) [arXiv:0804.0255].

[136] D. G. Boulware, L. S. Brown, R. N. Cahn, S. D. Ellis, and C. K. Lee, "Scattering on Magnetic Charge," *Physical Review* D 14, 2708 (1976).

[137] J. S. Schwinger, K. A. Milton, W.-Y. Tsai, L. L. DeRaad, and D. C. Clark, "Nonrelativistic Dyon-Dyon Scattering," *Annals of Physics* 101, 451 (1976).

Kinetic Quantities in QGP and Hadronic Matter

[138] C. Ratti and E. Shuryak, "The Role of Monopoles in a Gluon Plasma," *Physical Review* D 80, 034004 (2009) [arXiv:0811.4174].

[139] G. Policastro, D. T. Son, and A. O. Starinets, "The Shear Viscosity of Strongly Coupled $N = 4$ Supersymmetric Yang-Mills Plasma," *Physical Review Letters* 87, 081601 (2001) [hep-th/0104066].

[140] M. Prakash, M. Prakash, R. Venugopalan, and G. Welke, "Nonequilibrium Properties of Hadronic Mixtures," *Physics Reports* 227, 321 (1993).

[141] P. B. Arnold, G. D. Moore, and L. G. Yaffe, "Transport Coefficients in High Temperature Gauge Theories. 2. Beyond Leading Log," *Journal of High Energy Physics* 0305, 051 (2003) [hep-ph/0302165].

[142] B. A. Gelman, E. V. Shuryak, and I. Zahed, "Classical Strongly Coupled Quark-Gluon Plasma. I. Model and Molecular Dynamics Simulations," *Physical Review* C 74, 044908 (2006) [nucl-th/0601029].

[143] A. Nakamura and S. Sakai, "Transport Coefficients of Gluon Plasma," *Physical Review Letters* 94, 072305 (2005) [hep-lat/0406009].

[144] J. Liao and E. Shuryak, "Angular Dependence of Jet Quenching Indicates Its Strong Enhancement Near the QCD Phase Transition," *Physical Review Letters* 102, 202302 (2009) [arXiv:0810.4116 [nucl-th]].

[145] J. Xu, J. Liao, and M. Gyulassy, "Bridging Soft-Hard Transport Properties of Quark-Gluon Plasmas with CUJET 3.0," *Journal of High Energy Physics* 1602, 169 [arXiv:1508.00552 [hep-ph]].

QGP in Experiments and Cosmology

[146] M. A. Stephanov, K. Rajagopal, and E. V. Shuryak, "Signatures of the Tricritical Point in QCD," *Physical Review Letters* 81, 4816 (1998) [hep-ph/9806219].

[147] E. Shuryak, "Strongly Coupled Quark-Gluon Plasma in Heavy Ion Collisions," *Reviews of Modern Physics* 89, 035001 (2017) [earlier version: arXiv:1412.8393 [hep-ph]].

[148] E. V. Shuryak and O. V. Zhirov, "Vacuum Pressure Effects in Low-$p_\perp$ Hadronic Spectra," *Physics Letters* B 89, 253 (1980).

[149] T. Kalaydzhyan and E. Shuryak, "Gravity Waves Generated by Sounds from Big Bang Phase Transitions," *Physical Review* D 91, 083502 (2015) [arXiv:1412.5147 [hep-ph]].

Cold Quark Matter and Color Superconductivity

[150] G. Baym and S. A. Chin, "Can a Neutron Star Be a Giant MIT Bag?" *Physics Letters* B 62, 241 (1976).

[151] P. D. Morley and M. B. Kislinger, "Relativistic Many-Body Theory, Quantum Chromodynamics and Neutron Stars/Supernova," *Physics Reports* 51, 63 (1979).

[152] R. Rapp, T. Schäfer, E. V. Shuryak, and M. Velkovsky, "High Density QCD and Instantons," *Annals of Physics* 280, 35 (2000) [hep-ph/9904353].

[153] M. G. Alford, A. Schmitt, K. Rajagopal, and T. Schäfer, "Color Superconductivity in Dense Quark Matter," *Reviews Modern of Physics* 80, 1455 (2008) [arXiv:0709.4635 [hep-ph]].

[154] M. Alford, K. Rajagopal, and F. Wilczek, "QCD at Finite Baryon Density: Nucleon Droplets and Color Superconductivity," *Physics Letters* B 422, 247 (1998) [hep-ph/9711395].

[155] M. Alford, K. Rajagopal, and F. Wilczek, "Color-Flavor Locking and Chiral Symmetry Breaking in High Density QCD," *Nuclear Physics* B 537, 443 (1999) [hep-ph/9804403].

[156] R. Rapp, T. Schäfer, E. V. Shuryak, and M. Velkovsky, "Diquark Bose Condensates in High Density Matter and Instantons," *Physical Review Letters* 81, 53 (1998) [hep-ph/9711396].

[157] D. T. Son, "Superconductivity by Long-Range Color Magnetic Interaction in High-Density Quark Matter," *Physical Review* D 59, 094019 (1999) [hep-ph/9812287].

[158] T. Schäfer and F. Wilczek, "Continuity of Quark and Hadron Matter," *Physical Review Letters* 82, 3956 (1999) [hep-ph/9811473].

[159] T. Schäfer and F. Wilczek, "Superconductivity from Perturbative One-Gluon Exchange in High Density Quark Matter," *Physical Review* D 60, 114033 (1999) [hep-ph/9906512].

Chiral Symmetries in the QCD Vacuum and Instantons

[160] T. Schaefer and E. V. Shuryak, "Instantons in QCD," *Reviews Modern Physics* 70, 323 (1998) [hep-ph/9610451].

[161] Y. Nambu, "Axial Vector Current Conservation in Weak Intractions," *Physical Review Letters* 4, 380 (1960); Y. Nambu, "Quasi-Particles and Gauge Invariance in the Theory of Superconductivity," *Physical Review* 117, 648 (1960); Y. Nambu and G. Jona-Lasinio, "Dynamical Model of Elementary Particles Based on an Analogy with Superconductivity," *Physical Review* 122, 345 (1961).

[162] Vaks V. G. and A. I. Larkin, "On the Application of the Methods of Superconductivity Theory to the Problem of the Masses of Elementary Particles," *Soviet Physics JETP* 40, 282 (1961).

[163] S. P. Klevansky, "The Nambu-Jona-Lasinio Model of Quantum Chromodynamics," *Reviews of Modern Physics* 64, 649 (1992).

[164] T. M. Schwarz, S. P. Klevansky, and G. Papp, "The Phase Diagram and Bulk Thermodynamical Quantities in the NJL Model at Finite Temperature and Density," *Physical Review* C 60, 055205 (1999) [nucl-th/9903048].

Confinement, Monopoles, and Dual Superconductivity

[165] Y. Nambu, "Strings, Monopoles, and Gauge Fields," *Physical Review* D 10, 4262 (1974); S. Mandelstam, "Vortices and Quark Confinement in non-Abelian Gauge Theories," *Physics Reports* 23, 245 (1976); G. 't Hooft, "Gauge Theories with Unified Weak, Electromagnetic and Strong Interactions," in A. Zichichi, ed., *Proceedings of the European Physical Society International Conference on High Energy Physics.* Bologna: Editrice Compositori, 1976.

[166] G. 't Hooft, "Magnetic Monopoles in Unified Gauge Theories," *Nuclear Physics* B 79, 276 (1974).

[167] A. M. Polyakov, "Particle Spectrum in Quantum Field Theory," *JETP Letters* 20, 194 (1974).

[168] Y. M. Shnir, *Magnetic Monopoles.* Berlin: Springer, 2005.

[169] N. Seiberg and E. Witten, "Electric-Magnetic Duality, Monopole Condensation, and Confinement in $N = 2$ Supersymmetric Yang-Mills Theory," *Nuclear Physics* B 426, 19 (1994). Erratum: *Nuclear Physics* B 430, 485 (1994) [hep-th/9407087].

[170] G. S. Bali, "The Mechanism of Quark Confinement" [hep-ph/9809351].

[171] T. Kalaydzhyan and E. Shuryak, "Collective Interaction of QCD Strings and Early Stages of High Multiplicity pA Collisions," *Physical Review* C 90, 014901 (2014) [arXiv:1404.1888 [hep-ph]].

The Anomalies

[172] J. Steinberger, "Learning about Particles—50 Privileged Years," *Physical Review* 76, 1180 (1949).

[173] J. Schwinger, "On Gauge Invariance and Vacuum Polarization," *Physical Review* 82, 664 (1951).

[174] S. L. Adler, "Axial-Vector Vertex in Spinor Electrodynamics," *Physical Review* 177, 2426 (1969).

[175] J. S. Bell and R. Jackiw, "A PCAC Puzzle," *Nuovo Cim.* A 60, 47 (1969).

[176] J. Ambjorn, J. Greensite, and C. Petersson, "The Axial Anomaly and the Lattice Dirac Sea," *Nuclear Physics* B 221, 381 (1983).

[177] H. B. Nielsen and M. Ninomiya, "No Go Theorem for Regularizing Chiral Fermions," *Physics Letters* B 105, 219 (1981).

[178] K. Fujikawa, "Path-Integral Measure for Gauge-Invariant Fermion Theories," *Physical Review Letters* 42, 1195 (1979); K. Fujikawa, "Comment on Chiral and Conformal Anomalies," *Physical Review Letters* 44, 1733 (1980).

The Instantons

[179] A. A. Belavin, A. M. Polyakov, A. S. Schwartz, and Yu. S. Tyupkin, "Pseudoparticle Solutions of the Yang-Mills Equations," *Physics Letters* B 59, 85 (1975).

[180] G. 't Hooft, "Computation of the Quantum Effects Due to a Four-Dimensional Pseudoparticle," *Physical Review* D 14, 3432 (1976). Erratum: *Physical Review* D 18, 2199 (1978).

[181] E. V. Shuryak, "The Role of Instantons in Quantum Chromodynamics. 1. Physical Vacuum," *Nuclear Physics* B 203, 93 (1982).

[182] M. A. Escobar-Ruiz, E. Shuryak, and A. V. Turbiner, "Three-loop Correction to the Instanton Density. I. The Quartic Double Well Potential," *Physical Review* D 92, 025046 (2015). Erratum: *Physical Review* D 92, 089902 (2015) [arXiv:1501.03993 [hep-th]].

[183] M. A. Escobar-Ruiz, E. Shuryak, and A. V. Turbiner, "Three-Loop Correction to the Instanton Density. II. The Sine-Gordon potential," *Physical Review* D 92, 025047 (2015). [arXiv:1505.05115 [hep-th]].

[184] L. Y. Glozman, C. B. Lang, and M. Schrock, "Symmetries of Hadrons after Unbreaking the Chiral Symmetry," *Physical Review* D 86, 014507 (2012) [arXiv:1205.4887 [hep-lat]].

The Instanton-Dyons

[185] T. C. Kraan and P. van Baal, "Monopole Constituents inside $SU(n)$ Calorons," *Physics Letters* B 435, 389 (1998) [arXiv:hep-th/9806034].

[186] K.-M. Lee and C.-h. Lu, "$SU(2)$ Calorons and Magnetic Monopoles," *Physical Review* D 58, 025011 (1998) [hep-th/9802108].

[187] D. Diakonov, "Topology and Confinement," *Nuclear Physics B – Proceedings Supplements* 195, 5 (2009) [arXiv:0906.2456v1 [hep-ph]].

[188] R. Larsen and E. Shuryak, "Instanton-Dyon Ensemble with Two Dynamical Quarks: The Chiral Symmetry Breaking," *Physical Review* D 93, 054029 (2016) [arXiv:1511.02237 [hep-ph]].

[189] R. Larsen and E. Shuryak, "Instanton-dyon Ensembles III: Exotic Quark Flavors," *Physical Review* D 94, 094009 (2006) [arXiv:1605.07474 [hep-ph]].

[190] H. Kouno, Y. Sakai, T. Makiyama, K. Tokunaga, T. Sasaki and M. Yahiro, "Quark-Gluon Thermodynamics with the Z_{N_c} Symmetry," *Journal of Physics* G 39, 085010 (2012).

[191] T. Misumi, T. Iritani, and E. Itou, "Finite-Temperature Phase Transition of $N_f = 3$ QCD with Exact Center Symmetry," (2015) [arXiv:1510.07227 [hep-lat]].

General Parting Remarks

[192] F. J. Dyson, "Divergence of Perturbation Theory in Quantum Electrodynamics," *Physical Review* 85, 631 (1952).

[193] J. M. Maldacena, "Wilson Loops in Large *N* Field Theories," *Physical Review Letters* 80, 4859 (1998) [hep-th/9803002].

[194] S. Lin and E. Shuryak, "Toward the AdS/CFT Gravity Dual for High Energy Collisions. 3. Gravitationally Collapsing Shell and Quasiequilibrium," *Physical Review* D 78, 125018 (2008) [arXiv:0808.0910 [hep-th]].

[195] U. Gursoy and E. Kiritsis, "Exploring Improved Holographic Theories for QCD: Part I," *Journal of High-Energy Physics* 0802, 032 (2008) [arXiv:0707.1324 [hep-th]].

[196] U. Gursoy, E. Kiritsis, and F. Nitti, "Exploring Improved Holographic Theories for QCD: Part II," *Journal of High-Energy Physics* 0802, 019 (2008) [arXiv:0707.1349 [hep-th]].

[197] S. S. Gubser, I. R. Klebanov, and A. A. Tseytlin, "Coupling Constant Dependence in the Thermodynamics of $N = 4$ Supersymmetric Yang-Mills Theory," *Nuclear Physics* B 534, 202 (1998) [hep-th/9805156].

[198] P. Staig and E. Shuryak, "Fate of the Initial State Perturbations in Heavy Ion Collisions. III. The Second Act of Hydrodynamics," *Physical Review* C 84, 044912 (2011) [arXiv:1105.0676 [nucl-th]].

[199] R. A. Lacey, Y. Gu, X. Gong, D. Reynolds, N. N. Ajitanand, J. M. Alexander, A. Mwai, and A. Taranenko, "Is Anisotropic Flow Really Acoustic?" arXiv:1301.0165 [nucl-ex], 2013.

[200] P. A. R. Ade et al. [Planck Collaboration], "Planck 2013 Results. XV. CMB Power Spectra and Likelihood," *Astronomy and Astrophysics* 571, A15 (2014) [arXiv:1303.5075 [astro-ph.CO]].

[201] Y. Liu, E. Shuryak, and I. Zahed, "Confining Dyon-Antidyon Coulomb Liquid Model. I.," *Physical Review* D 92, 085006 (2015) [arXiv:1503.03058 [hep-ph]].

[202] R. Larsen and E. Shuryak, "Interacting Ensemble of the Instanton-Dyons and the Deconfinement Phase Transition in the $SU(2)$ Gauge Theory," *Physical Review* D 92, 094022 (2015) [arXiv:1504.03341 [hep-ph]].

Index